J[illegible]ollard
Adrian Schmit

HODDER
EDUCATION
AN HACHETTE UK COMPANY

Although every effort has been made to ensure that website addresses are correct at time of going to press, [illegible]ion cannot be held responsible for the content of any website mentioned. It is sometime[illegible]o find a relocated web page by typing in the address of the home page for a website in the [illegible] window of your browser.

Risk Assessment

As a service to our users, a risk assessment for this text has been carried out by CLEAPSS and is available on request to the publishers. However, the publishers accept no legal responsibility on any issue arising from this risk assessment. Whilst every effort has been made to check the instructions of practical work in this book, it is still the duty and legal obligation of schools to carry out their own risk assessments.

Orders: please contact Bookpoint Ltd, 130 Milton Park, Abingdon, Oxon OX14 4SB. Telephone: (44) 01235 827720. Fax: (44) 01235 400454. Lines are open 9.00–17.00, Monday to Saturday, with a 24-hour message answering service. Visit our website at www.hoddereducation.co.uk

First published in 2011 by
Hodder Education
An Hachette UK Company,
338 Euston Road
London NW1 3BH

Impression number	5	4	3	2	1
Year	2015	2014	2013	2012	2011

Cover photo © PASIEKA/Science Photo Library

Illustrations by DC Graphic Design Limited and Barking Dog Art

Typeset in England by DC Graphic Design Limited.

Printed in Italy

A catalogue record for this title is available from the British Library.

ISBN 978 1 444 124330

Contents

Acknowledgements

The Publisher would like to thank the following for permission to reproduce copyright material:

Text credits

p.177 Tim Webb, from 'At least 1 in 10 new homes fail the energy efficiency test' from *The Guardian* (14 April, 2010), copyright Guardian News & Media Ltd 2010, reproduced by permission of the publisher; **p.221** Robin McKie, from 'Is this the most radioactive place on the planet?', adapted from *The Observer* (19 April, 2009), copyright Guardian News & Media Ltd 2009, reproduced by permission of the publisher; **p.238** D. Shiga, 'Found: first rocky exoplanet that could host life', adapted from *New Scientist*, 23:34 (29 September 2010).

Photo credits

p.1 © Fotolia/Michael Jung; **p.5** © Fotolia/Elen; **p.13** © Mary Evans Picture Library; **p.14** *t* © Fotolia/Mark Penny, *b* © Fotolia/Siloto; **p.16** *l* © Fotolia/Outdoorsman, *m* © Fotolia/EcoView, *r* © Fotolia/Brad Sauter; **p.18** *t* © Fotolia/Maksym Gorpenyuk, *b* © Fotolia/Daniel Mortell; **p.20** © Fotolia/Stephen Meese; **p.21** *t* © Photolibrary Wales/Margaret Price, *b* © Photolibrary Wales/Jeff Morgan; **p.23** *tl* © British Lichen Society/Mike Sutcliffe, *ml* © British Lichen Society/Mike Sutcliffe, *mm* © British Lichen Society/Mike Sutcliffe, *mr* © Fotolia/Tatjana Gupalo, *bl* © British Lichen Society/Robin Crump, *bm* © British Lichen Society/Mike Sutcliffe, *br* © Irish Lichens/Jenny Seawright; **p.26** © Rex Features/Eye Ubiquitous; **p.27** © Fotolia/TDPhotos; **p.34** © Fotolia/Christian Musat; **p.35** *t* © Fotolia/Anderson Rise, *b* © Newspix/News Ltd/Gregg Porteous; **p.36** *t* © Alamy/Asperra Images, *b* © Science Photo Library; **p.39** *t* © Science Photo Library/Cordelia Molloy, *b* © Science Museum/SSPL; **p.42** © Science Photo Library/Philippe Psaila; **p.47** © Fotolia/Witold Kaszkin; **p.48** © The Library of Congress/Bain Collection; **p.49** *l* © Alamy/George Chambers, *r* © Fotolia/Paul Moore; **p.51** © Science Photo Library/Photo Researchers; **p.52** © Daily Mail Syndication/John Frost Newspapers; **p.54** © Alamy/FR Sport Photography; **p.59** © Photolibrary/Peter Arnold Images/Ed Reschke; **p.60** © Alamy/Radius Images; **p.62** © Fotolia/Alexander Raths; **p.64** © SPL/Biofoto Associates; **p.76** © Getty Images/Adam Gault; **p.77** © Science Photo Library/Ria Novosti; **p.81** *t* © The Art Archive/Bibliotheque des Arts Decoratifs Paris/Dagli Orti, *b* © Getty Images/Stock Montage; **p.84** © Science Photo Library; **p.86** *t* © The Trustees of the British Museum, *b* © Mary Evans Picture Library/GROSV; **p.87** *t* © Alamy/ The Photo Library Wales/David Williams, *m* © Alamy/E C Photography, *bl* © Pictures Collection, State Library of Victoria, Australia. Unearthing the Welcome Stranger Nugget by W.Parker – Accession number: H13298-a14416-A400, *br* © The Art Archive/Bibliotheque des Arts Descoratifs Paris/Gianne Dagli Orti; **p.88** *tl* © Science Photo Library/E.R.Degginger, *tm* © Fotolia/Bruce, *tr* © Fotolia/Jim Mills, *bl* © Science Photo Library/Dirk Wiersma, *bm* © Science Photo Library/Cordelia Molloy, *br* © Science Photo Library/Arnold Fisher; **p.89** *from top to bottom* © Science Photo Library/Biophoto Associates, © Science Photo Library/Scientifica, Visuals Unlimited, © Science Photo Library/Edward Kinsman, © Science Photo Library/Scientifica, Visuals Unlimited, © Science Photo Library/Charles D. Winters, © Alamy/Arco Images GmbH, © Science Photo Library/Scientifica, Visuals Unlimited; **p.91** © Milepost/Railphotolibrary.com; **p.95** © Alamy/The Photolibrary Wales; **p.97** *l* © Science Photo Library/Robert Brook, *r* © Photolibrary/John Phillips; **p.99** *t* © Fotolia/Ionescu Bogdan, *m* © Fotolia/Demarco, *b* © Fotolia/Scanrail; **p.100** *lt* © Alamy/David J. Green, *lb* © Fotolia/Small Tom, *mt* © Fotolia/Robyn Mac, *mb* © Fotolia/Dmitry Vereshchagin, *rt* © Fotolia/Ian Holland, *rb* © Science Photo Library/Shelia Terry; **p.101** © Science Photo Library/Maximilian Stock Ltd; **p.103** *tl* © Fotolia/Elen, *tr* © Alamy/Graficart.net, *ml* © Alamy/John James, *mm* © Science Photo Library/Charles D. Winters, *mr* © Fotolia/Philippe Devanne, *bl* © Alamy/Phil Crean A, *bl* © Rex Features/ AGB Photo Library; **p.104** © Alamy/Peter Brogden; **p.105** *t* © Science Photo Library/BSIP,Laurent/B.Hop Ame, *b* Fotolia/GordonSaunders; **p.106** *l* © Alamy/studiomode, *r* © Fotolia/Studiotouch; **p.107** © Science Photo Library/Trevor Clifford Photography; **p.108** © NASA Kennedy Space Center (NASA-KSC); **p.110** © Alamy/sciencephotos; **p.112** *t* © Fotolia/Bsilvia, *b* © Fotolia/iggyphoto; **p.115** *both* © Science Photo Library/Ria Novosti; **p.116** *both* © Science Photo Library/Andrew Lambert Photography; **p.122** *l* © Science Photo Library/Nasa, *r* © Still Pictures/Peter Arnold/Ray Pfortner; **p.126** © Science Photo Library/Ria Novosti; **p.127** © Science Photo Library/Andrew Lambert Photography; **p.129** © NASA Jet Propulsion Laboratory; **p.130** *l* © Photo courtesy of Karl Bruun., *r* © Image courtesy of Sapphire Energy, Inc.; **p.133** © fotoLIBRA/Kevin Fitzmaurice-Brown; **p.140** *l* © Science Photo Library/Dr Steve Gull& Dr John Fielden, *r* © Science Photo Library/Tom Van Sant/Geosphere Project, Santa Monica; **p.141** © Topfoto/The Granger Collection; **p.147** © Science Photo Library/Adam Hart-Davis; **p.150** *lt* © Fototla/Vladstar, *lb* © Alamy/Richard Levine, *m* © Science Photo Library/Victor de Schwanberg, *rt* © Science Photo Library/Cordelia Molloy, *rb* © Alamy/Malcolm Park; **p.151** © Getty Images/Bo Tornvig; **p.152** © Alamy/Paul Glendell; **p.153** © Alamy/dbphots; **p.154** © International Power/Jeff Jones; **p.155** © Alamy/Paul Glendell; **p.156** © Corbis/Andrew Aitchison/In Pictures.; **p.160** © Fotolia/Deanm1974; **p.161** © Mike Vetterlein; **p.165** *t* © Fotolia/Richard Cote, *b* © Alamy/Rami Aapasuo; **p.166** © Alamy/Ange; **p.167** *all* photograph by courtesy of Ampair Energy Ltd.; **p.169** *t* © Institution of Mechanical Engineers, *b* © Fotolia/Richard Cote; **p.170** © Getty Images/Francois Durand; **p.171** © Alamy/Phil Broom; **p.172** *tl* © Fotolia/Tyler Olson, *tr* © Fotolia/Cpauschert, *bl* © Alamy/Ron Chapple Stock, *br* © Fotolia/Sreedhar Yedlapati; **p.175** © Science Photo Library/Tony McConnell; **p.176** © Topfoto/Topham Picturepoint; **p.183** © Fotolia/Magann; **p.188** © Science Photo Library/Martyn F. Chillmaid; **p.190** © Alamy/Stocktrek Images; **p.191** *t* © Alamy/redbrickstock.com/Andrew Sheild, *b* © Alamy/Simon Burt; **p.192** © Science Photo Library/Lionel Bret/Eurelios; **p.197** *lt* © Science Photo Library/NOAO, *lm* © NASA Jet Propulsion Laboratory, *lb* © Science Photo Library/Dr Leon Golub, *r* © Fotolia/SpaceHiker; **p.198** © ESA/SPIRE/PACS/P. AndrŽ; **p.199** © Science Photo Library/Julian Baum; **p.200** *t* © ESA/AOES Medialab, *m* © Science Photo Library/NASA/JPL-Caltech/J.Hora(Harvard-Smithsonian CfA), *b* © Alamy/Ashley Cooper; **p.201** *m* © Alamy/Dirk v. Mallinckrodt, *bl* © Science Photo Library/Royal Observatory, Edinburgh/AAO, *br* © Science Photo Library/NASA/JPL/CALTECH; **p.202** *tl* © Science Photo Library/NASA, *tm* © Science Photo Library/Sinclair Stammers, *tr* © Science Photo Library/Dr P Marazzi, *b* © Science Photo Library/Detlev Van Ravenswaay; **p.203** *t* © Science Photo Library/Maximilian Stock Ltd., *ml* © NASA/Swift/Stefan Immler, *mr* © NASA E/PO, Sonoma State University, Aurore Simonnet; **p.204** *l* © Science Photo Library/Stevie Grand, *r* © Science Photo Library/Prof.J. Leveille; **p.206** *tl* © FLIR systems Inc., *tr* © FLIR systems Inc., *b* © Getty Images/Keith D McGrew/US Army; **p.207** *l* © Science Photo Library/Tony McConnell, *r* © Getty Images/Sonny Tumbelaka; **p.208** © Empics Sports Photo Agency; **p.211** *t* © Photolibrary/Cultura/Tim Hall, *b* © Photolibrary/Andrew Watson; **p.212** © Science Photo Library/Mikki Rain; **p.215** © Fotolia/Goldenangel; **p.216** © Rex Features; **p.217** © Fotolia/Jane Songhurst; **p.219** *l* © Alamy/Robert Brook, *r* © Getty Images/Steve Allen; **p.221** © Getty Images/Timur Grib; **p.223** © Science Photo Library/Martin Dohrn; **p.225** © Science Photo Library/Gustoimages; **p.231** © Getty Images/Odd Andersen/AFP; **p.234** © Touchstone/Spyglass/The Kobal Collection; **p.235** © Science Photo Library/NASA; **p.237** *t* © Fotolia/Stephane Benito, *b* © Fotolia/The Supe87; **p.241** © Bibliothèque de l'Observatoire de Paris.; **p.244** © Science@NASA; **p.245** *t* © NASA/ESA/S. Beckwith(STScI) and The HUDF Team, *b* © NASA; **p.246** © Science Photo Library/Emilio Serge Visual Archives/American Institute of Physics; **p.249** *t* © Science Photo Library, *l* © Science Photo Library/Science Source, *r* © Science Photo Library/N.A. Sharp,NOAO/NSO/Kitt Peak FTS/AURA/NSF; **p.250** *t* © Getty Images/Science & Society Picture Library, *b* © Science Photo Library/Royal Astronomical Society; **p.253** © Science Photo Library/Physics Today Collection/ American Institute of Physics; **p.254** © NASA/WMAP Science Team.

t = top, *b* = bottom, *l* = left, *r* = right, *m* = middle

Every effort has been made to trace all copyright holders, but if any have been inadvertantly overlooked, the Publisher will be pleased to make the necessary arrangements at the first opportunity.

Introduction

This book has been produced to complement the new GCSE Science specification which is to be taught from September 2011. Although much of the content remains the same, the new course, and therefore this book, places much greater emphasis on 'How Science Works', and exam candidates will also be expected to be able to clearly explain scientific concepts. Many of the exercises in the book are designed to develop and test these abilities.

'How Science Works'

This term includes an understanding of the methods scientists use to investigate problems, involving experimental design, risk assessment, careful measurement, presentation and analysis of results, and the evaluation of methods used. It also involves an understanding of how original ideas become accepted theories, or are rejected as a result of new evidence. Ethical issues need to be considered, too, as science becomes capable of doing things that some people find unacceptable for moral rather than scientific reasons. The exercises and questions in this book focus very much on scientific enquiry skills and general aspects of 'How Science Works'.

Communication skills

Communication is vitally important in the scientific community. It is no good being able to understand the nature of (or answer to) a scientific problem if you cannot clearly explain it to others. Public misunderstandings of some scientific issues have resulted from the facts being poorly explained either by scientists themselves or by the media. In the new Science GCSE, candidates are to be tested on their communication skills and it will no longer be enough to just know the facts. The exams will contain questions that require extended writing, and there are exercises in the book which provide opportunities to develop and practise such skills.

How to use this book

The subject content of the new WJEC Science GCSE is fully covered, both at Foundation and Higher level. In addition, there are numerous exercises which fall into the following categories:

Practical work

Practical exercises have been chosen to help the understanding of concepts, but also include questions which focus on the scientific enquiry skills which are so important for success at GCSE.

Tasks

The tasks are exercises that usually involve the use of second hand information and data that could not be obtained in a school laboratory, along with questions that consider such things as experimental design, analysis of data and judgements about the strength of evidence.

Questions

Questions are scattered throughout the book to test understanding and application of concepts. We have not included past exam paper questions, for reasons of space but also because the exam papers for the new specification will look rather different from those of the past. It is possible to download past papers, if required, from the WJEC website.

Discussion points

Discussion points are questions that could be answered by individuals, but that benefit from discussion with others, either in peer groups or led by the teacher. In such cases there are usually a variety of opinions or possible answers.

For teachers

Guidance and support materials for teachers whose classes are using this textbook can be found on the WJEC website at http://www.wjec.co.uk/sciencegcse.

Tiering

Higher-tier material in this book is indicated by the presence of a green bar alongside text, questions and figures. All material without a green bar is required for Foundation-tier students.

1 Scientific enquiry skills

Isn't science all about learning facts?

Science is not just a load of facts that need to be learnt. Science is asking questions about the world around us and trying to come up with answers. Sometimes these answers can be found by careful observation. Sometimes we need to test out a possible answer (a '**hypothesis**') by doing **experiments**. Facts are useful, though. We need to know if someone else has already found the answer that we are looking for (if so, there would be no point in doing an experiment to find it out again, unless we wanted to check the answer was right). Scientific facts may also help us to come up with a hypothesis.

The people we call 'scientists' don't sit around learning facts. They use the facts they already know, or that they can find out by research, to ask questions, suggest answers and design experiments. Science is a process of enquiry, and to be good at science you have to understand and develop certain enquiry skills.

How does science work?

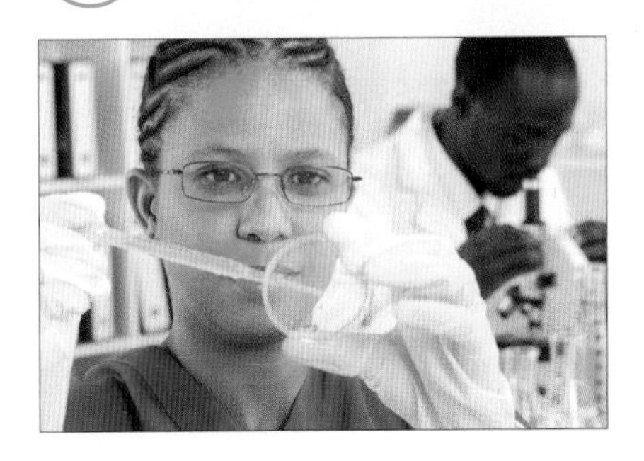

Figure 1.1 Science is all about doing experiments and making observations to find the answers to questions.

The way science is carried out and questions are answered is quite complicated and varied. A flowchart showing the ways in which scientists investigate things is shown on the following page. Not every question will involve *all* of these steps. The flowchart shows six skill areas that scientists need to develop:

1. an ability to ask scientific questions and to suggest hypotheses
2. experimental design skills
3. practical skills in handling apparatus
4. risk assessment
5. an ability to present data clearly and analyse it accurately (data handling)
6. communication skills.

This chapter will deal with these skills, which are essential for scientists to master.

How do I ask a 'scientific' question?

Sometimes you can ask a question, but there is no hope of getting a definite and undisputed answer. Look at these questions:

- Is there a God?
- What would be the best way to spend a lottery win of £10 000 000?
- Who is the greatest painter that ever lived?
- Is Paris a nicer place to live than London?

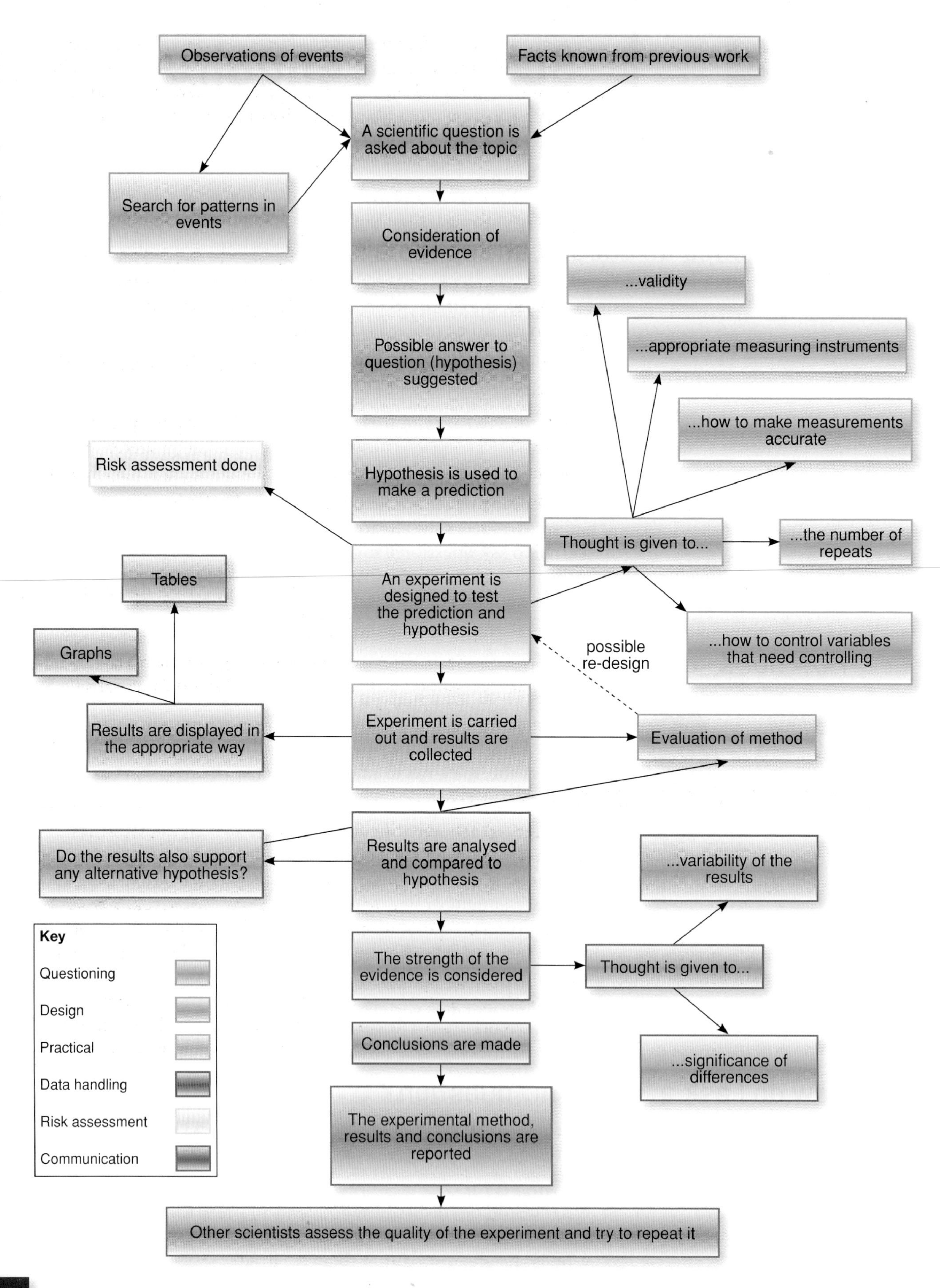

Figure 1.2 A model of how scientists work.

These are not scientific questions. Whether or not there is a God is a matter of faith and cannot be proved in a scientific sense. The other questions are all complex and open to different opinions. Scientific questions have the possibility of being answered by experiments.

Let's consider another question:

- How can I make the plants in my greenhouse grow better?

This is a scientific question but it's not a very good one. It is possible to answer it by experimenting, but you would have to do a lot of experiments because there are many factors (and combinations of them) that could affect plant growth. A better question would be more specific, such as:

- What effect does the temperature in my greenhouse have on the growth of the plants inside?

It is possible to find the answer to this by exposing the plants to different temperatures. It would be even better to specify one particular type of plant, because temperature may not have exactly the same effect on all the plants in the greenhouse.

QUESTION

1 Try to think up *one* scientific question about something in the world around you. There are two rules about this question:
- You must not already know the answer.
- You should not be able to find the answer by asking friends or family or looking it up in books or on the internet. You would only be able to find out the answer by doing some sort of experiment.

How do I design an experiment?

A good experiment will provide an answer to your question, or at least give information that will get you nearer to an answer. If you have a hypothesis, it will provide evidence as to whether the hypothesis is right or wrong (although it may not actually *prove* the hypothesis). We call experiments like this **valid** experiments. If the experiment design has any major faults, it will probably not be valid.

Two of the most important things that make an experiment valid are **accuracy** and **fairness**. If it's a fair test, and your results are accurate, then you are more likely to get the 'correct' answer.

TASK HOW DO I ENSURE A TEST IS FAIR?

This activity helps you with:
★ designing fair tests
★ judging whether not being able to control a variable makes an experiment invalid.

This is best shown by an example. Imagine a test to see how the speed of a car affects its braking distance. Let's think of all the variables apart from speed that could affect braking distance. It could be affected by:

- the car (some may have more efficient brakes than others)
- the tyres (some have a better grip)
- the person driving (some people may have quicker reaction times)
- the condition of the road (wet or icy roads are more slippery)
- the signal used to tell the driver to brake (one signal may be easier to notice than another)
- the slope of the road (you'll stop quicker going uphill)
- the time of day/light levels (in the dark, the signal to stop may be less noticeable unless it's a sound)
- how many times the driver has done the test (he or she might react quicker with more practice).

If the test is to be fair, we have to try to ensure that none of these things affect the experiment – so we would use the same car (with the same tyres), the same driver, the same stretch of road in the same conditions, and the same signal to stop each time.

The last two variables in the list are interesting. If the signal *was* a sound, then the light level becomes irrelevant and need not be controlled – you never need to control a variable that will not have any effect. You only control variables that are likely to affect the outcome. The last variable in the list is one that cannot be controlled. If you are using the same driver each time, the driver cannot help but get 'practice' at stopping. In this case it is unlikely to have much effect, but you would bear it in mind when you analyse the results, and look to see if there were any noticeable effects.

Imagine you wanted to find out if boys or girls were better at catching a ball. You would need to make sure that only the gender could be the cause of any effect.

1 What variables, apart from gender, could have an effect and would need to be controlled?
2 Are there any of these that cannot be controlled? If so, does this make the experiment a complete waste of time?

How do I make my measurements accurate?

Accurate measurements are defined as ones that are as near as possible to the 'true' value. The problem is, we don't know exactly what the true value is! So, it is impossible to be certain that a measurement is accurate. All we can do is make sure there are not any obvious inaccuracies.

Any instrument used for measurement should be as accurate as possible. It is usually best to use an instrument with a high **resolution**.

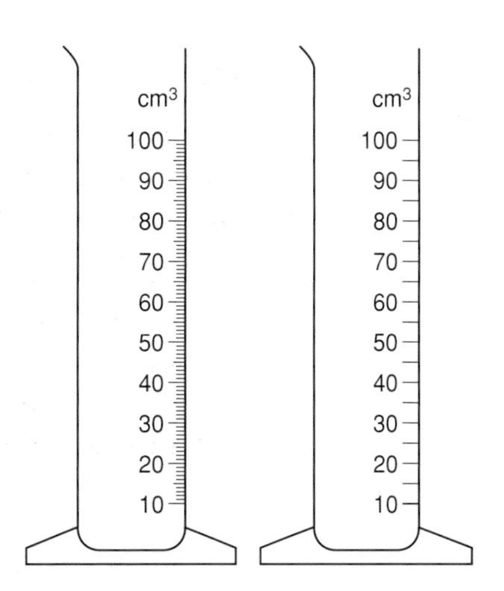

Figure 1.3 The measuring cylinder on the left has a higher resolution (i.e. smaller divisions) than the one on the right.

Inaccuracy can result from the units of measurement being imprecise. When measuring gas, for instance, counting bubbles is going to be inaccurate because the bubbles will not all be the same size, so that 25 bubbles in one case may contain more gas than 30 bubbles in another, if the first set of bubbles contains more big bubbles.

Inaccuracy can also result from human error caused by the means of measurement. If you are timing a colour change, for instance, it is often difficult to judge *exactly* when the colour changes, because it will be a gradual process.

Most measurements are not 100% accurate. This is acceptable as long as the inaccuracy is not so large that it makes comparisons of different measurements invalid.

In the 'counting bubbles' scenario above, for instance, if you have two readings of 86 bubbles and 43 bubbles then, even though there is inaccuracy, the difference is so large that the inaccuracy doesn't matter. If you had two readings of 27 and 32 bubbles, though, you could not confidently say that the second reading was actually larger than the first.

Figure 1.4 This photo shows how much bubbles vary in size. Therefore, 'one bubble' cannot be an accurate measurement of a volume of gas.

Why do scientists repeat experiments?

Scientific results always vary to some extent, and sometimes they vary a lot. Look at the two sets of results below.

Table 1.1 Number of paperclips lifted by an electromagnet.

	Trial 1	Trial 2	Trial 3	Trial 4	Trial 5	Average
Number of clips lifted	4	5	4	4	5	4.4

Table 1.2 Volume of gas produced in one minute by respiring yeast cells.

	Trial 1	Trial 2	Trial 3	Trial 4	Trial 5	Average
Volume of gas (cm^3)	8	25	13	28	21	19.0

The first set of results hardly varies at all, but the second set is very variable. In the yeast experiment, the first result appears to be quite unusual (a 'rogue'). If the readings were not repeated, the result would be highly inaccurate. Even after five readings, we still cannot be sure the average is accurate because the results are so variable.

In the electromagnet experiment, however, even if we took the first result and did not repeat it, it would be quite accurate, because the results hardly vary at all (but of course we need to do a few repeats to actually discover that).

So, consistent results require only a few repeats, and the more variable the results are, the more repeats are necessary.

What is the best way of presenting results?

When scientists record their results, it is important that these are done in a way that is clear to anyone reading their report. Generally, results are presented in tables and usually in some form of graph or chart.

Tables

Tables are ways of organising data so that they are clear and the reader does not have to search for them in the text. If you need to look at the method to see what the table means, the table is not doing its job.

- Tables must have clear headings.
- If the measurements have units, these should be shown in the column headings.
- Tables must have a logical sequence to the rows and columns.

QUESTION

2 This table shows the amount of carbon dioxide produced per minute when hydrochloric acid is reacted with a carbonate at different temperatures. It does not fully meet the criteria listed above and can be improved.

Temperature	Trial 1	Trial 2	Trial 3
20	12 cm^3	14 cm^3	16 cm^3
40	43 cm^3	49 cm^3	41 cm^3
50	79 cm^3	75 cm^3	82 cm^3
30	22 cm^3	22 cm^3	27 cm^3
60	142 cm^3	138 cm^3	150 cm^3

Redraw this table so that it meets the criteria for a good quality table.

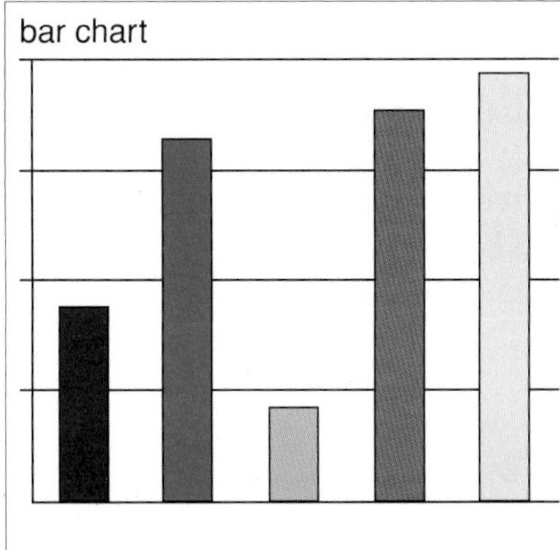

line graph

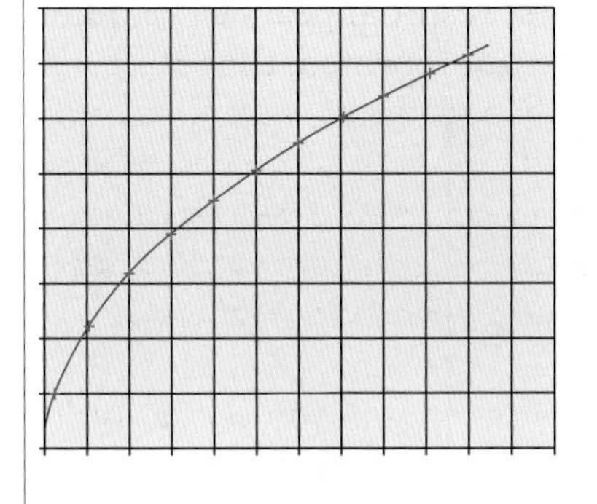

pie chart

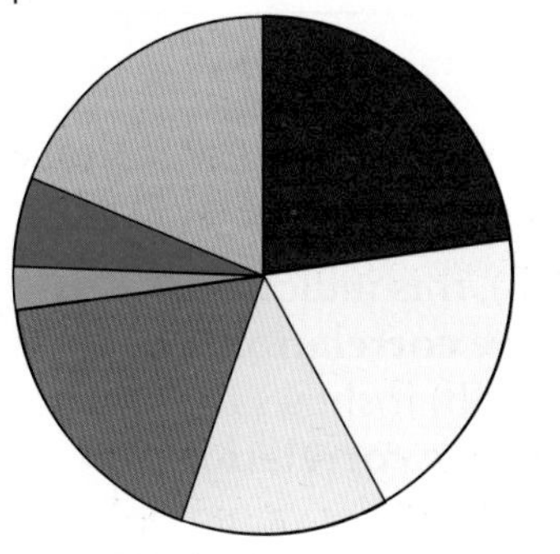

Figure 1.5 Data can be displayed in a variety of ways.

Graphs and charts

Graphs and charts are of several types, but the three types most often used are bar charts, line graphs and pie charts.

- **Bar charts** are used when the values on the *x* axis are a **discontinuous variable** (no intermediate values), e.g. months of the year, eye colour, etc.
- **Line graphs** are drawn when the *x* axis is a **continuous variable** (any value is possible), e.g. time, pH, concentration, etc.
- **Pie charts** are used to show the make-up of something. Each section represents a percentage of the whole.

Bar charts and line graphs are drawn to show patterns or trends more clearly than a table would. Once again, the graph should display everything necessary to identify the trend, without any need to read through the method.

A good quality bar chart or line graph must have the following:

- a title
- both axes clearly labelled with units where appropriate
- a 'sensible' and easy-to-read scale for each axis
- use of as much of the space available as possible for the scale (without making it awkward to read)
- axes the correct way round. If one factor is a 'cause' and the other an 'effect' then the cause (the **independent variable**) should be on the *x* axis and the effect (the **dependent variable**) should be on the *y* axis. Sometimes, the relationship is not 'cause and effect' and the axes can be either way round
- accurately plotted data
- clearly distinguished sets, if more than one set of data is plotted, plus a key to show which set is which
- in a line graph, if the data follow a clear trend, this should be indicated with a **line of best fit**. If there is no clear trend, the points should be joined by straight lines, or left un-joined.

How do I analyse results?

Results are usually analysed for one of three purposes:

1 To identify relationships between two or more factors.
2 To decide if a hypothesis is likely to be correct.
3 To help to create a hypothesis.

Relationships

Relationships are most clearly shown by line graphs. The direction the line slopes (or doesn't) indicates the type of relationship. There may be two or more different types of slope on some graphs.

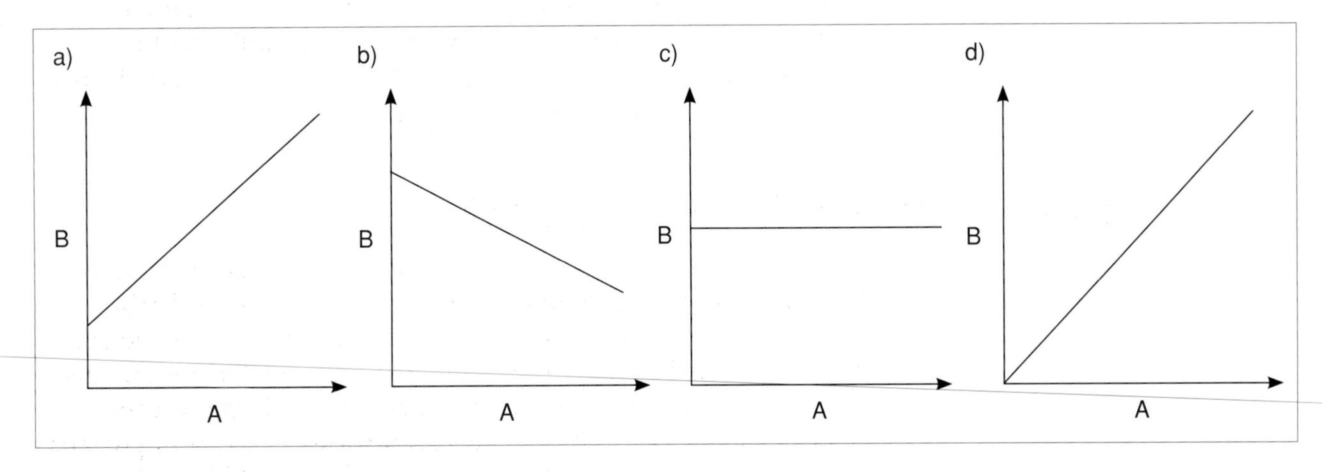

Figure 1.6 Line graphs can show different relationships between the variables.

When the line slopes upwards (Figure 1.6a), this indicates that as A increases, so does B. This is called a **positive correlation**.
When the line slopes downwards (Figure 1.6b) it shows that when A increases, B decreases. This is called a **negative correlation.**

If the line is horizontal (Figure 1.6c), it means that values of B are unrelated to A, and there is **no correlation** between the variables.
If the graph forms a straight line and it goes through the origin (Figure 1.6d), this is called a **proportional relationship**.

Just because two factors may show a relationship, it does not mean that either of them actually *causes* that relationship. If B goes up when A increases, it does not mean that the increase in A is what *makes* B go up.

A relationship may suggest or support a hypothesis, but to draw a firm conclusion you need to judge the strength of the evidence provided by the data.

How do I judge the strength of evidence?

To be confident that any conclusion you make is correct, you need strong evidence. Weak evidence does not mean your conclusion is wrong, it simply means you can be less sure it is right.

To judge the strength of evidence, you need to ask certain questions.

1 **How variable were the results?** The more variation in repeats, the weaker the evidence will be.

2 **Were enough repeats done? Was the sample big enough?** Variable results can still provide good evidence if the number of repeats or the sample size was big enough. You need to be certain that the results you have got were not 'freak' results. Freak results don't happen very often, so lots of repeats, or a big sample, mean that you will get a more accurate overall picture of what is happening.
3 **Were any differences significant?** Small differences may be just due to chance, as scientific measurements often cannot be perfectly accurate. Sometimes it is obvious that differences are or are not significant. If not, scientists may do statistical tests to measure how significant a difference is.
4 **Was the method valid?** Faults in the method (for example, inaccurate ways of measuring, variables that could not be controlled) reduce the strength of the evidence. Major faults may mean that the conclusions are completely unreliable.

Why do scientists publish their results?

However good an experiment might be, and however conclusive the results, scientists do not accept any conclusions until other scientists have had a chance to study what has been done and the results obtained to check the experiment and conclusions are valid. Others also need to repeat the experiment to see if they get the same results (this is known as **corroboration**). It is very important that scientists have good written communication skills so that others can understand their experiments.

Why are risk assessments necessary?

Every science experiment needs a risk assessment. Many use substances or equipment that can cause harm to the person doing the experiment. Risk assessments are done so that the experimenter is aware of any dangers before starting. They also offer advice on ways to avoid risks.

A good risk assessment does the following:

- It identifies all the **hazards**.
- It identifies the **risks** associated with those hazards.
- It gives some indication of the likelihood of each risk occurring.
- It advises **precautions** that will reduce or avoid the risk.

A hazard is something that is potentially harmful. This may be a chemical or a piece of apparatus. When identifying a hazard, the nature of the hazard should also be made clear, for example, not just 'hydrochloric acid', but 'hydrochloric acid is corrosive and can damage the eyes and skin'.

A risk is the action that might result in the hazard causing damage. Dilute hydrochloric acid in a bottle on the bench is a hazard but it is not a risk. Pouring it into a beaker is unlikely to be a risk as there is no significant likelihood of it getting in your eyes, and even if dilute acid gets on your skin it will not cause any damage. If the acid was concentrated, pouring it into a beaker would be a risk because it could burn your skin if it ran down the side of the bottle onto your hands. There is a risk of getting dilute hydrochloric acid into your eyes via your hands, though.

When using dilute hydrochloric acid in this context, then, the precaution should be: 'Avoid getting acid on your hands. If you do, do not touch your eyes and wash your hands immediately'.

QUESTIONS

3 For each of the statements below, say if it is a hazard or a risk:
  - a Powdered amylase is an irritant if breathed in.
  - b Iodine is poisonous if swallowed.
  - c When you pour the acid into the burette, it could splash into your eyes.
  - d Sodium hydroxide produces a lot of heat when dissolved in water.
  - e If you handle the test tube after heating it with the fuel, you could burn your hand.
  - f If you open the Petri dish after incubating it, you might breathe in harmful bacteria.
  - g Scalpels are sharp and can easily cut the skin.
  - h Thin wires can get very hot when a current is passed through them.

4 Look at the warning sign from a railway bridge.
Suggest reasons why this sign is less useful than it could be, and suggest some improvements.

PRACTICAL HOW GOOD A SCIENTIST AM I?

This activity helps you with:
★ practising a wide range of scientific enquiry skills.

This chapter outlines some of the skills that are needed to be a good investigative scientist. Listed below are some scientific questions. Pick one of them. Design and carry out an investigation to try and find the answer.

- Some people say that cut flowers last longer in lemonade than in water. Is this true?
- When asked, more people say they like Coke™ better than Pepsi™. However, it may be that they are just saying that because they think Coke™ has a better 'image'. Find out which brand people in your class say they prefer, and then test if that is really the case.
- In hot weather, is it best to wear black or white clothing to keep cool?
- Is there any difference between boys and girls in their ability to do Sudoku puzzles?
- How acidic does rainwater need to be to affect limestone buildings?
- If you want to design a sign to attract people's attention, what would be the best colour combination of background and text to use?
- Does anyone in your class have extra-sensory perception? Test if they can accurately predict what geometric shape you are looking at without seeing it themselves.
- It has been suggested that if you drop food on the floor, it is safe to eat provided you pick it up within five seconds, as bacteria will not have had time to get onto it in large numbers. Is this correct?

Chapter summary

- Scientists investigate the world around them by a complex process of enquiry.
- Not all questions can be answered by science.
- To be of any use, an experiment must be fair and valid, and measurements must be as accurate as possible.
- If a variable cannot be controlled, the likely effect of not controlling it must be taken into account when analysing results.
- The more variable the results are, the more often the experiment needs to be repeated.
- Presenting data in tables makes it clearer than in text. The table should be constructed so that it is clear and the reader does not need to refer back to the method to see what it means.
- Line graphs and bar charts are used to make trends and patterns in the data clearer.
- Line graphs are used if both variables are continuous. Bar charts are used when the independent variable is discontinuous.
- The shape of a line graph indicates the nature and the strength of any trend or pattern.
- Evidence varies in strength. Stronger evidence makes the conclusion more certain.
- Scientists publish their results so that others can check their methods and conclusions. For this to happen, scientists need to write precisely and clearly.
- All experiments must have a risk assessment before they can be carried out.
- A good risk assessment identifies hazards and risks, and clearly explains how the risks can be reduced or avoided.

Variety of life, adaptation and competition

Biodiversity

In any small area of land, you will find many different kinds of living things. In a garden, for example, you may come across earthworms, slugs, snails, centipedes, insects, spiders, birds, and even moles and voles. There are very obvious differences between these creatures. The same is true of plants. Your garden may provide a home for mosses, ferns and plants with flowers, for example. Millions of types of living things are already known to science, and many others are being discovered each year. At the moment, about 1.75 million species have been found, but scientists estimate the total number of species on the planet to be somewhere between 3 million and 100 million!

Biodiversity is the name given to the variety of life on Earth. It includes the different species that are found, the genetic variations within those species, and the different ecosystems, such as those occurring in woodland, grassland, desert, the sea, rivers, etc.

Living things vary in a number of ways. A bacterium has very little in common with an elephant, for instance – it is much simpler, much smaller and has completely different features. On the other hand, horses and donkeys are similar in many ways, even though they have a number of differences.

People have always tended to group similar living things together and give that group a name. Some of the groups you are likely to have heard of are shown in Figure 2.1.

QUESTION

1 'Invertebrates' include all animals (such as worms, snails, crabs, insects, starfish, and spiders) that do not have a backbone. It is not a group that is used by scientists to classify animals. Which of the two possible reasons below do you think is the *best* explanation of why the group is not used for classification? Justify your answer.
   - a Invertebrates don't have a common feature. The lack of a backbone is not a feature, it is just the *absence* of a feature.
   - b The group contains too much variety – the animals in it are very different.

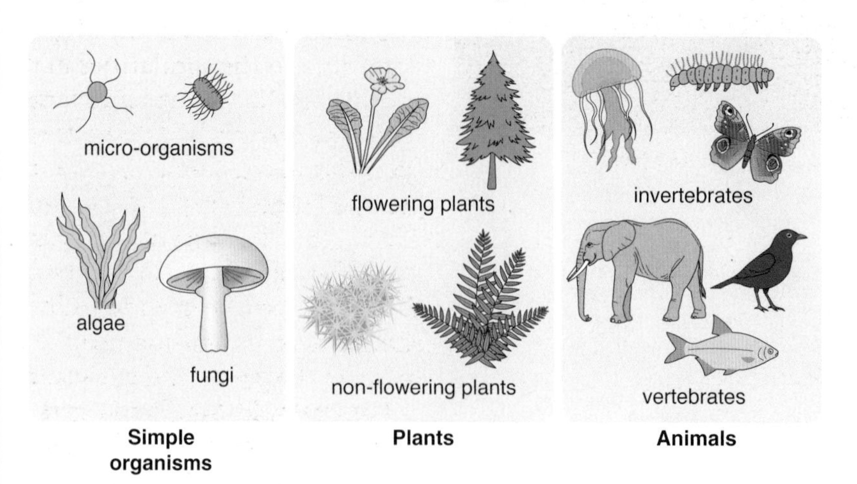

Figure 2.1 The variety of living things.

The principle of classification

Long ago, scientists decided that it was necessary to group living things into groups that had features in common, to make them easier to study. Things discovered about one member of a group might apply to the others, for instance. It was also necessary to give each living thing a name that would be recognised across the whole world, instead of 'common' names that vary in different places.

A good example of why agreed names are needed is the woodlouse. There are 42 species of woodlice in Britain, and in different places they have different names, such as monkey pea (Kent), grammersow (Cornwall), chucky pig (SW England), cheeselog (Berkshire), Granny Grey (S.Wales), parson's pig (Isle of Man). Across the world, many other names are used. It is possible that two people from different regions could be talking about the same species but using completely different names, and within one region several different species of woodlice could all be given the same name.

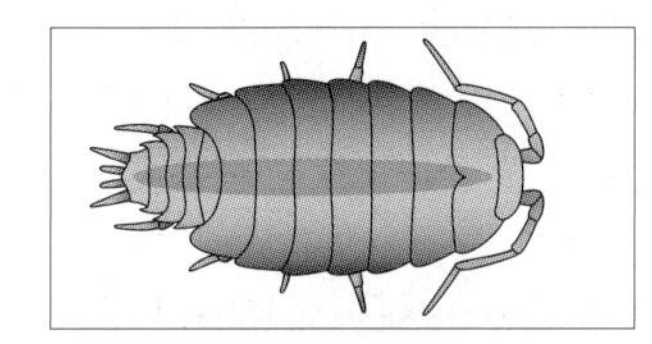

Figure 2.2 The common striped woodlouse, *Philoscia muscorum*.

Figure 2.2 shows one species of woodlouse, called the common striped woodlouse, but known across the world by its scientific name (in Latin) of *Philoscia muscorum*.

The question is, how did this way of naming living things come about, and who decides on what a new species should be called?

Carolus Linnaeus

Figure 2.3 Carolus Linnaeus.

Carl von Linné (1707–78) a Swedish scientist, was the first to devise a logical way of naming living things. He had a love of Latin and called himself Carolus Linnaeus.

Scientists had already decided that each animal or plant should have a scientific name that was agreed on, and that the name should be in Latin, which was the language used in universities at the time. However, the names were given individually and were not logically organised so that similar things had similar names. Linnaeus was a doctor, and at that time a lot of plants were used in medicines. It was very important that doctors used the right plants, and having a scientific system for naming them was clearly important.

Linnaeus devised a way of grouping (classifying) plants according to the similarities in their reproductive organs, and put the smaller groups into larger groups to produce a logical classification method. Similar species were grouped into a **genus** (plural genera), similar genera formed an **order**, orders were grouped together in a **class**, and similar classes made a **kingdom**. This is still the basis of the classification system that scientists use today, although some extra groups like **families** and **phyla/ divisions** have been added. Linnaeus also decided that every species should have a name made up of two words. The first word would indicate the genus and the second the species.

For example, lions and tigers are separate but closely related species. They belong to the same genus – *Panthera*. The scientific name for the lion is *Panthera leo* and that of the tiger is *Panthera tigris*.

Scientific ideas change as new information becomes available. Linnaeus's method of classifying plants according to their reproductive organs alone has been replaced by methods involving many more features. New groups have been added to the ones he suggested, but his system of naming living things has remained unchanged. Modern classification does not link organisms just by features, but also takes into account how the animals or plants evolved. Linnaeus thought that species always remained the same, and it was only when Charles Darwin proposed his theory of natural selection that scientists realised that physical similarities can indicate how closely related species are in terms of their evolution.

QUESTIONS

2 Here are the scientific names of three animal species:
Common frog – *Rana temporaria*
Painted frog – *Discoglossus pictus*
Stream frog – *Rana graeca*
  a Which two of these frogs are the most closely related? Give a reason for your answer.
  b Do any of these frogs belong to the same species? Give a reason for your answer.

3 a Based on the information above, which do you think was the most successful of Linnaeus's ideas?
    A his system for naming organisms
    B his way of classifying plants
    C his idea of grouping living things into gradually smaller and smaller groups.
  b Justify your choice.

4 Why did Linnaeus decide to give species Latin names?

5 Use the internet to find out when Charles Darwin proposed his theory of natural selection.

Figure 2.4 Can you tell the difference between a hoverfly and a wasp? Which is which?

Carolus Linnaeus came up with many names of species that have remained the same, but new species are constantly being discovered and have to be named. Nowadays, the names for animals are decided by an organisation called the International Commission on Zoological Nomenclature (ICZN). Plant names are not decided by one organisation but are named according to an internationally agreed code.

DNA and classification

Linnaeus and many later scientists had to rely on visible features to classify living organisms. Nowadays, scientists are able to study the DNA and genes of different species and look for similarities there. If two organisms share a high percentage of their genes it is likely that they are closely related, sharing a common ancestor in the not too distant past. This can be a better way of deciding relationships than physical features, because sometimes an animal evolves to 'mimic' another one and so looks quite similar even though there is no close relationship. An example of this is seen in hoverflies, which have evolved to look like wasps (because predators are wary of wasps!) but, according to their DNA, they are not closely related to them.

TASK LOOKING AT EVIDENCE

This activity helps you with:
- ★ interpreting data
- ★ drawing conclusions.

Studies have been done on the genomes (genetic make-up) of humans, chimpanzees, orang-utans and gorillas to see how closely related humans are to these species. The results are shown in Table 2.1.

Table 2.1 Comparison of the genetic make-up of different primates with humans.

Species	% of genes the same as humans
Gorilla	98.48
Orang-utan	98.47
Chimpanzee	98.76

From these data it is possible to create an 'evolutionary tree' to show how closely related humans and these great apes are. The branches show how long ago the species separated in evolution. The more similar the genes of two species are, the more recently they separated. Several different versions of these trees are shown in Figure 2.5.

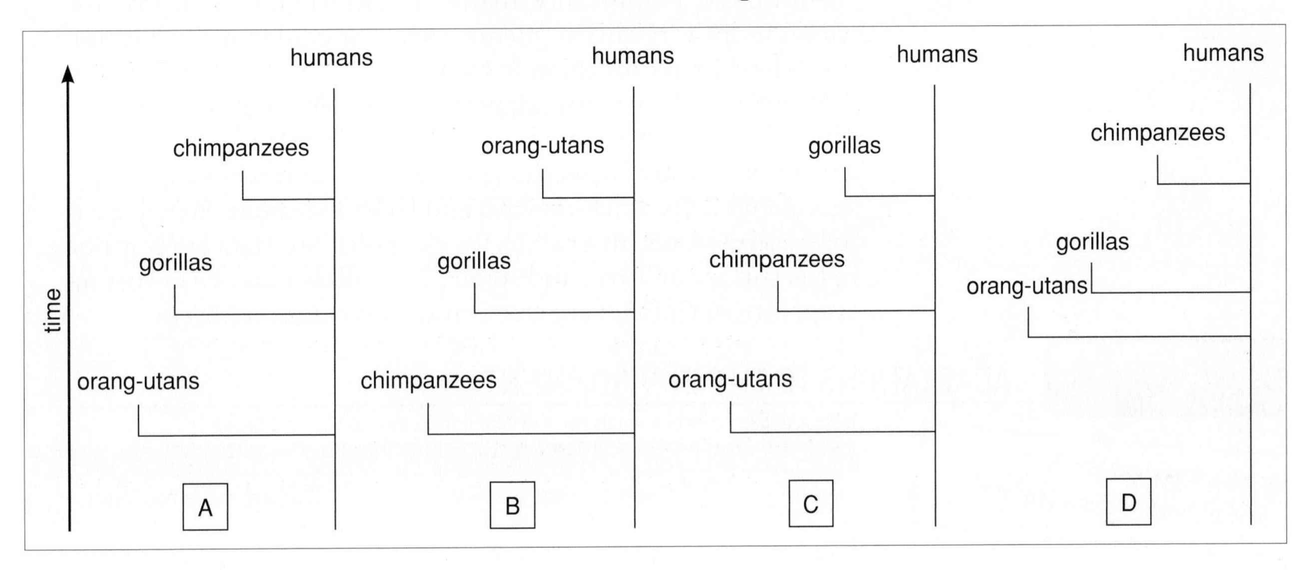

Figure 2.5 Different evolutionary trees.

1. For each tree, say whether the evidence in the table supports the shape of the tree.
2. On the basis of the evidence, which tree do you think is the most accurate?

Adaptations to the environment

All organisms living today have become **adapted** to their environment in many specialised ways. Each part of the Earth has its own special animals and plants. For instance, the arctic fox is closely related to the desert fox but they have both developed adaptations to help them survive. The large ears of the desert fox help to radiate heat away from the animal. In contrast, the very small ears of the arctic fox allow it to retain as much heat as possible. The red fox is found in Wales and other European countries. It has ears that are not obviously very large or very small because it does not have to survive at extreme temperatures.

Figure 2.6 a) Arctic, b) desert and c) red foxes are closely related but have different adaptations according to where they live.

As well as adaptations to their bodies, called **morphological** adaptations, animals often adapt their behaviour to suit their environment. For instance, many animals will hibernate over the winter months in environments where the cold weather means that it will be difficult for them to feed.

Plants too show many adaptations to their environment. The ability to survive in dry conditions depends on how good a plant is at retaining water. The reduction of leaves to spines, succulent stems, thick **cuticles** and lack of **stomata** are all used to conserve water (in a cactus for example). Stomata are tiny holes in the surface of leaves and stems. They allow water to be lost by evaporation. Cuticles are waterproof coverings on leaves.

TASK ADAPTATIONS TO THE ENVIRONMENT

This activity helps you with:
★ researching information.

Table 2.2 Examples of animals living in different environmental conditions.

Environment	Conditions	Examples
Desert	Extreme heat; shortage of water	Kangaroo rat; camel; desert iguana
Arctic	Extreme cold; covering of snow and ice	Polar bear; caribou; arctic hare
Savannah	Hot; open, so easy for predators and prey to be seen by each other; dry for most of the year except for wet season; food shortage in the dry season	Zebra; lion; meerkat

For each environment listed in Table 2.2, research *one* animal (either from the examples provided in the table or choose your own). For each animal, find and record *one* **morphological adaptation** and *one* **behavioural adaptation** to its environment.

The need for resources

In order to survive, all organisms need a supply of energy. Plants get their energy directly from sunlight by photosynthesis; animals have to get their energy from food (i.e. from other living organisms). Energy is needed to carry out all living processes but a supply of raw materials is also needed for chemical processes and to build bodies. Any environment which is to sustain life needs an adequate supply of these materials. Energy constantly enters most ecosystems in the form of sunlight, but the raw materials are limited and have to be reused over and over again.

The resources needed by living things are summarised in Table 2.3.

Table 2.3 Resources needed by living things.

Resource	Needed
Light	by plants to make food for energy
Food	by animals for energy
Water	for all the chemical reactions in the bodies of plants and animals
Oxygen	to break down food to release its energy
Carbon dioxide	by plants for photosynthesis
Minerals	different minerals are needed for certain chemical reactions in the bodies of plants and animals

Competition

There is always a limited amount of energy coming into an ecosystem and a limit to the other resources available. This puts a limit on the number of living things an ecosystem can support. The organisms have to compete with each other for the resources and those that are better at this competition will survive better than the rest. Competition always takes place between members of the same species, because they all require the same things (e.g. they eat the same food) but also occurs between different species with similar needs.

TASK COMPETITION IN FLOUR BEETLES

This activity helps you with:

- ★ understanding control experiments and fair testing
- ★ drawing conclusions
- ★ judging the strength of evidence.

Scientists kept two very similar species of flour beetle (the red flour beetle and the confused flour beetle) in some flour, which provides them with both food and an environment in which to live. They found that after about 350 days, the red flour beetle had died out, leaving only the confused flour beetle. They also kept each beetle on their own in separate samples of flour. Some of their results are shown in Figure 2.7.

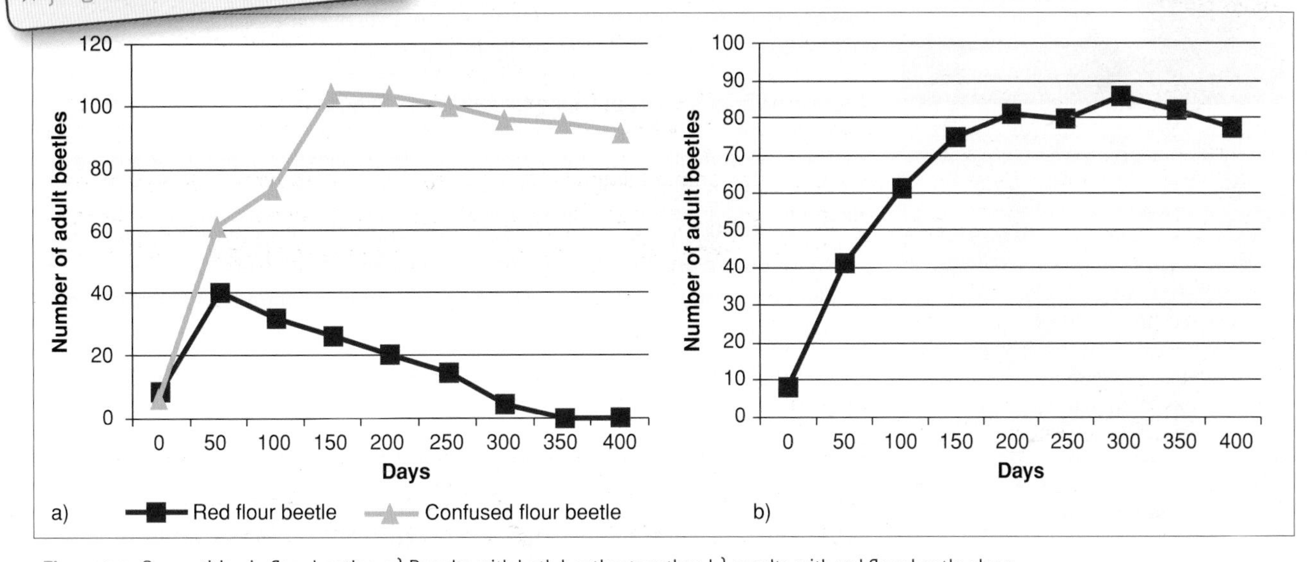

Figure 2.7 Competition in flour beetles. a) Results with both beetles together; b) results with red flour beetle alone.

The scientists concluded that the flour beetles were competing for a resource, and that the confused flour beetle was more successful in this competition than the red flour beetle.

continued...

1. Why was it important for the scientists to study how the red flour beetle population grew when it was on its own?
2. What resource(s) might the two beetles be competing for?
3. How strong is the evidence that there is competition between the two species of beetle? Explain your answer.
4. Suggest a reason why *both* species increased in the first 50 days.
5. The scientists kept the temperature of the flour the same throughout the experiment. Suggest why they did that.

TASK

COMPETITION IN SQUIRRELS

This activity helps you with:
- ★ judging ethical issues
- ★ communicating ideas clearly
- ★ considering opposing views and forming an opinion.

There are two species of squirrel in Britain – the red and the grey squirrel. Both live in broadleaved deciduous woodland (made up of trees like hazel, oak, ash and beech). The red squirrel is native to this country but the grey squirrel was introduced into Britain in the late nineteenth century. It is able to out-compete the red squirrel in most habitats, although the red squirrel does do slightly better in coniferous woodland, where it feeds on pine cones.
The reasons the grey squirrel does so well are:

- The grey squirrel can eat acorns which are very common in broadleaved woodland, but the red squirrel cannot. This means the grey squirrel has access to much more food.
- The grey squirrel seems to be able to store more fat, and so is better able to survive periods of food shortage.
- The grey squirrel feeds mainly on the forest floor and the red squirrel mostly in the trees. As seeds and nuts often fall from the trees when they are ripe, this means that again the grey squirrel has more food available.
- The grey squirrel carries a disease which it is resistant to, but which kills the red squirrel.

Figure 2.8 a) red squirrel; b) grey squirrel.

There are some areas of the UK where the red squirrel is still found, and efforts are being made to conserve it. If the grey squirrel becomes established in these areas, the red form is likely to die out.

There are various ways of trying to conserve red squirrels:

1. Trapping and killing as many grey squirrels as possible in areas where red squirrels are found.
2. Removing all the oak trees in woodlands where the red squirrels are found, so that there are no acorns for the grey squirrels to feed on.
3. Managing coniferous forests where red squirrels are found to ensure that no oak trees start to grow there.
4. Studying the virus that grey squirrels carry to see if there is a way to stop it being spread so easily to red squirrels.

Each of these methods has advantages and disadvantages, and no one method is likely to succeed on its own.

Hold a class debate to discuss the pros and cons of each of these conservation measures.

PRACTICAL COMPETITION IN CRESS SEEDS

This activity helps you with:
★ following instructions
★ presenting results
★ drawing conclusions
★ understanding about repetition of experiments.

This experiment looks at the effect of competition for water on the germination of cress seeds.

Apparatus
* 2 sheets of filter paper
* 2 Petri dishes
* syringe
* cress seeds
* forceps

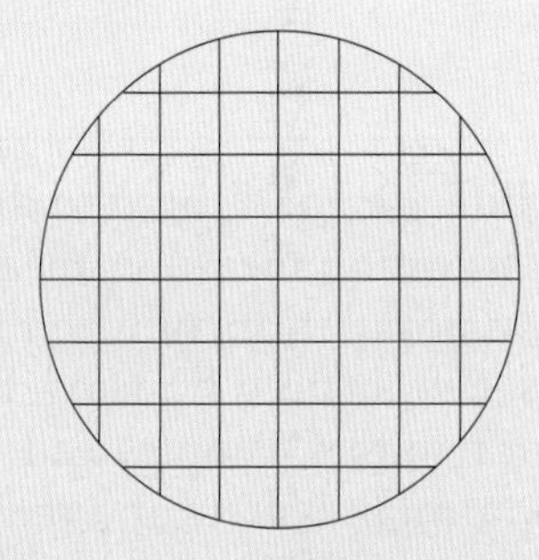

Figure 2.9 A grid of lines 1 cm apart, drawn on a filter paper circle (not to scale).

Procedure

1. Prepare two sheets of filter paper. Cut each so that it fits inside the lid of a Petri dish. Draw a grid on each piece of paper, as shown in Figure 2.9.
2. Put each piece of filter paper into a Petri dish lid.
3. Using a syringe, add enough water to dampen all of the filter paper in one Petri dish. Note how much water you used.
4. Add the same amount of water to the second Petri dish.
5. Using forceps, carefully place one cress seed inside each square on the filter paper in the first Petri dish. Try to put the seed in the middle of the square. Count and record the total number of seeds used.
6. Put three seeds into each square in the second Petri dish. Count and record the total number of seeds used.
7. Cover the lids of the Petri dishes and leave for 3–4 days to germinate.
8. Record the percentage germination of the seeds in each dish.

$$\%\text{ germination} = \frac{\text{number of seeds germinated}}{\text{total number of seeds}} \times 100$$

9. Draw conclusions and explain your results.
10. Stephanie and Ivan did the experiment. Stephanie thought the experiment should be repeated, but Ivan said that because they had used lots of seeds, they had already repeated it. Discuss the merits of each idea.

Chapter summary

- Living things vary from one another in many ways.
- Organisms that have similar features and characteristics can be classified together in a logical way.
- Similarities in the DNA of organisms can help scientists to classify them.
- Scientists give every species a scientific name that is used worldwide.
- These names are made up of two words; the first indicates the organism's genus, the second its species.
- An international committee decides on the scientific names of animals; the scientific names of plants are decided using an internationally agreed code.
- Organisms adapt and develop structural and behavioural adaptations that help them survive in their environment.
- All living things need energy.
- Energy enters ecosystems in the form of sunlight, which can be used directly by plants.
- Animals get their energy by consuming other living organisms (either plants or animals).
- A variety of chemical 'raw materials' is also needed for life processes.
- Organisms of the same species will always compete with each other for sunlight, food or other raw materials, because their needs will be the same.
- Competition also occurs between different species if they have similar needs.
- If a species cannot compete successfully for a resource that is in short supply, it may die out.

The environment

Are you concerned about the environment?

Nowadays, many people are concerned about 'the environment'. Before we think about whether we're concerned about it, we need to be certain what the environment is. For biologists, the environment is the place where living things live (known as a habitat) together with all the living things in it, and the physical factors that affect it (its 'climate'). All of these things are interlinked, and changes in one aspect can have effects on the others. In this chapter we will look at some of these interactions.

Should we always conserve wildlife?

Most people think that conserving wildlife is a good thing to do. The issue can be complicated when conservation measures mean a significant disadvantage for the human population, or when conserving one species can harm others. People need housing, but building it can destroy animal and plant habitats. People need food, but converting wild habitats into farmland alters and reduces the numbers of species that live there. The world needs to build up alternative sources of energy but the development of things like tidal barrages and hydroelectric power stations can destroy or completely alter natural habitats. It is also worth remembering that over the whole of history, environments have constantly changed – it is not 'natural' for everything to stay the same.

In order to make any decisions about the effect of anything humans do on the environment, we need to have scientific information about the state of the environment and any changes that are happening to it. In the UK, much of the environmental monitoring is done by the **Environment Agency**.

Figure 3.1 Are you happy with your environment?

TASK WAS THE CARDIFF BAY BARRAGE A GOOD IDEA?

This activity helps you with:
- ★ research skills
- ★ considering the strength of evidence
- ★ considering ethical issues in science.

Between 1994 and 1999, a barrage was constructed across Cardiff Bay, making a freshwater lake 2 km^2 in area. When it was proposed, there was a lot of opposition from environmentalists because it would have a big effect on the wildlife of Cardiff Bay.

Look at the information below, which summarises the good and bad effects of the barrage.

Figure 3.2 Cardiff Bay before and after the barrage was constructed.

Benefits

- The freshwater lake looks a lot better than the mudflats that were there before.
- Because the area is attractive, a huge amount of development has occurred around the bay. This has commercial benefits for Cardiff because of tourism, and has created a lot of jobs.
- The freshwater lake is a valuable leisure facility, where people can enjoy a variety of water sports.
- Alternative feeding grounds have been created in the Usk estuary for the birds that used to feed on the mudflats.
- A 'fish pass' has been installed to allow salmon to travel up the river from Cardiff Bay to breed, as they did before the barrage was built.

Disadvantages

- Many of the shore birds have not used the alternative site and have left the area. Although many of these birds are common in other areas of the UK, Cardiff Bay was considered to be a 'site of international importance' for migrating birdlife.
- One bird, the common redshank, has set up a colony in the nearby Rhymney estuary, but the feeding does not seem to be so good there, and fewer of them survive the winter.
- The barrage was very expensive to build (about £200 million) and costs nearly £20 million a year to maintain.
- The creation of the alternative bird feeding site required 1000 acres of farmland to be flooded.

1 Hold a class debate on the topic: 'It was right to build a barrage across Cardiff Bay'.

2 Since moving their feeding grounds, the survival rate of the common redshank has gone down from 85% to 78%. It is possible that this is because the food supply on the new feeding ground is not as good as in Cardiff Bay. What information would scientists need to collect before they could confirm this hypothesis?

What pollutants are in our environment?

A **pollutant** is something that has been added to the environment and which damages it in some way. There are many types of pollutant. Pollutants are not necessarily 'unnatural', as some natural substances can be harmful if introduced in the wrong place or in large quantities.

Some common pollutants are:

- solid or liquid chemicals, such as oil, detergents, fertilisers, pesticides and heavy metals
- gaseous chemicals such as carbon dioxide, methane, chlorofluorocarbons (CFCs), sulfur dioxide and nitrous oxides
- human and animal sewage
- noise
- heat
- non-recyclable household waste.

It is impossible to prevent pollution but we must try to limit the levels of pollution so that they do little or no permanent damage to the environment.

TASK

HOW CAN WE TELL HOW POLLUTED THE ENVIRONMENT IS?

This activity helps you with:
- ★ analysing data
- ★ judging the strength of evidence
- ★ drawing conclusions.

Some pollutants can be measured directly, and pollution can also be detected by a *fall* in the **oxygen level** or **pH** of the water in streams and rivers. Scientists can often judge the overall level of pollution by using **indicator species** (Table 3.1). Some plants and animals are more tolerant of pollution than others. In an environment there are certain species you would expect to find. If some expected species are absent, this can give an idea of how polluted the environment is.

Table 3.1 Insect indicators of pollution levels in streams.

Water quality	Insects present
Clean water	Stonefly nymph, mayfly nymph
Low level of pollution	Freshwater shrimp, caddis fly larva
Moderate pollution	Water louse, bloodworm
High pollution	Sludge worm, rat-tailed maggot

Figure 3.3 shows a stream that scientists were studying for pollution. Two farms, Mill Farm and Tipton Farm, were near the stream and it was thought that sewage from one or both farms might be getting in to the stream. The stream was sampled at five places, labelled A–E on the diagram.

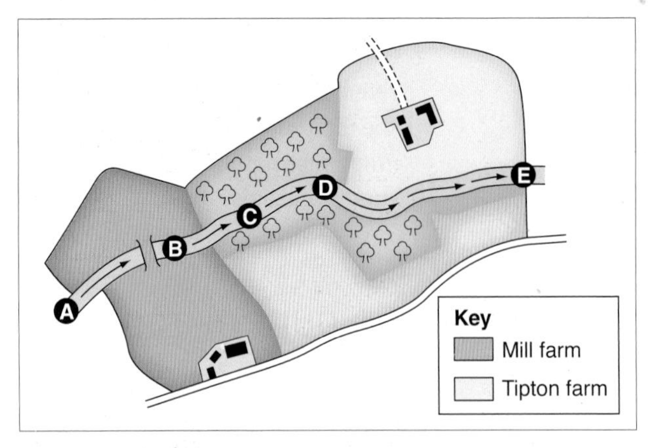

Figure 3.3

The results are shown in Table 3.2.

continued...

Table 3.2 Results of the stream sampling.

Sample point	Species found (numbers per m²)							
	Stonefly nymph	Mayfly nymph	Freshwater shrimps	Caddis fly larvae	Bloodworm	Water louse	Rat-tailed maggot	Sludgeworm
A	11	15	5	12	0	2	0	0
B	0	0	3	4	6	16	12	3
C	0	0	3	8	8	14	2	0
D	0	0	4	10	4	6	0	0
E	0	0	0	4	12	20	2	0

From the results, write a report of the pollution of the stream and its likely causes. Make sure your report is detailed and use evidence from the results to justify your conclusion.

TASK

HOW POLLUTED IS THE AIR *YOU* ARE BREATHING?

This activity helps you with:
- ★ understanding fair testing
- ★ designing experiments.

Lichens are used as indicators of air pollution. Figure 3.4 shows lichens that are found in clean air and lichens that are found in areas polluted with nitrous oxides or sulfur dioxide. It also shows examples of plants that are not lichens, but are sometimes mistaken for them.

All these lichens grow on the bark of trees. Carry out a survey of the trees around your school to see if the air in your area is polluted.

1. If you were comparing your area with another one, explain why it would be necessary to look at just one species of tree in both areas in order to make the test fair.
2. If you wanted to compare your area with others, it would be useful if you could get quantitative data (numbers) to compare. How could you design your survey so that it gave some sort of figure for the air pollution that could be reliably compared with another area?

Nitrogen-loving

Xanthoria parietina

Physcia tenella

Usnea cornuta

Sulfur-tolerant

Xanthovia polycarpa

Indicator of clean air

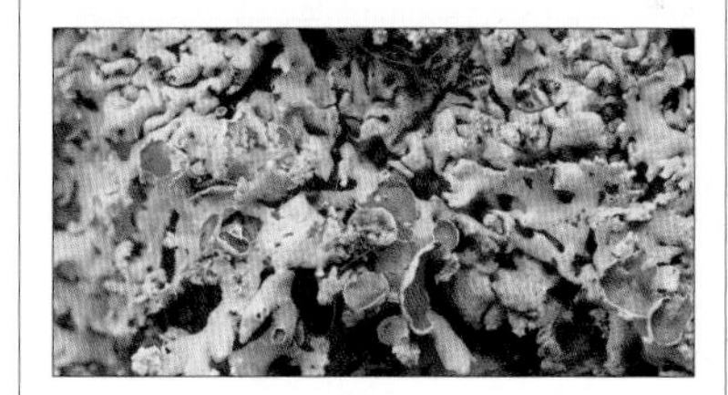

Hypogymnia physodes

These are not lichens!

Moss

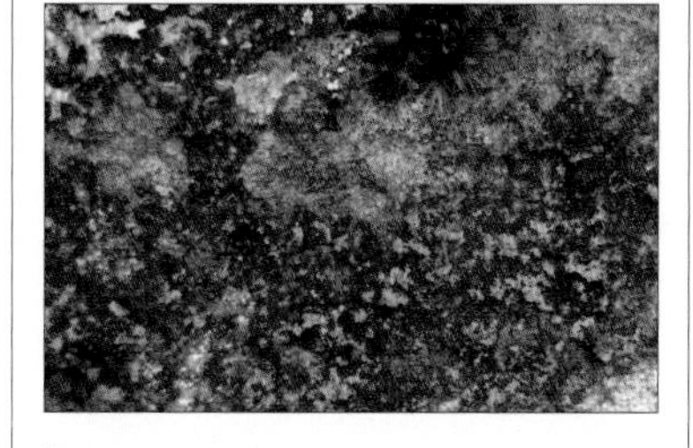

Desmococcus

Figure 3.4 Lichens and air pollution indicator species.

Why do we worry about chemicals 'entering the food chain'?

The main chemicals to be concerned about here are **heavy metals** and **pesticides**. Living things need small quantities of metals but too much can harm them. Some metals, such as lead and mercury, are poisonous even in small quantities. Most heavy metal pollution is caused by industrial processes. There used to be a lot of lead pollution from vehicles burning petrol, but nowadays petrol is lead-free, although leaded petrol is still used in some developing countries.

Pesticides are poisonous chemicals that are used to kill agricultural pests, usually by spraying on crops. Some of them take time to break down, so traces of them can be found on fruits and vegetables in the shops. Pesticides left in the soil can also be washed into rivers and streams by rainfall, and pesticide sprays can drift in the air beyond the area being sprayed.

In the UK there are controls on the use of these polluting chemicals. Some are banned in certain situations and the Environment Agency monitors the environment to look for any signs of harmful levels of pollutants. Although there are occasional accidents, causing high levels of pollution, the levels of these pollutants released into the environment are usually small and controlled. Problems arise, though, if these chemicals 'enter the food chain'.

A famous case of this happening occurred in Minemata, Japan in the 1950s. The city is on the shore of Minemata Bay, and the population lived almost entirely on fish caught in the Bay. Many people in Minemata suddenly started showing the symptoms of mercury poisoning, and 20 died. Mercury was used in a factory on the edge of the bay, but there had been no large spillage. However, the mercury had been absorbed by microscopic plants in the bay, which were part of the human food chain. Part of the food web in the bay is shown in Figure 3.5.

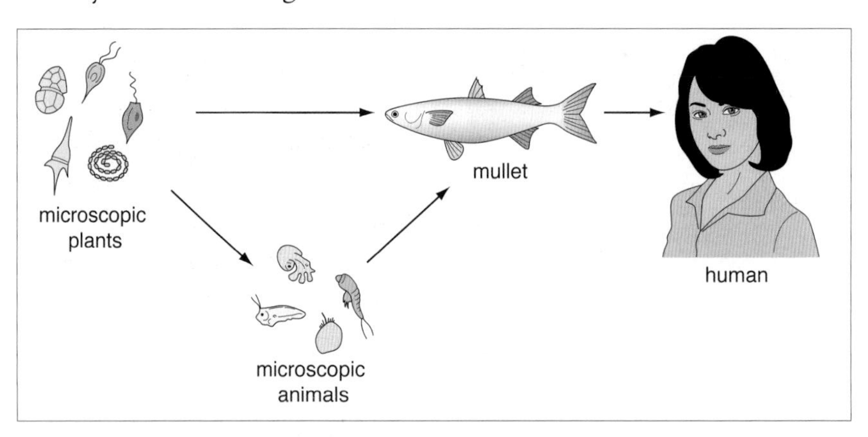

Figure 3.5 Part of the food web around Minemata Bay. The humans ate a variety of other fish and shellfish apart from mullet. These in turn all ate the microscopic plants and animals.

The poisoning had happened in this way:

- The microscopic plants absorbed mercury in the water.
- The microscopic animals ate large quantities of the plants, and the mercury built up inside them.
- The fish ate very large quantities of the microscopic plants and animals, and so the mercury built up to even higher levels in them.
- The fish were, in effect, poisonous because of the levels of mercury they contained. When the humans ate a lot of these fish, the mercury levels in them became so high that it made them very ill or killed them.

QUESTION

1 Why are humans possibly more at risk than other organisms if poisons enter the food web of a habitat?

Why do sewage and fertilisers kill fish?

Sewage and fertilisers sometimes get into streams and rivers from farmland, washed from the soil by rain. This starts a process called **eutrophication**, which can kill fish and other animals. It happens in the following way.

- The sewage/fertiliser causes an increase in the growth of microscopic plants.
- These plants have short lives so, soon afterwards, the number of dead plants in the water goes up.
- Bacteria rot the bodies of the plants, and because there are so many dead plants the population of bacteria goes up sharply.
- These bacteria use oxygen for respiration, and the oxygen level of the water goes down.
- Animals such as fish, which need a lot of oxygen, die because there is now not enough oxygen in the water.

PRACTICAL

HOW CAN YOU TEST THE LEVEL OF ORGANIC POLLUTANTS IN WATER?

This activity helps you with:
★ developing skills in handling scientific apparatus
★ choosing appropriate apparatus
★ making decisions about types of graph to display data
★ drawing graphs.

As stated above, sewage or fertiliser pollution causes an increase in the number of bacteria in water. These bacteria produce an enzyme, catalase, which breaks down hydrogen peroxide in their cells into water and oxygen.

hydrogen peroxide → oxygen + water

Polluted water, with more bacteria, will break down hydrogen peroxide faster than clean water will, and the breakdown can be measured by the oxygen produced.

Apparatus

* pond water
* tap water
* hydrogen peroxide solution
* 2 × 1 cm^3 syringes
* 2 × 20 cm lengths of capillary tubing
* rubber tubing
* 2 clamp stands
* stopwatch
* ruler
* marker pens
* latex gloves
* eye protection

continued...

PRACTICAL *contd.*

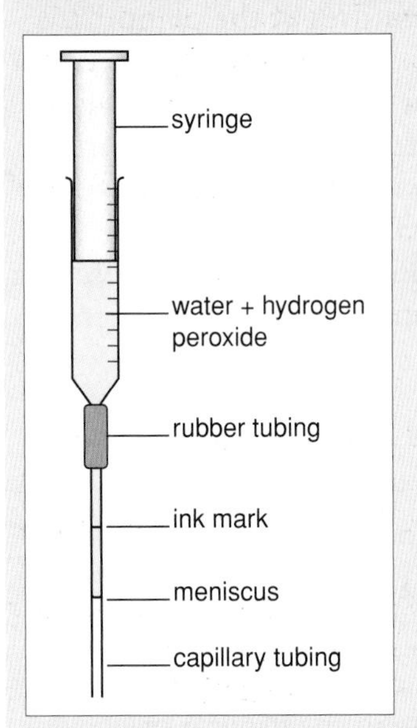

Figure 3.6 Apparatus for the practical.

Risk assessment

- **Wear eye protection.**
- **You will be supplied with a risk assessment by your teacher.**

Procedure

1. Draw 0.5 cm^3 of hydrogen peroxide into the syringe, then draw pond water in to fill the syringe up to the 1.0 cm^3 mark. Shake the syringe gently to mix.
2. Attach the rubber tubing and capillary tubing as shown in the diagram.
3. Place the apparatus in a clamp stand.
4. Gently squeeze the syringe until the solution is visible in the capillary tube. Mark the position of the meniscus with a marker pen.
5. Set up another identical set of apparatus, except that tap water should be used instead of pond water.
6. If oxygen forms, it will collect at the top of the syringe and force the liquid down the capillary tube. Every minute, for 5 minutes, mark the new position of the meniscus on each capillary tube.
7. At the end of 5 minutes, use a ruler to measure the distance travelled each minute in each tube.
8. Record your readings in a suitable table.

Analysing your results

1. Draw a graph of your results. Decide whether this should be a bar graph or a line graph, and explain your choice.
2. What are your conclusions from your results?
3. Why is it better to use capillary tubing in this experiment rather than normal glass tubing?

Discussion Points

1. This experiment cannot prove that the pond water is polluted. Why not?
2. How could you modify the experiment so that it *would* give information about possible pollution of the pond water?

Do we want cheap food or happy farm animals?

Figure 3.7 Battery-farmed chickens are kept in very confined spaces.

One environmental issue that some people have strong opinions about is battery farming, where animals are kept in huge numbers in a small space. This is just one example of **intensive farming** methods. Intensive farming is an agricultural system that aims to produce a maximum yield from the land available. It applies to both animals and plants, and as well as battery farming it involves the use of chemicals such as pesticides and fertilisers to increase yield and to control diseases.

There are advantages and disadvantages, and scientific evidence can clarify these, but in the end people have to decide their own opinions about it. Scientific data cannot decide an ethical issue, but can make sure that the information people have to make their decision is accurate.

The advantages of intensive farming are:

- The yield is high because more animals can be kept and conditions can be controlled. Therefore the food is cheaper to produce. The farmer makes more money, which can sometimes keep him or her in business. If farms go out of business, the UK becomes less self-sufficient in food.
- The food is cheaper in the shops, allowing poor people to have a healthier diet.
- By increasing the yield, it allows the UK to grow more food, to meet the needs of a growing population.

The disadvantages are:

- The chemicals used can enter the food chain and get into our bodies.
- The chemicals can cause pollution and harm wildlife other than pests.
- Natural environments are destroyed. For example, hedgerows are uprooted to make large fields suitable for intensive farming.
- Although no one can really know what an animal 'feels', it is likely that intensive farming causes animals stress and discomfort. Their quality of life is very poor.

Should we kill badgers to stop bovine TB?

Figure 3.8 Badgers are believed to transmit bovine TB to cattle.

Another environmental issue where science can play a role is whether or not to kill ('cull') badgers to reduce cattle deaths from **bovine tuberculosis** (TB). There is an ethical issue here, because although nobody really wants to kill badgers, some people think that it would be acceptable if it was effective in stopping cattle dying from TB. What science can do is to collect evidence about how effective badger culling is. Unfortunately, the evidence is not straightforward. It is summarised below:

- Badgers definitely carry bovine TB and pass it on to cattle.
- If every badger in an area is killed in a very short period of time, culling works, and bovine TB is significantly reduced.
- It is extremely difficult to kill every badger in an area.
- If there is a partial cull and some badgers survive, they tend to move away. TB in the culled area is reduced, but badgers carry the infection to surrounding areas, and TB there goes up.

Farmers cannot kill badgers unless they get a licence from the Government. In 2010 farmers in some areas of Wales were licenced, and there was a proposal to allow English farmers to cull badgers, too.

QUESTION

2 On the basis of the scientific evidence (not how much you like badgers), what do you think the UK Government's policy on badger culling should be?

Where do we get our energy from?

Energy enters the planet constantly in the form of sunlight. This energy passes from organism to organism by means of **food chains**. Plants are the first links in all food chains because they are **producers** – they change energy in sunlight into stored chemical energy. At this point, quite a lot of the energy is wasted – plants only manage to capture about 5% of the energy in sunlight. When plants are eaten by herbivores, some of the energy is passed to this next link, the **consumers** in the food chain. When the herbivore is eaten by a carnivore, the process of energy transfer is repeated. Energy passes in this way from carnivores to scavengers and decomposers, which feed on dead organisms. However, not all the energy stored by a herbivore is stored by a carnivore that feeds on it. Much is used in life processes such as movement, growth, cell repair and reproduction. Some is also wasted as heat during respiration. Only leftover energy is stored by the carnivore.

Consider the food chain through which energy flows when we eat fish such as tuna. Energy from the Sun is first used by plant plankton (microscopic algae). It then passes to animal plankton, then to small fish, then to larger fish, then to tuna and then to us. There is usually no predator to eat us, so we are the top carnivores in this food chain.

plant plankton → animal plankton → small fish → large fish → tuna → humans

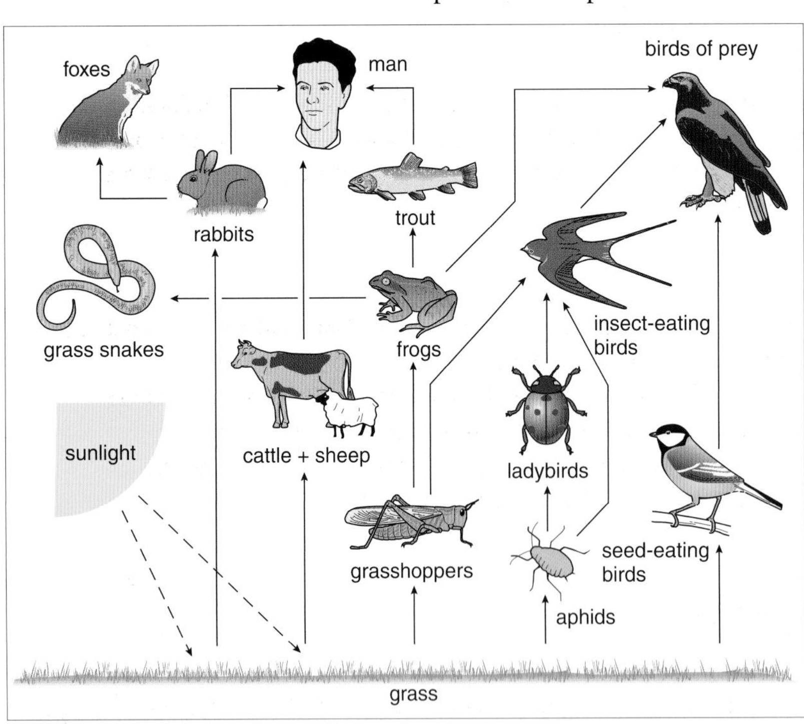

Figure 3.9 A typical food web.

In nature, the food chains often interlink, because most organisms eat a lot of different things and are eaten by many different animals, too. Interlinked food chains are called **food webs** (Figure 3.9). Even this food web is an over-simplification of all the feeding relationships that would exist in this environment.

Feeding relationships can be illustrated as pyramids (see Figures 3.10 and 3.11).The width of each block in the pyramid is an indication of the number of that type of organism (or its mass). These pyramids can tell us more about the energy that is available to organisms living in a measured area or volume. The pyramids can be drawn in different ways:

- A **pyramid of numbers** shows the number of organisms per unit area or volume in each feeding level.
- A **pyramid of biomass** shows the dry mass of organic material per unit area or volume at each feeding level.

QUESTION

3 Food webs link together the lives of the animals and plants in them, and so what happens to one organism can affect many others. In the food web in Figure 3.9, suggest what might happen to the population of birds of prey if the population of ladybirds was drastically reduced.

The number of links in a food chain are limited by the chemical energy available. As the quantity of chemical energy at each feeding level becomes smaller, so does the amount of living material that can be supported by that level. When the chemical energy decreases to nothing, the food chain (and the pyramid) ends.

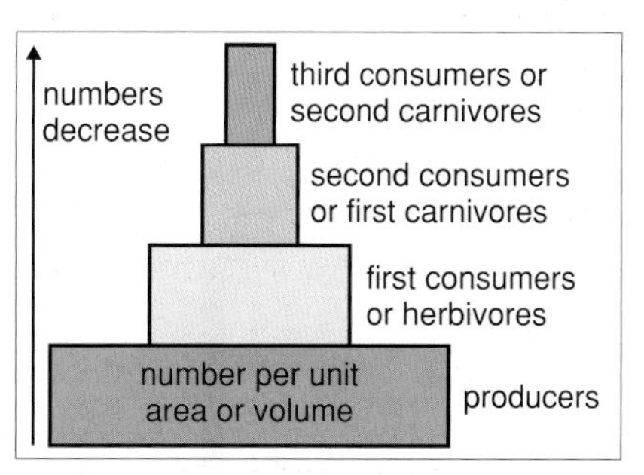

Figure 3.10 A general pyramid of numbers.

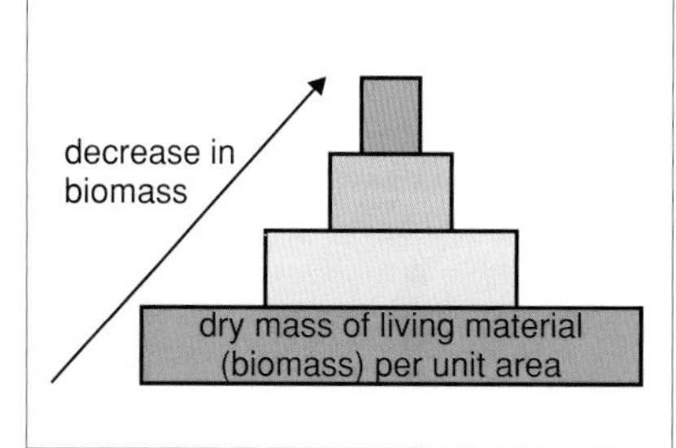

Figure 3.11 A pyramid of biomass for a woodland food chain.

TASK

WHAT'S WRONG WITH MY PYRAMID?

This activity helps you with:
- ★ plotting graphs
- ★ explaining results
- ★ evaluating experimental methods.

Some data have been collected from an area of woodland (Table 3.3).

Table 3.3 Numbers of organisms at different feeding levels.

Organisms	Number
Plants (trees and ground plants)	55
First consumers (herbivores)	120
Second consumers (small carnivores)	36
Third consumers (large carnivores)	2

1 Use graph paper to construct a pyramid of numbers for this data.
2 Your 'pyramid' is not pyramid shaped! Explain the likely reason for this.
3 Why would it be better to draw a pyramid of biomass if the data were available?

Discussion Points

A number of factors influenced the collection of information. How might each of these have affected the data?

a The plants were easier to find and count because they did not move.
b Some of the first consumers were very small and difficult to see.
c The third consumers moved around a lot – the two seen were just the ones that happened to be there at the time.

Am I going to be recycled?

Many people like to recycle things to help conserve the environment, but did you know that you will be recycled when you die? All living things are made of chemicals containing carbon, and this carbon has to be recycled otherwise supplies would run out. The **carbon cycle** is shown in Figure 3.12.

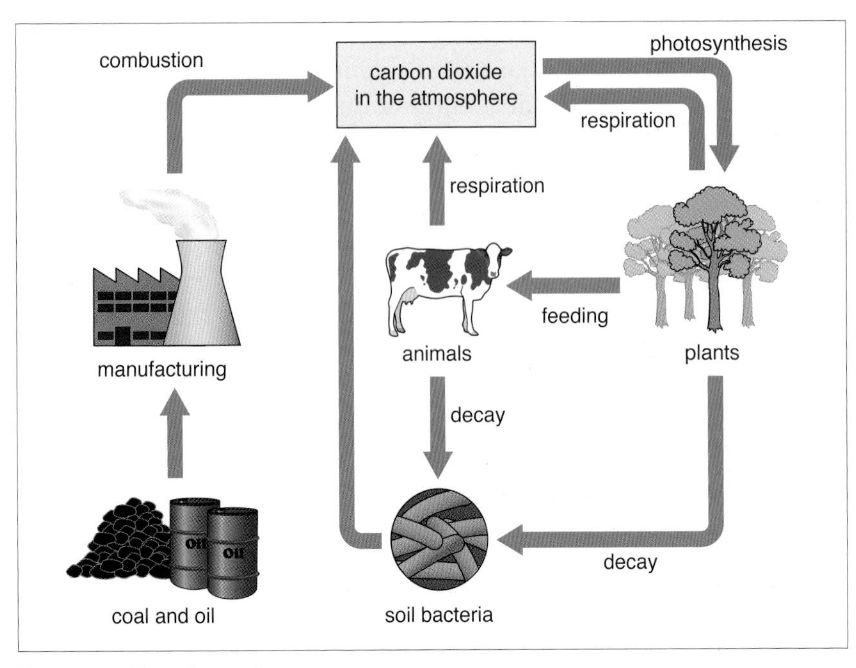

Figure 3.12 The carbon cycle.

Carbon dioxide in the air is made into food by green plants in a process called **photosynthesis**. Animals get their carbon by eating plants (or other animals). The carbon in dead animals and plants is released back into the atmosphere by the process of decay. The bacteria involved release carbon dioxide when they **respire**. Living animals and plants also respire, and so put carbon dioxide back into the atmosphere. Extra carbon dioxide is added to the atmosphere by the burning of **fossil fuels**. Fossil fuels were made from the dead bodies of plants and animals, millions of years ago. They were not completely decomposed, so a lot of the carbon in them was 'locked' into the fossil fuels. When fossil fuels are burnt, the carbon in them is released as carbon dioxide, adding to the levels in the atmosphere. Humans have only started extracting and burning fossil fuels in large quantities in the last 200 years. This has disrupted the balance and led to an increase in the levels of carbon dioxide in the atmosphere.

What else is recycled?

All the minerals in living things are recycled, not just carbon. There is a limited supply of all minerals on the Earth and if they were not recycled, this supply would run out. Plants take in these substances from their environment and animals take them in as food. They are released again when dead bodies decay. This maintains a balance between the living things and the atmosphere. Two important minerals are nitrogen and phosphorus. The '**nitrogen cycle**' and '**phosphorus cycle**' ensure they are put back into the environment for reuse. The nitrogen cycle is shown in Figure 3.13.

Nitrogen in the air is changed by bacteria in the soil into nitrates, which plants can absorb and use. Other bacteria can change this nitrate back into nitrogen. The nitrates absorbed by the plants are passed onto animals that eat the plants, and the nitrogen is eventually returned to the soil in urine and faeces from animals, and when dead animals and plants decay.

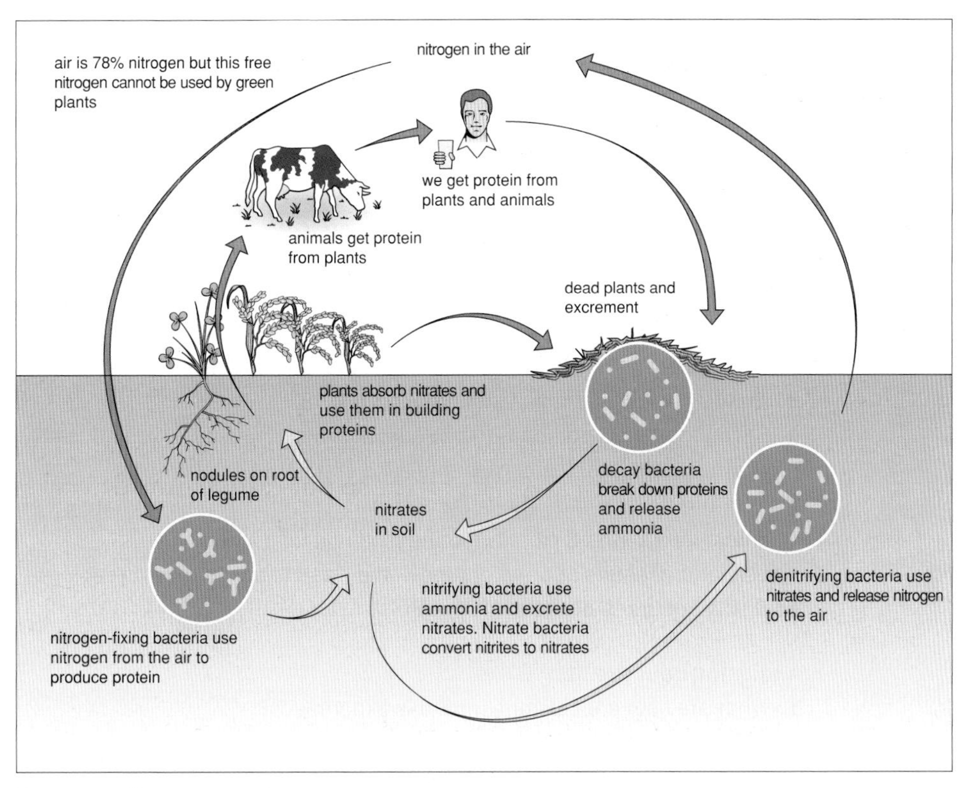

Figure 3.13 The nitrogen cycle.

PRACTICAL HOW DOES URINE HELP KEEP THE NITROGEN CYCLE GOING?

This activity helps you with:
- ★ drawing conclusions
- ★ judging the strength of evidence
- ★ understanding control experiments.

Decomposers, as part of recycling, secrete enzymes into the soil that break down waste such as urea. One of these enzymes is called **urease**. This experiment investigates the effect of the concentration of urease on its reaction with urea. Urease catalyses (speeds up) the breakdown of urea to release ammonia. The ammonia dissolves in water to form an alkaline solution, which can be converted in the soil into nitrates.

Risk assessment

- **Wear eye protection.**
- **You will be supplied with a risk assessment by your teacher.**

Apparatus

- 4 test tubes
- test tube rack
- labels
- 3 syringes
- dropping pipette
- Bunsen, tripod and gauze
- large beaker (water bath)
- pH colour chart
- 0.1 mol/dm^3 urea solution
- fresh urease solution
- boiled and cooled urease solution
- ethanoic acid
- Universal Indicator solution
- distilled water
- eye protection

Procedure

1. You will need a boiling water bath for this experiment. Half fill the large beaker with water and start to heat it now.
2. Label four test tubes A, B, C and D.
3. Using a syringe, add 3 cm^3 of urease solution to tube D and place the tube in the boiling water for 4 minutes. Remove the tube and allow it to cool. (You can speed up the cooling process by running the bottom of the tube under cold water.)

continued...

PRACTICAL *contd.*

4 Using the second syringe, add the following to each of the tubes:
   a 5 cm^3 of urea solution
   b 3 drops of ethanoic acid
5 Using a dropping pipette, add 10 drops of Universal Indicator solution to each tube.
6 Using the first syringe, add urease solution to tubes A, B and C as follows:
   Tube A – 1 cm^3 urease
   Tube B – 3 cm^3 urease
   Tube C – 5 cm^3 urease
7 Tube D already contains 3 cm^3 of boiled and cooled urease.
8 Draw up a suitable table to record your results.
9 Shake the tubes, compare the colour with the pH chart and record the approximate pH of the contents of each of the tubes in the table.
9 At 2-minute intervals, shake the tubes and record the approximate pH of each tube (A, B, C, D). Continue the readings for 12 minutes.
10 Collect the results of other groups in the class to act as repeat readings.
11 Display your results as a line graph.

Analysing your results

1 What are your conclusions from this experiment?
2 How certain are you of these conclusions? Give reasons for your answer (hint: look at variations between the different groups' results).
3 Why do you think ethanoic acid was added to the tubes?
4 What was the point of tube D?

Discussion Points

In this experiment, the urease used was the same *concentration*, but different *volumes* were used. Does this mean the experiment is not valid as a means of testing the effect of concentration?

Chapter summary

- The needs of wildlife have to be balanced with human requirements for food and economic development.
- To reduce the impact of development on the environment, it is important to collect detailed and reliable biological data and to monitor any changes.
- Pollution in streams and rivers can be monitored by measuring oxygen levels or pH, or by looking for the presence of indicator species.
- Heavy metals and pesticides entering food chains can cause particular health problems for animals higher in the food chain because they get more concentrated as the food chain progresses.
- Untreated sewage and fertilisers cause rapid growth of algae in fresh water. When these plants die, the bacteria that decay them reduce the oxygen levels in the water, causing fish and other animals to die as a result.
- Food chains and webs show the transfer of energy between organisms. This energy enters the environment in the form of sunlight.
- Much of the sunlight energy is not absorbed or used effectively by plants.
- Some energy is used or wasted at each stage of a food chain.
- Pyramids of number and biomass can be used to represent the amount of organisms at different feeding levels. Pyramids of biomass are more accurate than pyramids of number.
- Microbes bring about the decay of dead bodies and waste materials. In doing this, the microbes respire and release carbon dioxide into the atmosphere.
- Carbon, nitrogen and phosphorus are constantly reused in nature because they are 'cycled'.
- Photosynthesis and respiration are important processes in the carbon cycle.
- Burning fossil fuels results in the release of large quantities of carbon dioxide.

Variation and inheritance

Am I unique?

Yes, you are. No human being is identical to any other, living now or at any time in the past. Even 'identical' twins, though they look very similar, are not actually identical in every way. There are many ways in which humans vary.

We will look at variation in your class. Try and find at least 20 ways in which individuals in your class vary from one another. Do not include things like clothes and jewellery, but just variations in their bodies.

All the people in your class are of similar age. Yet they will vary quite a lot in size. This is not just their height, but the size of different parts of their bodies, too.

PRACTICAL VARIATION IN FINGER LENGTH

This activity helps you with:
- ★ choosing how to plot data
- ★ constructing bar graphs
- ★ judging hypotheses
- ★ considering the strength of evidence.

Procedure

1. Collect data on the length of the middle finger of everyone in your class. Measure the finger length as shown in Figure 4.1.
2. In each case, record whether the person is male or female – we can then ask the question 'Is there any difference between the length of fingers in boys and girls?'
3. The data can be plotted as a bar chart in several different ways, as shown in Figure 4.2.
4. Which would be the best way of plotting the data:
   - **a** if your main interest was in the data for the group as a whole, but you had some interest in gender differences
   - **b** if you were interested in differences in the pattern in boys and girls?

   Give reasons for your answers in each case.

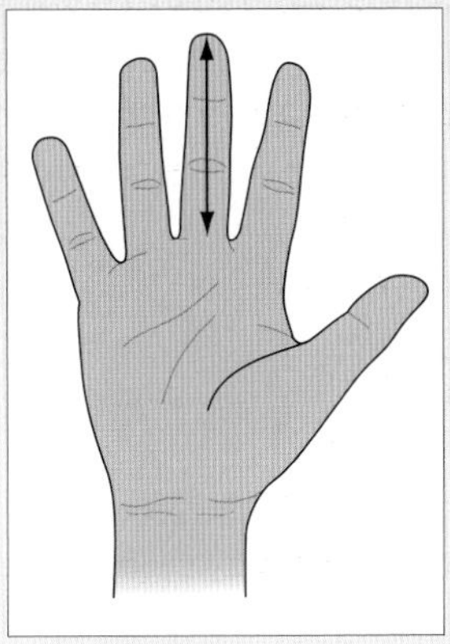

Figure 4.1 Measure the middle finger from the tip to the bottom of the crease where the finger joins the palm.

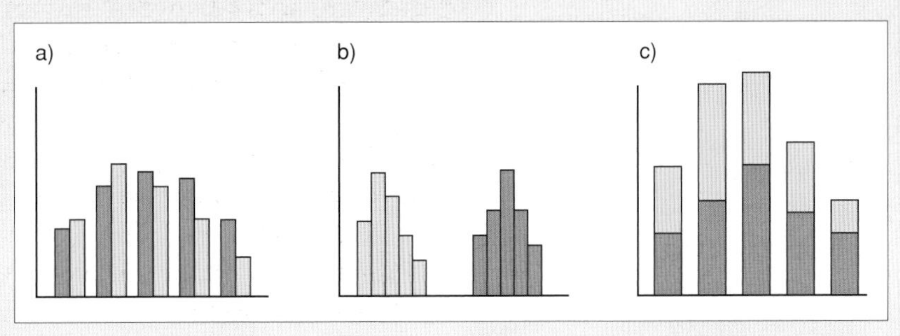

Figure 4.2 In each of these bar charts, data from boys are in orange, and those from girls are in yellow. a) Boy and girl data plotted alongside each other; b) Boy and girl data plotted separately (could also be on two graphs); c) Boy and girl data combined but distinguished.

continued...

PRACTICAL contd.

Analysing your results

1 There are three possible hypotheses in connection to our question, 'Is there any difference between the length of fingers in boys and girls?'
   i There is no difference in finger length between boys and girls.
   ii Boys tend to have longer fingers than girls.
   iii Girls tend to have longer fingers than boys.
   a From your data, which hypothesis is supported by the evidence?
   b How conclusive do you think this evidence is?

   Give reasons for your answers.

PRACTICAL DO *ALL* LIVING THINGS VARY?

This activity helps you with:
★ observing closely
★ describing features clearly.

Apparatus
* 2 garden snails
* hand lens

Procedure

You will be provided with two snails of the same species.

1 Look at the snails carefully (with the naked eye and with a hand lens). Try to find at least ten ways in which one snail differs from the other. Do *not* include overall size, because this will be at least partly related to the age of the snail, and we cannot tell how old they are.
2 From your observations, would you conclude that all living things vary?

Figure 4.3 The garden snail – are all garden snails the same?

What are continuous and discontinuous variation?

Scientists describe variation as either **continuous** or **discontinuous**. Continuous variation is where there is a continuous range with no 'categories' (e.g. height in humans; people can be any height within a certain range) whereas in discontinuous variation there are distinct groups (e.g. fingerprint types; a persons fingerprint can be one of an arch, a whorl or a loop – there are no 'in-between' fingerprints).

QUESTION

1 For each of these human variations, decide whether they are continuous or discontinuous:
   a weight
   b eye colour
   c right- or left-handed
   d blood group
   e short sight/normal vision/long sight.

a)

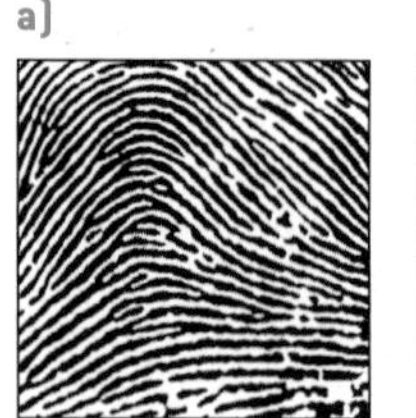

b)

c)

Figure 4.4 Finger print groups: a) an arch; b) a loop; c) a whorl. This is an example of discontinuous variation.

What causes variation?

Variation is caused by two factors. The genes of an organism control its characteristics and different sets of genes will result in **genetic variation**. In humans, the only people with identical genes are identical twins (or triplets, etc.) because they are formed from the splitting of a single fertilised egg cell. Yet there are variations even between identical twins. These variations are **environmental variations**, and are caused by the environment (mainly things that result from events in their lives, sometimes by choice).

Figure 4.5 Identical twins have identical genes but there are still some differences between them.

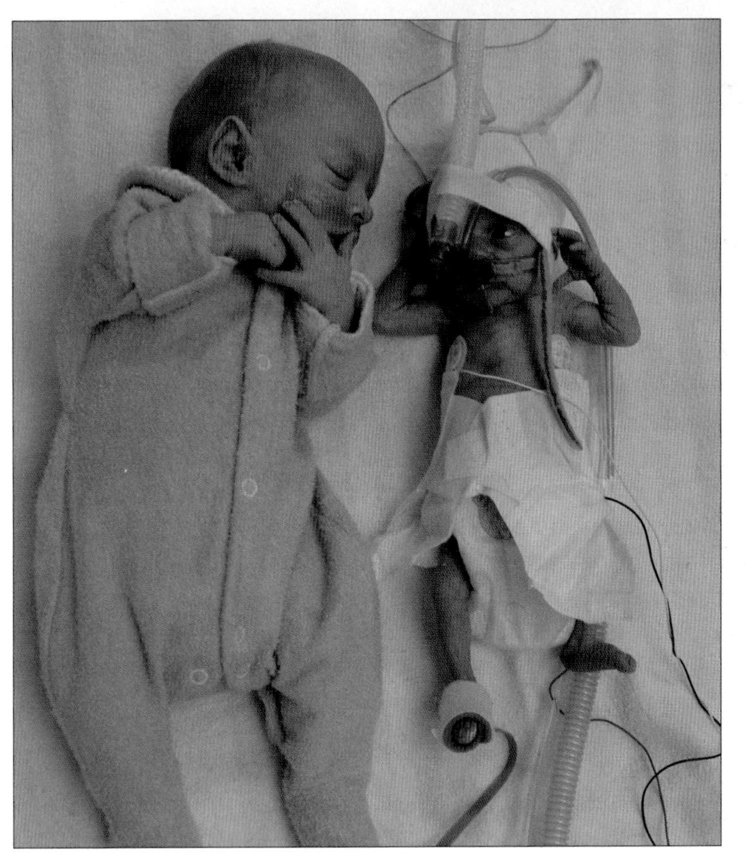

Sometimes one twin is born much smaller than the other, because for some reason it was starved of oxygen in the mother's womb (see Figure 4.6). This is an 'environmental' variation.

Other examples of environmental variation in humans include body piercings, scars, hairstyles, tattoos and facial hair. Some variations might be a combination of genetic and environmental factors – height and weight, for example, have genetic components but are also affected by diet.

Figure 4.6 These two babies are identical twins but there was a complication in pregnancy which resulted in blood flowing from the smaller one to the larger one, so the smaller twin was starved of oxygen and nourishment.

Why don't we look exactly like either parent?

Offspring become genetically different from their parents as a result of **sexual reproduction** which involves an **egg** fusing with a **sperm** in the process of **fertilisation**. The genes from the mother in the egg are mixed with different genes from the father in the sperm. The cell formed as a result of fertilisation (the **zygote**) has one set of genes from the father and one from the mother. The 'set' of genes represents only half of the mother's or father's total number of genes, and the combination of genes making up the 'set' varies, which is why brothers and sisters are similar but different.

Organisms that reproduce by **asexual reproduction** do not mix their genes because fertilisation does not take place. One individual produces offspring that are genetically identical to each other and to the parent. These are called **clones**.

Figure 4.7 A set of identical ('clone') plants produced from cuttings taken from a single parent plant.

How do we know about genes?

The first person to use the term 'gene' was the Danish scientist Wilhelm Ludwig Johannsen in 1902, but the idea of the gene had been established 40 years earlier by an Austrian monk, **Gregor Mendel**, who is now regarded as one of the most famous scientists of all time.

Figure 4.8 Gregor Mendel, the father of genetics.

When Mendel did his experiments on pea plants, scientists knew that characteristics were inherited from parents, but thought that the inherited character was a sort of 'blend' of the parents' characters. Mendel used 'tall' and 'dwarf' pea plants that were 'pure breeding', i.e. always produced the same type of offspring. Both types varied in size a bit, but there was a clear difference in the range of heights of the two categories. He crossed tall plants with dwarf ones by careful pollination, but he did not get the expected 'medium size' plants. In fact, all the new plants were tall.

When Mendel crossed two of these plants together, he got mostly tall plants, but about a quarter were dwarf. He now knew that whatever caused the 'dwarf' feature *had* been inherited in the first cross, but for some reason had been masked. The real genius of

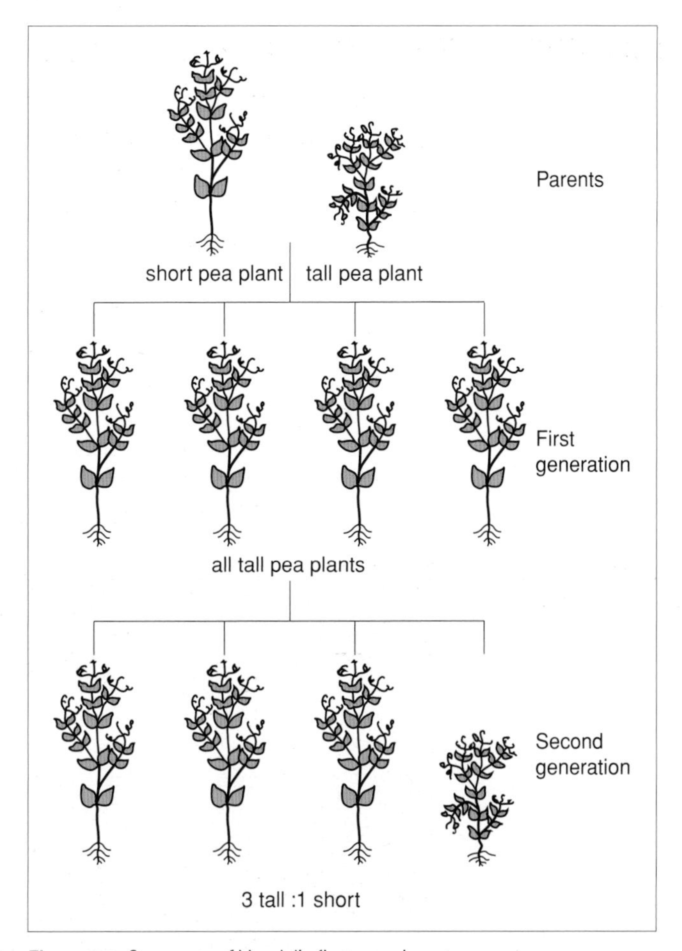

Figure 4.9 Summary of Mendel's first experiments.

Mendel was how he investigated and explained what had happened.

Mendel knew that, if the dwarf character had been masked, another (tall) factor must be masking it. Therefore, the pea plants had two factors that controlled height, and the tall factor seemed to be in some way stronger than the dwarf one. We now call these different factors **alleles**, and we refer to 'stronger' ones as **dominant** and 'weaker' ones as **recessive**.

Mendel worked out how this pattern of inheritance had come about. It can be shown by a diagram, in which the dominant allele (tall) is shown by the symbol **T**, and the recessive dwarf allele is shown by the symbol **t** (Figure 4.10). Geneticists always show dominant alleles in capital letters and recessive one in lower case. The genetic make up of an organism is called its **genotype** and what it 'looks' like is called its **phenotype**.

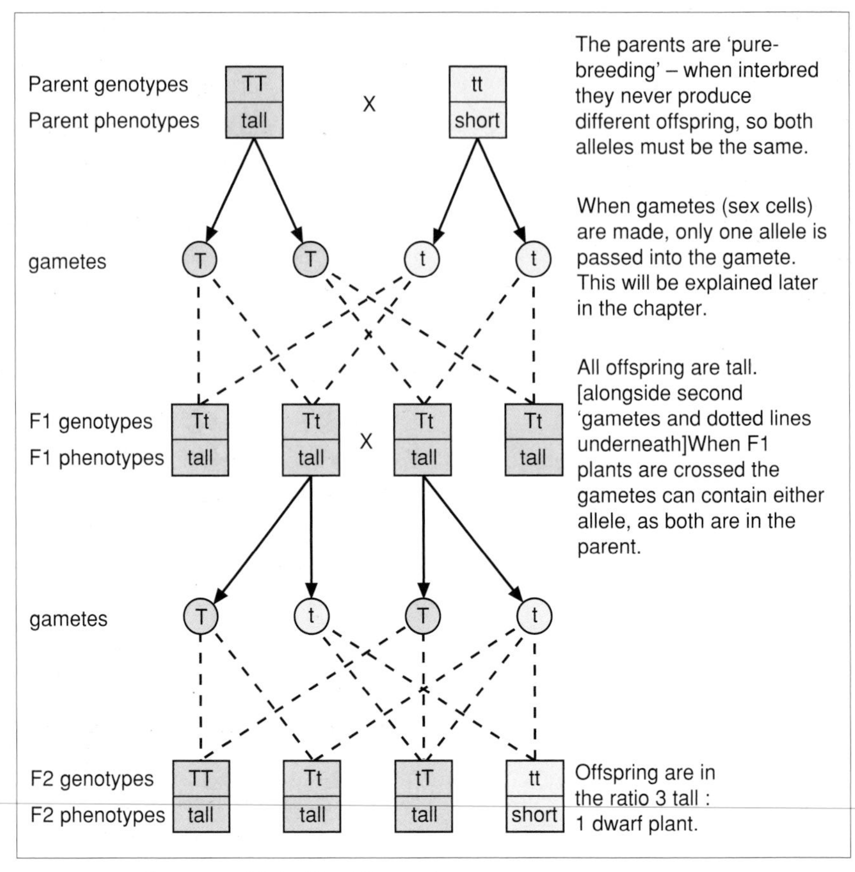

Figure 4.10 A diagram explaining Mendel's experiments.

QUESTIONS

2 Why was it important that Mendel looked at a variety of characteristics, not just tall × dwarf?

3 Mendel's scheme predicts a 3 : 1 ratio in F2 plants. His results all showed a roughly 3 : 1 ratio, but never an *exact* 3 : 1 ratio. Does this matter?

4 In humans, the ability to roll the tongue is dominant (Figure 4.11) and an inability to roll the tongue is recessive. We can give the alleles the symbol **R** for roller and **r** for non-roller. A man with the genotype **Rr** has children with a woman with the genotype **rr**.
   a What would be the phenotypes of the parents?
   b Use a Punnett square to predict the genotypes and phenotypes of any possible children.

In crosses like the second one in Figure 4.10 the use of arrows can be a bit confusing. Biologists tend to put the gametes in a table called a **Punnett square**, as shown here.

		Male gametes	
		B	b
Female gametes	B	BB	Bb
	b	Bb	bb

Mendel tried his experiments with a number of characteristics apart from height. In each case, he found that there was a dominant and a recessive characteristic. In the initial crosses, all the offspring showed the dominant phenotype. When two F1 plants were crossed, the ratio between dominant and recessive phenotypes was around 3:1.

Mendel's conclusions were:

- Characteristics are controlled by a pair of alleles, which may be the same as each other or different.
- One allele is dominant, the other is recessive. If both alleles are present, only the dominant one can be seen in the phenotype.
- Only one allele from each pair is passed into each gamete.
- If two pure-breeding parents (one for each allele) are crossed, the offspring all show the dominant characteristic.
- If two of the offspring are crossed, the new generation has approximately three individuals showing the dominant characteristic to every one showing the recessive characteristic.

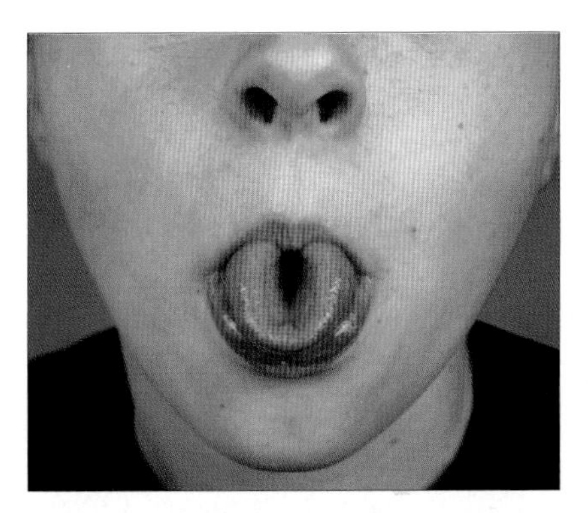

Figure 4.11 Not everyone can roll their tongue like this. It is a genetic trait.

Mendel was able to make his discoveries because inheritance in peas is relatively simple, and many characteristics are controlled by a single pair of alleles.

In humans this is rare; most of our features are controlled by a combination of genes.

Genetic terminology

You need to know the following genetic terms:

Allele: a variety of a gene

Genotype: the genetic make-up of an individual (e.g. **BB**, **Bb**, **bb**)

Phenotype: the description of the way the genotype 'shows itself' (e.g. blue eyes, curly hair, red flowers, etc.)

Dominant: the allele that shows in the phenotype whenever it is present (shown by a capital letter)

Recessive: the allele that is hidden when a dominant allele is present (shown by a lower case letter)

F1/F2: short for first generation (F1) and second generation (F2) in a genetic cross

Homozygous/homozygote: a homozygote contains two identical alleles for the gene concerned; it is homozygous

Heterozygous/heterozygote: a heterozygote contains two different alleles for the gene concerned; it is heterozygous.

Why wasn't Mendel famous (in his lifetime)?

Although Mendel is now regarded as one of the greatest scientists of all time, nobody really took much notice of his experiments in his lifetime. This was because he did not publish his results in an international science journal, as scientists do today. He did give a presentation at the Natural Science Society in his home town of Brno and a paper linked to the talk was printed in the magazine of that society, but not many people read it.

This is a good example of why scientists publish their results. It lets people know what they have discovered and how they discovered it, and allows other scientists to repeat the experiments to see if they get the same results. In Mendel's case, it was more than 30 years before his work became widely known, by which time Gregor Mendel was dead.

Versuche über Pflanzen-Hybriden von Gregor Mendel.

Einleitende Bemerkungen

Abb. 9

Faksimile der ersten Seite von Mendels Original-Manuskript (nach dem in Brünn vorhandenen Original)

22

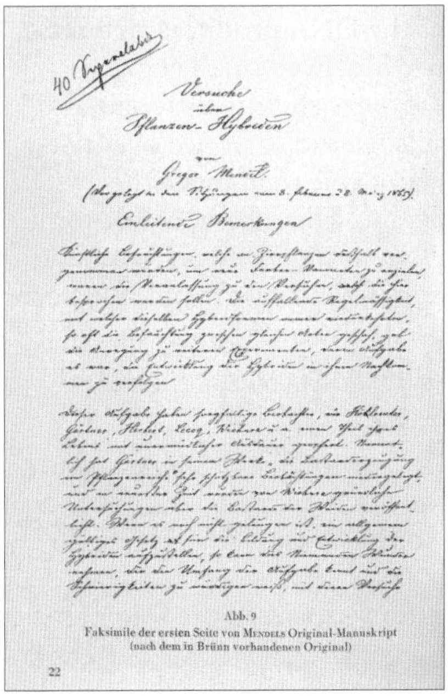

Figure 4.12 Mendel's handwritten manuscript of the paper about his work in the *Proceedings of the Natural Science Society* in Brno, 1866.

What is a gene?

Mendel called the things that influence inheritance 'factors'. We now call them **genes**, which occur in different 'varieties' (alleles). However, it is only in the last 50 years that we have discovered exactly what a gene is.

The nucleus of a cell contains an amazing chemical called **DNA** (deoxyribonucleic acid), which controls what proteins the cell makes, and can make exact copies of itself. In the nucleus, the long DNA molecules are coiled up into structures called **chromosomes**. DNA is the raw material of genes – a gene is a short length of DNA. This is summarised in Figure 4.13.

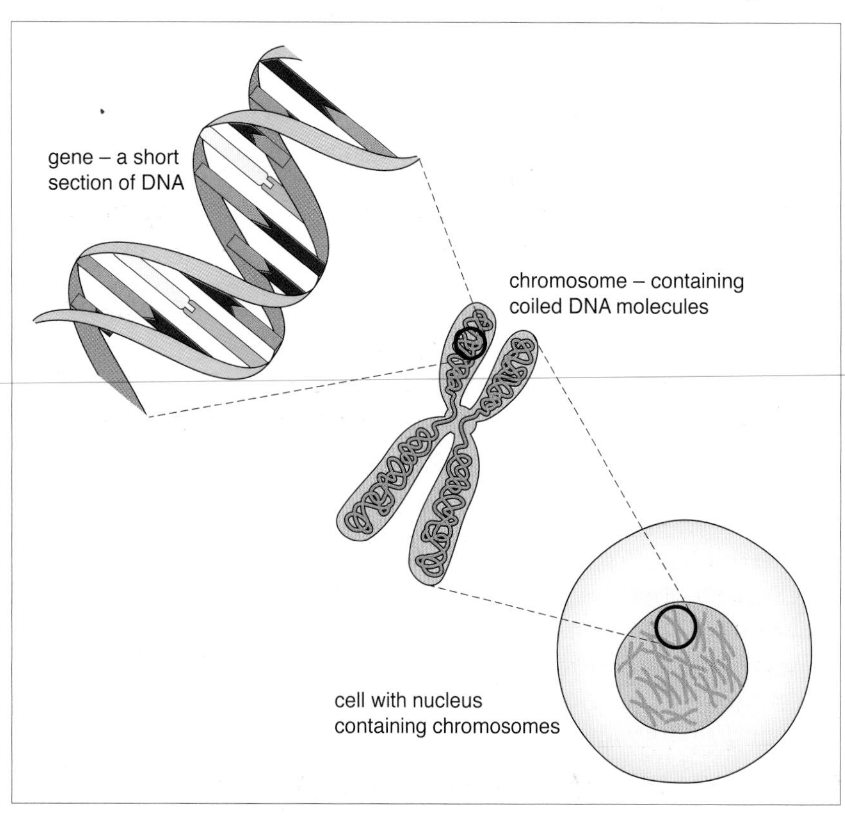

Figure 4.13 The structure of a gene in relation to DNA and chromosomes.

In Figure 4.13 the coloured bands seen running across the centre of the DNA molecule are pairs of chemicals called **bases**. The sequences of these bases in a DNA molecule form a sort of a 'code', which decides what proteins the cell makes. The proteins determine how cells function and so control characteristics of both the cell and the whole plant or animal.

In Chapter 1 we saw that scientists can look at the bases in a DNA molecule and see to what extent different DNA samples are similar. The analysis of the DNA produces a **genetic profile**. Everyone's genetic profile is slightly different and this has led to police scientists being able to identify the person who may have left some DNA at a crime scene ('genetic fingerprinting'). It has also been used to identify the father of a child, where this is not certain.

Genetic profiling consists of several steps:

1. A sample of cells is collected, e.g. from blood, hair, semen, skin, etc. The cells are broken up and the DNA extracted.
2. The DNA is 'cut up' by enzymes, so that it ends up in bits of different sizes.
3. A technique called 'gel electrophoresis' separates the fragments. The DNA segments are placed on a gel and an electric current is passed through the gel. The segments move across the gel, with the smaller ones moving the furthest. A pattern develops, which is the 'genetic profile'.

TASK IS GENETIC FINGERPRINTING A GOOD THING?

This activity helps you with:
★ considering the ethics of scientific investigations
★ making judgements about a complex argument
★ justifying conclusions using evidence
★ expressing ideas clearly.

DNA profiling has proved useful in solving crimes. It can also identify potential health problems that have not yet become obvious, so that they can be treated. However, some people think that indiscriminate genetic profiling can infringe personal liberty.

Use the internet to research the advantages and potential problems associated with genetic profiling. Write a report that considers both sides of the argument. Give your own opinion about genetic fingerprinting and justify it using evidence from your research.

PRACTICAL CAN WE SEE DNA?

This activity helps you with:
★ following instructions
★ measuring accurately
★ designing experiments.

It is relatively easy to extract DNA from cells. The experiment works particularly well with kiwi fruit because they contain a protease enzyme that helps to break down the cells and release the DNA.

Apparatus

* distilled water
* ice-cold alcohol
* kiwi fruit
* ice
* 2 g sodium chloride
* 5 g washing-up liquid
* top pan balance
* mortar and pestle
* filter funnel
* filter paper
* 250 cm^3 beaker
* 250 cm^3 glass measuring cylinder
* glass stirring rod
* scalpel
* tile
* wire 'hook'
* eye protection

Risk assessment

- **Wear eye protection.**
- **You will be supplied with a risk assessment by your teacher.**

Procedure

1. Peel the kiwi fruit and chop it into small pieces using a scalpel and white tile.
2. Mash the kiwi fruit in the pestle and mortar.
3. Add 5 g washing-up liquid and 2 g salt to 100 cm^3 distilled water in the 250 cm^3 beaker.
4. Stir *slowly* to dissolve the salt, trying to avoid creating bubbles.
5. Add the kiwi fruit and mix.
6. Place the beaker in a water bath at 6 °C for 15 minutes.
7. Stand the measuring cylinder in ice to cool it.
8. Filter the mixture (using filter paper and a funnel) into the measuring cylinder.
9. Carefully, and slowly, pour 100 cm^3 ice-cold ethanol down the side of the beaker.
10. A jelly-like material forms between the water and alcohol layers – this is the DNA.

continued...

PRACTICAL contd.

11 Using a wire hook, you can hook out the DNA and place it on your white tile.

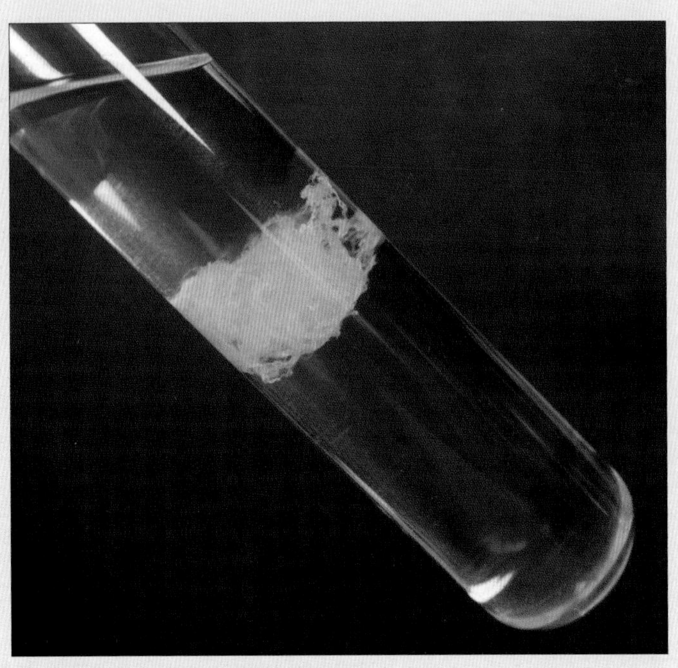

Figure 4.14 Jelly-like DNA forms where the water and alcohol layers meet.

Analysing your results

1 This experiment works to some extent with many fruits. If you wanted to compare the success of the experiment using kiwi fruit with the same experiment using strawberries, how could you *measure* the difference?
2 Why did you need to cut up and mash the kiwi fruit before using it in the experiment?

How are new genes made?

Figure 4.15 It is thought that prehistoric humans were mainly black haired and dark skinned (although new evidence indicates that some may have been redheads).

The genes of a species do not remain the same for all time. New genes and characteristics are constantly appearing. For example, it is likely that there were no blonde prehistoric humans. At some point in human history, a 'blonde' gene must have appeared. Completely new genes are rare, but when they occur they are caused by **mutation**. A mutation is a change in the structure of a gene. These changes occur naturally but can also be caused by ionising radiation and by certain chemicals in the environment. Most mutations make such small changes that no effect is seen. Some are harmful, but very rarely a mutation can arise that actually 'improves' the design of the organism and helps its survival.

QUESTION

5 X-rays can cause mutations. If you are X-rayed, small mutations may occur in a few of your cells but this will not harm you. However, doctors try to avoid X-raying pregnant women, especially if they are in the early stages of pregnancy. Suggest a reason for this.

How are genes inherited?

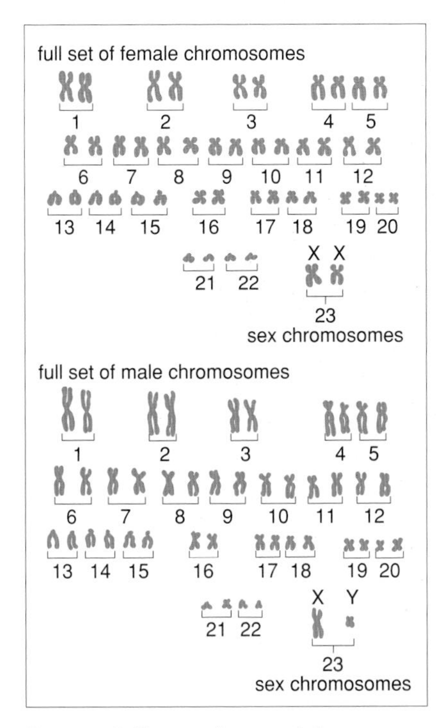

Figure 4.16 The complete set of chromosomes of a man and a woman.

Human families often have similar characteristics because they share a lot of the same genes. Parents pass on their genes to their children, who will then have a mixture of their mother's and father's genes and characteristics. The genes are passed on in the nuclei of the **gametes** (sperm and egg), which fuse to form a zygote that will develop into a child. Humans have 46 chromosomes in their body cells, which are actually made up of 23 pairs – one of each pair coming from the mother and one from the father. That is why, as we saw earlier with Mendel's work, there are two copies of every gene (which may be the same or different).

When sex cells are made, only one chromosome from each pair is passed into the sperm or egg cell. The number of chromosomes in a human gamete is therefore 23. This ensures that the body cells of the baby also have 46 chromosomes, not 92. The 23 chromosomes in a gamete are not always the same combination, so each sperm and egg cell is genetically different. The same principles operate in all living organisms although the number of chromosomes will be different.

QUESTIONS

6 A species of lily has 24 chromosomes in its body cells. How many chromosomes would you expect to find in an ovule (egg cell) of this lily?

7 Suggest a reason why sometimes two children of the same parents look very similar, but in other cases they look very different?

What determines whether a baby is a boy or girl?

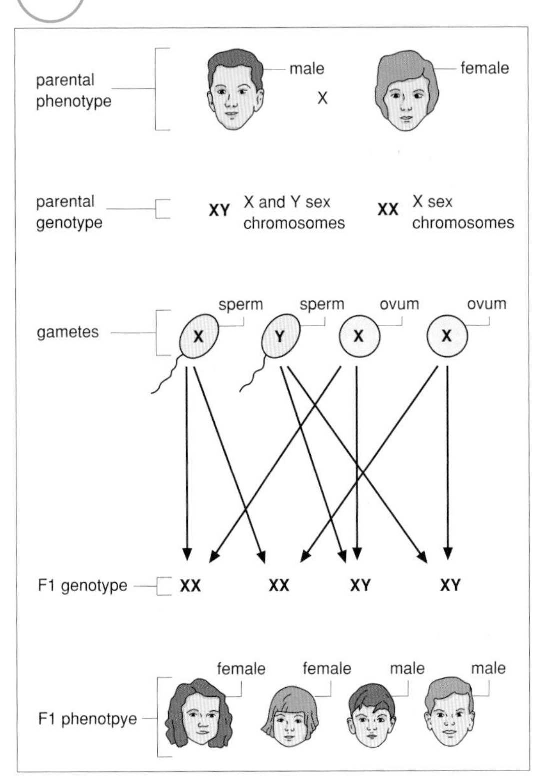

Figure 4.17 Human sex chromosomes.

In Figure 4.16 you will notice that the chromosomes in pair 23 of a female are the same, but in a male they look different. This pair of chromosomes determines whether the individual is male or female. Females have two 'X' chromosomes, but males have one 'X' and one 'Y'. When eggs are formed they all contain an X chromosome (because there is no alternative), but in sperm half the cells have an X chromosome and half a Y. The sperm cells carrying an X will make female babies, while those carrying a Y will make males. This creates a 50% chance of having either a boy or a girl, as shown in Figure 4.17, because sperm and egg cells combine at random, and roughly half of the sperm carry each chromosome.

How can diseases be inherited?

Sometimes a disease can be caused by a gene that makes the body function 'incorrectly' in some way, rather than by a micro-organism. This gene, and therefore the disease, can be inherited. An example of this is the disease **cystic fibrosis**, where the lungs and digestive system are clogged with thick mucus. This makes breathing and digesting food difficult, and leads to reduced life expectancy.

The cystic fibrosis allele is recessive, so the disease only appears when an individual has *both* alleles for cystic fibrosis. There are about 8500 cystic fibrosis sufferers in the UK but over 2 million people 'carry' the faulty allele. Someone who is heterozygous for a recessive trait, such as cystic fibrosis, won't suffer from the disease but may pass it on to any children. This person is called a **carrier**. If two people carrying the cystic fibrosis allele have children, there is a 1 in 4 chance that any child they have will suffer from the disease. This can be shown by using a Punnett square as used earlier in this chapter. Let us call the normal allele **C** and the cystic fibrosis allele **c**.

Parent genotype:	**Cc**	×	**Cc**
Parent phenotype:	normal		normal
Gametes:	**C** or **c**		**C** or **c**

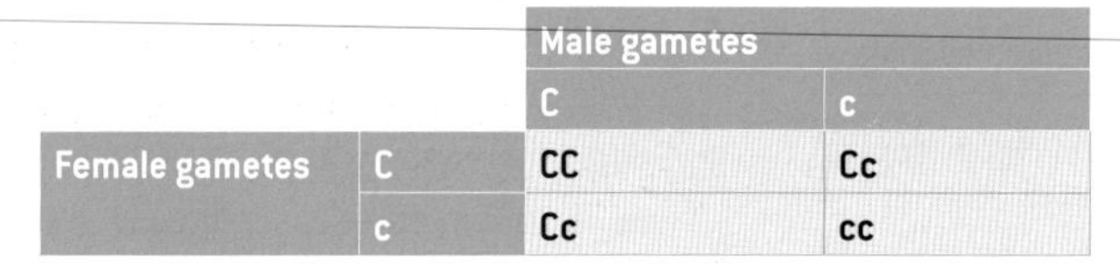

		Male gametes	
		C	c
Female gametes	C	CC	Cc
	c	Cc	cc

A child with alleles **cc** will suffer from cystic fibrosis.

It is likely that around half of the couple's children will carry the disease but not suffer from it.

QUESTION

8 If a sufferer lived long enough to have children, use Punnett squares to work out what the chances are of their children having the disease:

a if the other parent was a cystic fibrosis carrier

b if the other parent was not a carrier (i.e. genotype **CC**).

Is it a good thing to change genes artificially?

Scientists can now extract genes from one organism and put them into another, and can also 'swap' one gene for another. The introduction of genes into food plants is becoming more common and is known as **genetic modification**. In the 1980s, the first commercial genetically modified (**GM**) crop was developed that was resistant to insects and pests. It was the potato, and it was modified so that it made its own built-in **insecticide**. The insecticide was an insect poison normally produced by a type of bacterium that lives in the soil. The gene for the poison's production was transferred to potato plants, which then made the plant resistant to insect pests.

Resistance to **herbicide** is now common in genetically modified crops. By 2007, more than 50% of soya harvested across the world was genetically modified. In 2010, the European Commission approved a measure to allow different countries to choose for themselves whether or not to develop genetically modified crops.

Weeds compete with crops if left unchecked. For a great many years farmers have attempted to get rid of weeds by using chemicals called herbicides. However, selective herbicides that kill only weeds and not plants that are wanted are difficult to produce. A herbicide-resistance gene can be taken from a bacterium that normally grows in soil and transferred to a plant such as soya.

Unfortunately, there have been some potential problems with this technology to which some people have objected. For example, there is a possibility of herbicide-resistant plants escaping into the environment and flourishing. How could they be destroyed if herbicides cannot kill them? The answer is to ensure that the plants are sterile and can only reproduce asexually. Another unwanted side effect found in herbicide-resistant soya was that many plants split their stems in hot climates and could not support themselves.

The advantages of herbicide-resistant and insect-resistant plants are that far fewer chemicals need to be introduced into the environment to kill insects and weeds. Theoretically, high yields from crops can be maintained without affecting the environment. However, GM crops are new technology and more scientifically valid trials need to be carried out to decide if they are beneficial. There seem to be both advantages and disadvantages to GM crops.

The case for GM:

- Crops could be tailor-made to suit the varied farming conditions found throughout the world. In this way they could provide more nutritional value and a higher income.
- Energy-producing crops could save natural resources and so conserve the environment.

The case against GM:

- GM crops could reduce the developed countries' reliance on crops from developing countries. This could result in loss of trade and severe economic damage for the developing countries.
- Because of political reasons and mismanagement, there is doubt as to whether the populations of developing countries will actually receive the benefits from genetically modified crops in many cases.

These issues raise important political, ethical and trade questions that are not unique to modern biotechnology. They must be resolved at government and international level to maximise the benefits from gene technology.

Chapter summary

- Variation can be continuous or discontinuous.
- In asexual reproduction, offspring are genetically identical to each other and the parent, but sexual reproduction produces offspring that are different from the parents.
- New genes arise through the mutation of existing genes.
- Mutation rates can be increased by ionising radiation.
- Chromosomes are linear arrangements of genes and are made from DNA.
- Chromosomes occur in pairs and therefore so do genes.
- Different forms of genes are called alleles.
- DNA controls which proteins are made in a cell.
- Scientists can analyse an organism's DNA to produce a 'genetic profile'.
- In humans, one pair of chromosomes determine sex (XX = female, XY = male).
- When gametes are formed, the number of chromosomes is halved.
- Gregor Mendel's experiments provided the basis of modern genetics.
- Some diseases are caused by faulty alleles and can be inherited (e.g. cystic fibrosis).
- Genes can be transferred artificially from one organism to another.

Evolution

How do organisms become adapted to their environment?

We saw in Chapter 2 that organisms are well adapted to their particular environment. In Chapter 4 we looked at how variations in organisms occur. These two ideas are closely connected, because without variation, living things could never become adapted to their environment. How they do this is explained by the theory of **evolution**.

In a polar environment, nearly always snow-covered, it is best for animals to be white. They will be camouflaged and this will allow them to avoid being eaten or, if they are a predator, to be able to approach their prey without being noticed. Because of variation, any animal population will contain individuals of various 'shades', some darker than others. In a polar environment the paler animals will survive better than the darker ones because they will be better camouflaged. More of them will survive to breed, and they will then pass on their 'pale' genes to their offspring. In this way, there will be more of the paler individuals in the next generation. This process will continue in each generation and the population will become paler and paler until all the individuals will be basically white (although there will always be some variation in colour).

Figure 5.1 Polar bears have evolved to be perfectly suited to their frozen environment.

Over long periods of time, animal and plant populations change in ways that make them better adapted to their environment. This gradual change is called evolution. If the environment changes significantly for some reason then the process may have to be re-set, and new adaptations may need to evolve for the new conditions.

PRACTICAL CAN WE 'MODEL' EVOLUTION?

This activity helps you with:
- ★ understanding the use of models in science
- ★ judging the accuracy of models.

It is difficult to conduct experiments on evolution because the process often takes thousands of years. When scientists cannot carry out experiments directly on living organisms for some reason, they can sometimes 'model' the process they wish to investigate. We can do this to investigate the way in which camouflage evolves.

Apparatus
- * 100 plain cocktail sticks
- * 100 cocktail sticks, coloured green
- * stopwatch

Procedure
1. Count out 20 plain and 20 green cocktail sticks.
2. Mark out an area of long grass 1 m × 1 m.
3. One person in the group should now scatter the cocktail sticks across the grass in the area. It is better if the sticks are well spread, not clustered together.
4. A different person in the group now picks up cocktail sticks for 15 seconds. If 20 sticks are picked up before the time is up, stop. The person is acting as a predator and the collected cocktail sticks have been 'eaten'.

continued...

PRACTICAL *contd.*

Discussion Points

Models will only provide useful data if they are a fairly accurate reflection of reality (models are never 100% accurate). How accurate is this model of predation? Consider the following:

- How accurate are the cocktail sticks as prey?
- How accurately does the colouring represent variation in a natural population?
- How accurately does the person collecting the cocktail sticks behave as a predator?
- How accurate is the 'breeding' process in the model?
- Overall, do you think the model is accurate enough to provide a useful insight into real situations?

5 Draw a table to record your results in.
6 The remaining sticks in the area now 'breed'. Work out how many green and plain cocktail sticks remain, and double the number of each, by scattering new cocktail sticks back in the marked area. For example, if there are 12 green sticks left and 8 plain, then add another 12 green sticks and 8 plain sticks into the area. Record the new numbers of green and plain cocktail sticks.
7 Repeat steps **4** and **6** four more times, or until the population would exceed the number of cocktail sticks you have available.
8 Plot your results as a bar chart.

Analysing your results

1 What happens to the green and plain cocktail stick 'populations' over the 'generations' in this experiment?
2 Why did you plot the data as a bar chart rather than a line graph?

What is the 'theory of natural selection'?

The theory of natural selection gives a mechanism by which evolution is thought to occur. It is one of the most famous theories in science, and was originated by Charles Darwin.

In Darwin's time, many people believed that God had created every species separately and that one species never changed to give rise to new ones. Others believed in evolution but thought that changes came about by what the organism did, or what happened to it, in its lifetime. Darwin went on a 5-year scientific voyage of discovery on the ship *H.M.S. Beagle* between 1831 and 1836. He discovered many new species and noticed that, often, different species were variations on what seemed to be a common basic model. What was more, the variations all seemed to be linked with the organism's environment or lifestyle.

Darwin could not do experiments on evolution, which can take thousands of years. All he could do was to make careful observations and then try to devise hypotheses to explain them. He noticed that there were species of finch that were only found on certain islands in the Galapagos archipelago. They were similar to each other, and to a type of finch found on the South American mainland about 500 miles away, but each one had their own specific characteristics. In particular, their beak shape and size seemed to reflect the food they ate. This is shown in Figure 5.4.

Figure 5.2 Charles Darwin, the great British naturalist.

Figure 5.3 Variations in Galapagos tortoises. The one on the left has a domed front to its shell and a longer neck. It lives on an island with little ground vegetation so it has to reach up to feed on bushes. Its shell and neck adaptions allow it to do this.

tree finches	fruit eaters	parrot-like bills	*Camarhynchus pauper*		
	insect eaters	grasping bills	*Camarhynchus psittacula*	*Camarhynchus parvulus*	*Camarhynchus pallidus*
ground finches	cactus eaters	probing bills	*Geospiza scandens*		
	seed eaters	crushing bills	*Geospiza difficills*	*Geospiza fuliginosa*	*Geospiza magnirostris*

Figure 5.4 Galapagos finches, showing variations in beak shape and size according to diet.

Darwin considered his observations on the finches and tested them as evidence for or against the two existing hypotheses.

Hypothesis 1: God created all of the finches separately

It seemed very likely that finches on the Galapagos islands had originally come from the mainland, yet none of these exact varieties was present on the mainland. The Galapagos Islands were very different in many ways from the South American mainland, yet the finches were very similar to those on the mainland. Darwin wondered 'Why should the species which are supposed to have been created in the Galapagos Archipelago, and nowhere else, bear so plain a stamp of affinity to those created in America?' (taken from *On the Origin of Species* by Charles Darwin). However, science cannot disprove the existence or actions of a god, who is all-powerful and does not necessarily obey the laws of science. Any hypothesis involving the actions of a god cannot be a 'scientific' hypothesis.

Hypothesis 2: The characteristics had been acquired and then inherited

There is no known mechanism that would cause a beak to get bigger and stronger through trying to crush seeds, as the beak is not made of muscle. There is also no way that probing into bark for insects would cause a beak to get thinner and therefore better at probing. The observations do not fit this hypothesis.

For the next year or two, Darwin thought about what he had seen and eventually developed his **theory of natural selection** to explain the evidence. This was as follows:

- Most animals and plants have many more offspring than can possibly survive, therefore the offspring are in a sort of 'battle' for survival. This is the idea of **over-production**.
- The offspring are not all the same; they show **variation**.
- Some varieties must be better equipped for survival than others, because they are 'better fitted' to the environment. These will be more likely to survive (i.e. '**survival of the fittest**').
- Those that survive will **breed** and pass on their characteristics to the next generation (Darwin did not know the details of this, because Mendel had not yet done his experiments to discover genes).
- Over many generations, the best characters will become more common and eventually spread to all individuals. The species will have changed, or **evolved**.

Why do things become extinct?

Millions of species that existed in the past are no longer found on Earth – they have become **extinct**. This could happen for a variety of reasons.

1. The organism has failed to adapt successfully to its environment.
2. The organism has adapted to its environment to some extent, but another similar organism has adapted better. The less successful organism cannot compete and eventually dies out.
3. The organism has adapted to its environment well, but the environment suddenly changes and the organism cannot survive in the new conditions.

The first reason (complete failure to adapt) is virtually unknown. It might explain the extinction of a new mutation, but not a whole species.

The second reason is more common. For example, since the 1990s numbers of white-beaked dolphins have been declining around the Scottish coast, and this has been linked with an increase in numbers of another species, the short-beaked dolphin. Warmer sea temperatures have encouraged the short-beaked dolphin to move into the area from further south, and it is thought that it might be 'out-competing' the white-beaked form. The situation with red and grey squirrels we saw in Chapter 2 is another example of this.

QUESTIONS

1. Explain how Mendel's discoveries (see Chapter 4) helped to further explain the theory of natural selection.
2. Imagine a gene in an animal species that caused the animals to be very virile (successful at mating) but also caused an early death. Would such a gene eventually become common due to natural selection? Explain your answer.

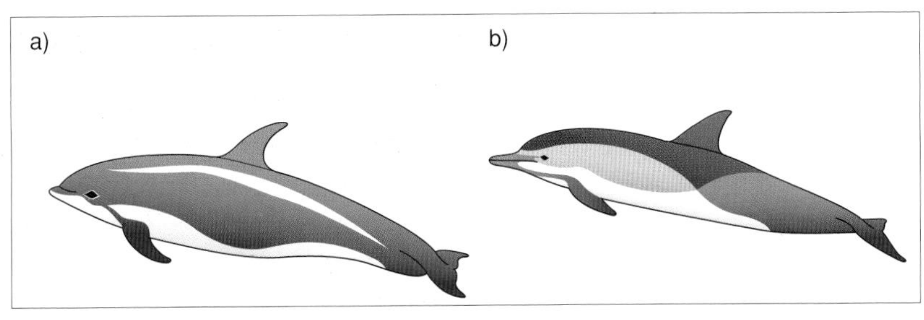

Figure 5.5 a) White-beaked dolphin; b) Short-beaked dolphin. The two species are competing in the seas around Scotland.

Figure 5.6 The dodo, which became extinct at the end of the seventeenth century – less than 100 years after its discovery.

The third reason is a common cause of extinction and is often linked with the activity of humans. A famous extinct animal is the dodo, a flightless bird that used to live on the island of Mauritius. It had adapted successfully to its environment and had no natural predators. When Dutch settlers colonised the island in 1638, they brought with them cats, dogs, rats (from the ships) and pigs. The dodos were easy prey for humans as they could not fly and had never needed to have evolved to be cautious. Humans ate quite a lot of them (although they didn't taste particularly good) and the cats, dogs and rats also fed on their eggs and young. Within a century the dodo was extinct, its environment having changed completely with the introduction of predators.

Why is natural selection a theory, not a hypothesis or a law?

A **theory** is an idea that seeks to explain scientific phenomena, which is generally accepted and has been supported by a range of evidence. A theory that has been disproved is discarded, but it is very difficult to conclusively prove that a theory is correct, as there is always the possibility that in the future new evidence may cause the theory to be changed or rejected. Darwin's theory of natural selection is a suggested explanation for how evolution works. There is lots of evidence that supports it, and no evidence that disproves it. It is accepted as true by the vast majority of scientists (but not all of them).

A **hypothesis** is like an educated guess. It is a suggested explanation for an event or observation, which is then tested by experimentation. In other words, a hypothesis does not have substantial evidence to back it up. If and when it does, and if it has widespread application, it will become a theory. Sometimes a theory involves a group of hypotheses. Natural selection is not a hypothesis because there is a lot of evidence that supports it.

A **law** is rather different. A law describes how something works or behaves and can be used to predict things, but it does not *explain* what happens. Newton's law of gravity describes how an object falls and can be used to predict what will happen when something is dropped, but it does not explain *why* there is an attraction between objects. There can never be any exceptions to a law – if there are, the law will need to be modified. Darwin's theory seeks to describe *how* evolution has happened, rather than describe it, and it cannot be used to make accurate predictions of exactly how an organism will evolve, so it is not a law.

What evidence is there for natural selection?

Now 100 die every week in Britain's filthy hospitals

SUPERBUGS: CRISIS GROWS

By Beezy Marsh
Health Correspondent

THE NHS is losing its battle against the deadly hospital superbug MRSA.

The number of cases has reached a new high, official figures revealed yesterday.

They showed that Britain has the worst record in Europe, with the risk of catching the infection more than 40 times as high here as in Denmark and the Netherlands.

The bug kills some 5,000 patients a year in the UK and costs the health service at least £1billion.

Its spread has been blamed on dirty wards and poor hygiene practices by doctors and nurses as well as the overuse of antibiotics.

Ministers reacted to the shameful figures by announcing a new cleanliness crackdown.

Health Secretary John Reid said 'infection czars' will be appointed in every hospital in England. He said he was declaring 'a call to arms' and demanding that hospitals put the problem at the top of the agenda.

'There should be teams going round hospitals, looking at the million little ways in which you can transfer bugs,' he said.

The Health Department also issued a league table naming and shaming the worst NHS trusts.

But critics said the initiative will simply create more [illegible] rather than making the NHS cleaner and safer for patients.

There are also fears that efforts to reduce MRSA cases will founder due to a lack of isolation wards and the pressure to treat growing numbers of patients. Ministers issued new

Turn to Page 4, Col. One

Figure 5.7 A variety of 'superbugs' have appeared that are very difficult to treat with antibiotics.

Natural selection usually takes a very long time, but in certain circumstances it can happen quite quickly. One example is the evolution of '**superbugs**'. These are micro-organisms that are resistant to the antibiotics which are normally used to treat infections.

Bacteria show variation like every living organism. Most of them are susceptible to antibiotics, but when you take these medicines there will always be a few bacteria that are naturally resistant and will survive. If there are only a few of them in you, they will not cause problems, but they could still be spread to others. If lots of people use the same antibiotics, after a while the susceptible bacteria are mostly killed off, leaving the resistant ones able to multiply. Given time, the whole population of bacteria will be resistant to the antibiotic (see Figure 5.8). The bacteria have not become a new species but in a fairly short period of time they have evolved to become resistant to certain antibiotics by a process of natural selection. To stop resistance building to new antibiotics, doctors try not to prescribe these drugs unless it is absolutely essential, and then try to use a variety of different antibiotics.

QUESTION

3 Explain why using different antibiotics to treat a given bacterial infection might stop the evolution of resistance.

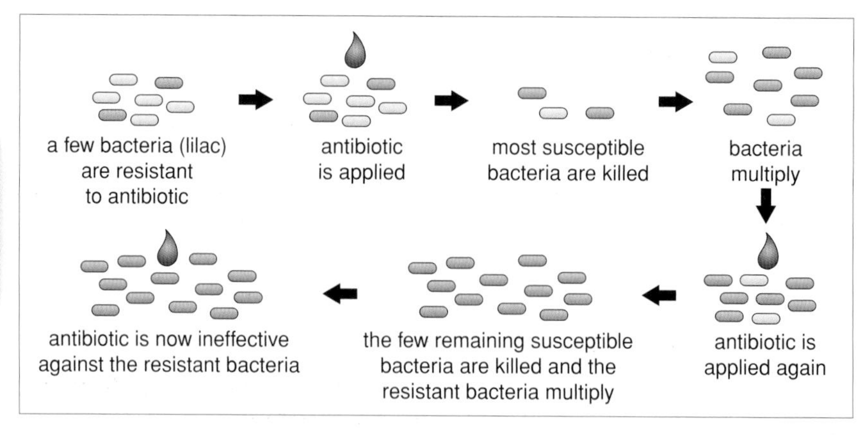

Figure 5.8 Evolution of resistance to antibiotics in bacteria. Note that in reality many more generations would be required before full resistance evolved.

Bacteria are useful organisms to use when studying evolution because they reproduce roughly every 20 minutes and so many generations can be studied in a short time.

TASK WHY HAS RAT POISON STOPPED WORKING?

This activity helps you with:
★ analysing data
★ drawing conclusions
★ judging the strength of evidence.

Warfarin is a common ingredient of rat poison but its use is now declining because rat populations are becoming resistant to it, so its effectiveness is decreasing. It is thought that this is because of natural selection, like the situation with antibiotics described above. Rats that have a natural resistance to warfarin have survived whereas those that are susceptible have been mostly killed off.

Different rat populations have differing levels of resistance to warfarin. Scientists sampled rat populations from five sites (A–E) in a fairly wide geographical area and tested them for warfarin resistance. The results are shown in Figure 5.9.

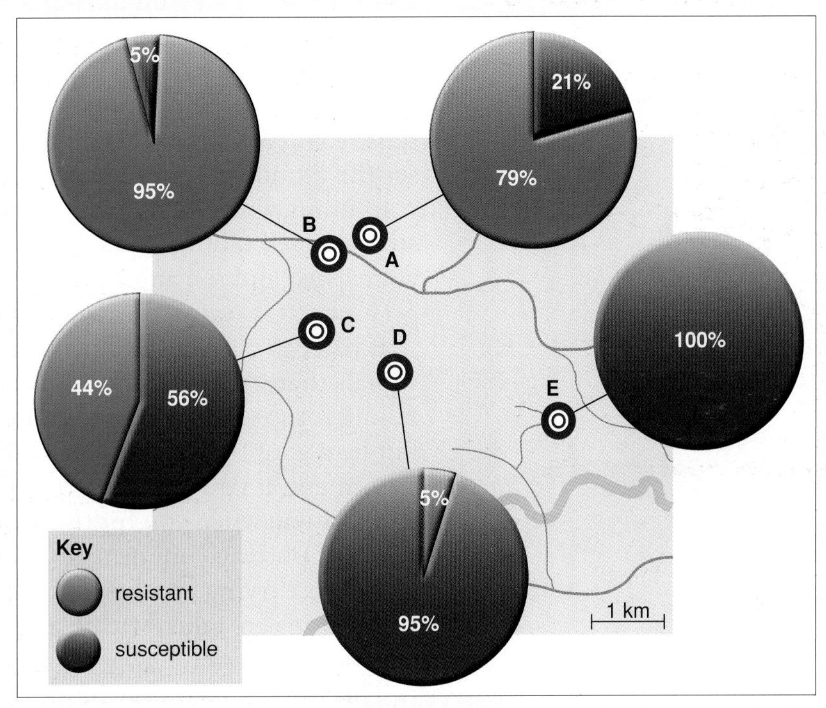

Figure 5.9 Results of warfarin-resistance tests on rat populations from five different sites (A–E).

Discussion Points

Look at the conclusions listed below. In each case, say whether you think the data support the conclusion and, if so, how strong the evidence is (with reasons).

1. Warfarin has never been used at site E.
2. The rat populations at the different sites are genetically different.
3. Resistance to warfarin first arose in site B.
4. Environmental conditions are different in sites D and E when compared with the other sites.
5. Resistant rats have migrated from areas A and B to area C.

Analysing the results

1. Look at Figure 5.9. Put sites A–E in order of resistance to warfarin, as shown by the rat populations.
2. Do you think the differences in resistance in the five sites are significant? Justify your answer.

Chapter summary

- Evolution depends on heritable variation in living organisms.
- Charles Darwin's theory of natural selection proposes a mechanism by which organisms can evolve to become adapted to their environment.
- Species that have not adapted, or have adapted to an environment that has now significantly changed, may become extinct.
- Although evolution can take many years, it can also sometimes happen rapidly (e.g. warfarin resistance in rats, antibiotic resistance in bacteria).
- Scientists may use 'models' to draw conclusions, which can be useful when experiments are difficult to carry out.

Response and regulation

If I stumble, will I fall?

We all trip up from time to time. Usually we will stumble and right ourselves, but occasionally we actually fall. Have you ever thought about what is going on during this process?

When you trip, the first stage in avoiding a fall is to realise that you've tripped. There are lots of signs – sense organs in your ears detect that you are no longer vertical; your muscles and skin may sense that you are not in contact with the ground; your eyes may see the ground coming towards you! All this information is sent to your brain, which then has to make a very rapid decision – can you stop yourself falling? This decision involves all sorts of factors like your speed and the angle of your body, and it's important the brain gets it right, because the decision will affect what it does next.

If you can correct the stumble, your brain must send out signals that will re-balance your body. For example, if you are falling forwards, it is best to put one leg forward to block any fall, but lean your body slightly backwards to balance your forward movement. If you are going to fall, you put your arms forward to support you when you hit the ground and protect your face and ribs, but putting your arms forward would be counter-productive if you were trying to stop yourself falling. All of these decisions and actions are taken within a split second, and your brain nearly always gets it right.

This is what your brain and nervous system do – they **control and co-ordinate** the senses and responses in your body. And they do it very well.

Figure 6.1 This footballer's brain has obviously decided he's going to hit the ground.

How does your brain get its information?

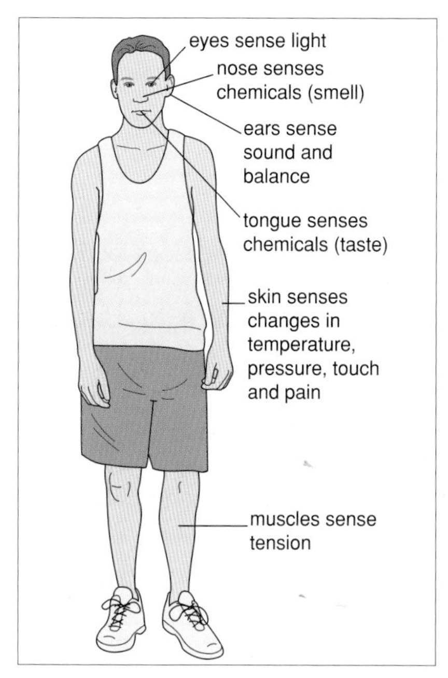

Figure 6.2 Some of the body's sense organs.

Information is fed to the brain constantly by a system of **sense organs** scattered around the body.

Sense organs are groups of special cells called **receptor cells**, which can detect changes around them, either internally or in the external environment. These changes are called **stimuli** (singular: stimulus) and include light, sound, chemicals, touch and temperature. Each group of receptor cells responds to a specific stimulus. The ears, for example, detect both sound and balance, but that is because they contain two different groups of receptor cells, one for sound and one for balance. The information from the sense organs travels to the brain and spinal cord (the **central nervous system**) along nerve cells (also called **neurones**).

PRACTICAL HOW SENSITIVE IS YOUR SKIN?

This activity helps you with:

- ★ analysing data
- ★ drawing conclusions
- ★ evaluating sources of error
- ★ improving experimental design
- ★ organising and displaying data.

One of the things your skin can detect is touch. Scattered around in the skin are touch receptors, which detect if the skin is touched near them. If the skin is touched in between the 'detection areas' the body will not detect it.

Some areas of your skin have these touch receptors more closely packed than others, so that the sensitivity of your skin varies on different parts of your body. We can see how sensitive different areas of the skin are by an experiment. We will test how good the skin is at distinguishing two separate contacts that are very close together. Sensitive areas will detect that they have been touched twice. Less sensitive areas will only detect one touch.

Figure.6.3 Diagram of the skin's surface showing areas covered by touch receptors. Any contact in the pink areas will be detected. Any contact in the spaces between the pink areas will not be detected.

Risk assessment

There are no significant risks associated with this experiment.

Standard procedure

Work in pairs. One person will be the test subject, the other will be the experimenter.

1. Bend the paper clip into a shape similar to that shown in Figure 6.4. By pressing on the sides, the gaps between the points can be altered.
2. Set the points so that they are 10 mm apart. Ask the test subject to look away.
3. Touch the paperclip to the skin on the test subject's fingertip 20 times. Sometimes, touch both points on the skin, sometimes just one point. Each time, the subject must say if he/she feels one point or two. Construct a suitable table to record your results. Record how many times the subject was right and how many times he/she was wrong.

Apparatus

* paperclip
* ruler

continued...

PRACTICAL contd.

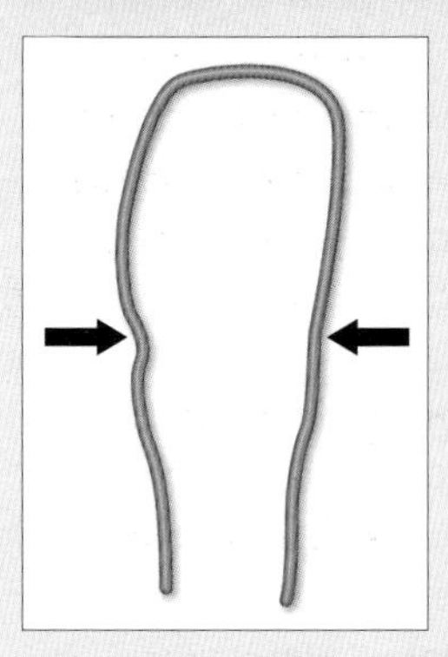

Figure 6.4 A paperclip unfolded for use in the experiment. Pressing as shown by the arrows can adjust the distance between the points.

4 Repeat the test on two further areas of skin – on the palm of the hand and on the back of the hand.
5 Test all three areas again with the paperclip points at different distances apart – 8 mm, 6 mm and 4 mm.
6 If time allows, the test subject and experimenter can swap roles and repeat the experiment.
7 Draw a suitable graph to display the data.

Analysing your data

1 What is your conclusion from the data? How strong is the evidence for this conclusion?
2 What are the possible sources of error in the experiment? Do the data indicate that any of these errors might have been significant? Is there anything that you could do to reduce these errors?
3 Why did you choose the type of graph you did? Would it have been possible to use any other sort of graph?

How does information travel in the body?

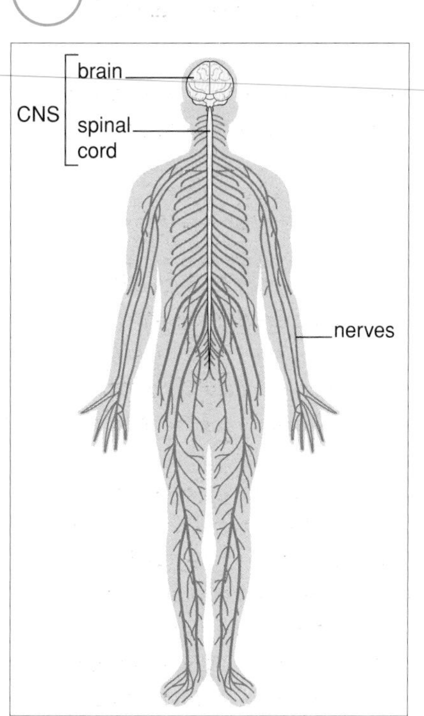

Figure 6.5 The CNS and nerves in the body.

Your brain and spinal cord make up the **central nervous system**, or **CNS**. Together they co-ordinate and control your body, but in order to do so they need to get information in from the sense organs and send out information to muscles in order to make things happen. The information travels as electrical signals along **nerves**.

Some nerves take signals to the CNS from sense organs and other nerves take information out from the CNS to the body. So, when your body senses a stimulus, the sense organ concerned sends a signal along a nerve to the CNS. Sometimes the message goes to the brain, which then decides what to do about it. If action is needed, the brain sends a signal along a nerve to the appropriate part of the body, which then reacts. This reaction is called a **response**. The time taken between the stimulus and the response is known as the **reaction time**.

PRACTICAL CAN YOU IMPROVE YOUR REACTION TIME?

This activity helps you with:
- ★ designing experiments (fair testing, repeats)
- ★ presenting data
- ★ analysing data (patterns and trends, significance of differences)
- ★ drawing conclusions (judging hypotheses, strength of evidence)
- ★ evaluating data and experiments (experimental error, repeatability).

Apparatus
* 30 cm ruler

This experiment looks at whether practice can improve reaction time. You can get a measure of reaction time by seeing how quickly someone catches a ruler that is dropped between their fingers. Our hypothesis is that **practice will reduce the reaction time**. Let's look at the evidence for this hypothesis, because a hypothesis is more than just a guess – it has to be built on scientific principles.

- It is known that the time taken for a signal to travel along a nerve is fixed. Every time you catch the ruler, the signals will travel along the same nerves, so the overall time taken will always be the same. This is evidence *against* our hypothesis.
- Catching the ruler does not just involve a simple pathway. You may be able to anticipate that the person is going to drop the ruler from tiny 'signals'; you may be able to develop greater powers of concentration. These are skills that could be improved by practice. This is evidence *for* our hypothesis.
- We know that similar things can be improved by practice. For example in cricket, fielders practise to improve their catching ability. Catching a cricket ball is a more complex action, yet this still provides evidence *for* our hypothesis.

The hypothesis suggested is a good one. There is evidence to back it up, yet we don't know for certain that it is true (if we did, there would be no point doing an experiment!).

Here is the basic method for working out reaction times.

Standard procedure

Work in pairs. One person is the experimenter, the other one is the subject.

1. Mark a pencil line down the centre of the subject's thumbnail on the right hand.

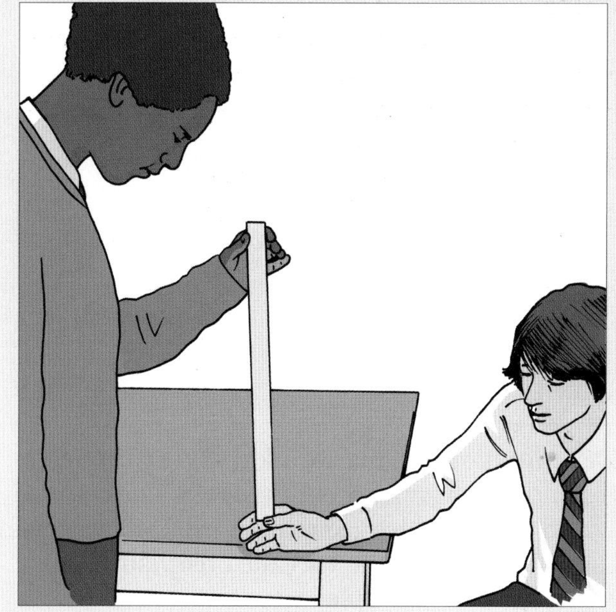

Figure 6.6 Carrying out the procedure.

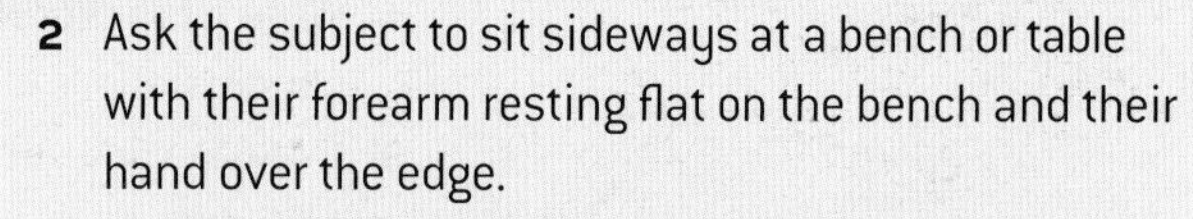

2. Ask the subject to sit sideways at a bench or table with their forearm resting flat on the bench and their hand over the edge.
3. Hold a ruler vertically between the subject's first finger and thumb with the zero opposite the line on the thumb but not quite touching the thumb or fingers. The distance between thumb and finger should be exactly the same for every trial.
4. Ask the subject to watch the zero mark and, as soon as you release the ruler, the subject must try to catch it between their finger and thumb to stop it falling any further. Construct a suitable table to record your results. Record the distance on the ruler opposite the mark on the thumb.
5. Repeat this four more times (to give five in total) and calculate the average distance. Convert this distance to a time using the graph in Figure 6.7.
6. Now use this method to design and carry out an experiment to test the hypothesis that practice improves the reaction time in this exercise.

continued...

PRACTICAL contd.

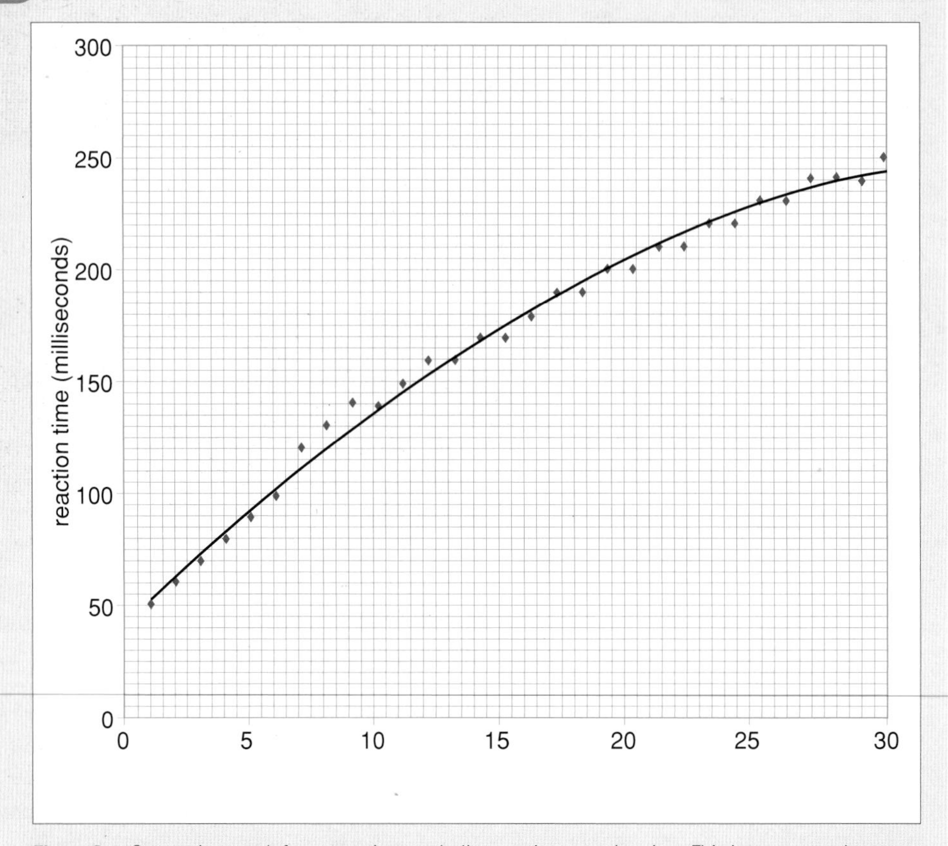

Figure 6.7 Conversion graph for converting catch distance into reaction time. This is necessary because the ruler accelerates as it falls.

Analysing your results

1. Write a full report of your experiment, describing your method, results and conclusions.
2. Evaluate your design. Could it be improved in any way?

Extension

A similar method could be used to investigate the variation between different people in their ability to catch the ruler.

1. Are some people naturally and consistently better at it?
2. If so, is this linked with any other factor (e.g. gender, playing of ball sports involving hand–eye co-ordination)?

Do plants respond to stimuli?

Plants do not move around, so you may think that they do not respond to changes in the environment, but they do. Their responses are slow and often involve growth towards, or away from, a stimulus. These growth 'movements' are called **tropisms**. There are several different kinds. Two examples are:

- **Phototropism.** This is growth in response to light. Plant shoots grow towards the light (*positive* phototropism) and roots grow away from light (*negative* phototropism).

QUESTIONS

1 Stems are positively phototropic and will grow towards a light source. Negative gravitropism causes stems to grow upwards away from the pull of gravity. How could we tell if a plant stem was negatively gravitropic *as well as* positively phototropic?

2 Look at Figure 6.8. What does this suggest about the strength of the phototropic and gravitropic responses in the shoots of this plant?

Figure 6.8 These seedlings have grown towards the light from a nearby window.

- **Gravitropism.** This is growth towards the pull of gravity. Roots of plants show positive gravitropism and stems show negative gravitropism.

Plants don't have nerves, so these responses are brought about by special chemicals called **hormones**. Hormones are produced in response to a stimulus and travel to another part of the plant, where they cause a response. Animals also have hormones, as we shall see later.

PRACTICAL: DOES THE COLOUR OF LIGHT MAKE ANY DIFFERENCE TO PHOTOTROPISM?

This activity helps you with:

★ understanding fair testing and controls

★ drawing conclusions.

Daylight is made up of white light, which is a mixture of all the possible colours of light. This experiment investigates whether plants are sensitive to specific colours of light. We will try out red and blue light.

To do this, you will need a special 'light box', as shown in Figure 6.9.

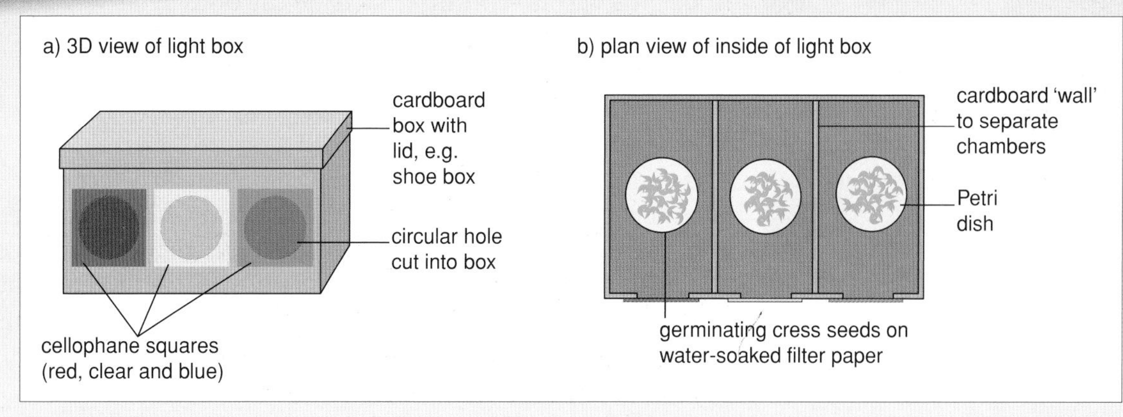

Figure 6.9 A light box.

continued...

PRACTICAL contd.

Apparatus
* for light box: rectangular cardboard box with lid, e.g. shoe box; extra cardboard for divisions; cellophane squares [red, blue and clear]; sticky tape
* sharp scissors
* desk lamp
* 3 Petri dishes
* 3 filter paper discs to fit the Petri dishes
* cress seeds

Risk assessment
- **You will be supplied with a risk assessment by your teacher.**

Procedure
1 First make the light box. Divide the box into three sections using the extra cardboard. Cut a circular hole at the front of each section and cover it with cellophane as shown in the diagram.
2 Place a disc of filter paper into each Petri dish and saturate it with water. There should be no excess water forming pools on the top of the paper.
3 Scatter cress seeds into each Petri dish.
4 Place one Petri dish into each chamber of the light box and replace the lid.
5 Place the lamp in front of the box so that it will shine through the cellophane sheets and switch it on.
6 Leave for several days. Each day, check for growth and water the filter paper if necessary.
6 Observe and record how the cress has grown in a suitable table.

Analysing your results
1 What do you conclude about the effects of different colours of light on phototropism?
2 Listed below are a number of features of the experiment. In each case, say whether you think they are likely to have significantly affected the result. Give reasons for your answers.
   a The amount of water added to the cress seeds was not measured.
   b The number of cress seeds in each dish was not identical.
   c The light was positioned more directly in front of [probably] the clear sheet than the red and blue ones.
3 We know that plants grow towards white light. Why did we use a sheet of clear cellophane when we knew what would happen in that chamber?

What do animal hormones do?

It was stated earlier that animals have hormones as well as plants. These chemicals are made in certain organs and travel around in the bloodstream, affecting various parts of the body. They are mainly used for medium- and long-term regulation, whereas nerves generally control quicker responses.

One of the most important things that hormones help with is keeping certain conditions inside the body relatively constant. This is very important for survival. The main conditions that are controlled by hormones in the body are described below.

Figure 6.10 If we lose water in sweat, we need to replace the lost water and maintain the balance in our body.

Water content

The concentration of chemicals in the cells can affect the essential chemical reactions going on inside the body. All of these reactions take place in water, which is therefore essential for life. Too little water (**dehydration**) may make the body fluids too concentrated and damage the body, but too much water can also be dangerous, as it dilutes the body fluids. The concentration of our bodily fluids is maintained within safe limits by hormones.

Glucose levels

One very important chemical in the body is the sugar, glucose. It is the main source of the body's energy, but can damage cells if present in high concentrations, so its level must be kept within a safe range. If blood sugar levels get too high following a meal, they can be reduced by the hormone **insulin**, which is released by the pancreas. Insulin is released into the bloodstream where it converts soluble glucose to insoluble **glycogen**, which is stored in the liver.

Some people have a condition where their body produces little or no insulin. If untreated, their blood sugar levels become dangerously high. This condition is called **diabetes**. Activities also influence blood glucose levels. Eating carbohydrates raises blood glucose, while exercise lowers it.

QUESTIONS

3 There is a condition called adipsia in which the patient loses the sense of 'thirst'. Why might this be dangerous?

4 The graphs in Figure 6.11 show the levels of glucose and insulin in a healthy person's blood over a 24 hour period.

a The graphs stop around midnight. What do you think will happen to the person's glucose and insulin levels over the next few hours? Give reasons for your answer.

b Insulin is known to decrease glucose levels, yet the graphs show several steep increases in insulin without any corresponding drop in glucose. Explain why this happens.

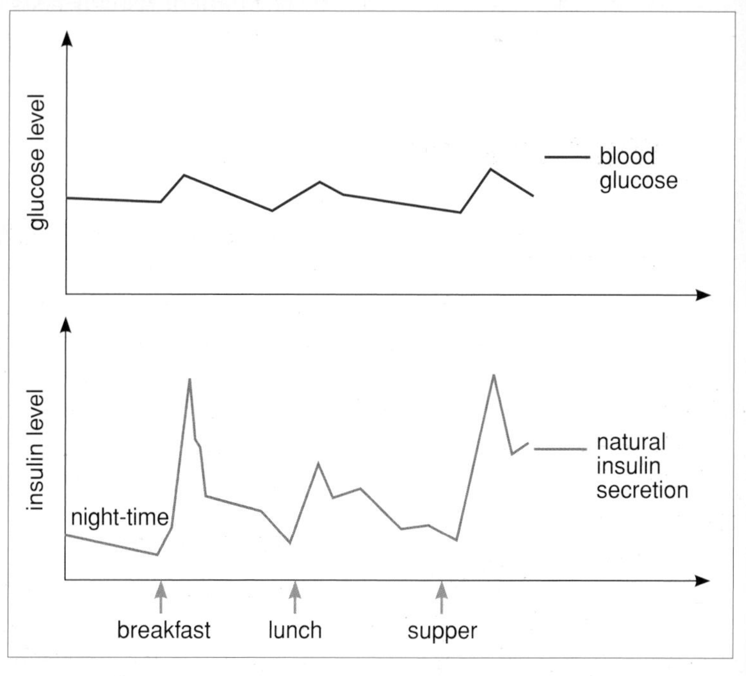

Figure 6.11 Comparison of glucose and insulin levels.

What happens if you get diabetes?

In **Type 1 diabetes** (the most common in young people), your body stops producing insulin. The blood sugar levels go up and up and your body tries to get rid of the excess sugar in the urine. The following symptoms are noticed:

- Sugar appears in the urine (a doctor has to test for this).
- The patient passes a lot of urine because the body is trying to 'flush out' the glucose.
- Because of all the water that is also being lost in the urine, the patient gets very thirsty.
- The body cannot actually use the glucose in the blood without insulin, so the patient feels very tired.

Doctors diagnose diabetes by the presence of **sugar in the urine**, because all of the other symptoms can be caused by other things apart from diabetes.

If diabetes is not treated, the blood sugar level will become so high that the patient will die. It cannot be cured but it can be managed so that the sufferer remains otherwise healthy. The treatment of Type 1 diabetes consists of three things:

- The patient has to inject themselves with insulin (usually before every meal) to replace the natural insulin that is no longer being produced.
- The diet has to be carefully managed. The patient has to eat the right amount of carbohydrate (which is the source of glucose) to match the amount of insulin injected.
- The patient usually tests his or her blood glucose levels several times a day, to make sure the level has not gone too high or too low.

There is another type of diabetes (Type 2), which is more common in older people. It is milder and can usually be controlled by tablets, or even by just being careful with the diet.

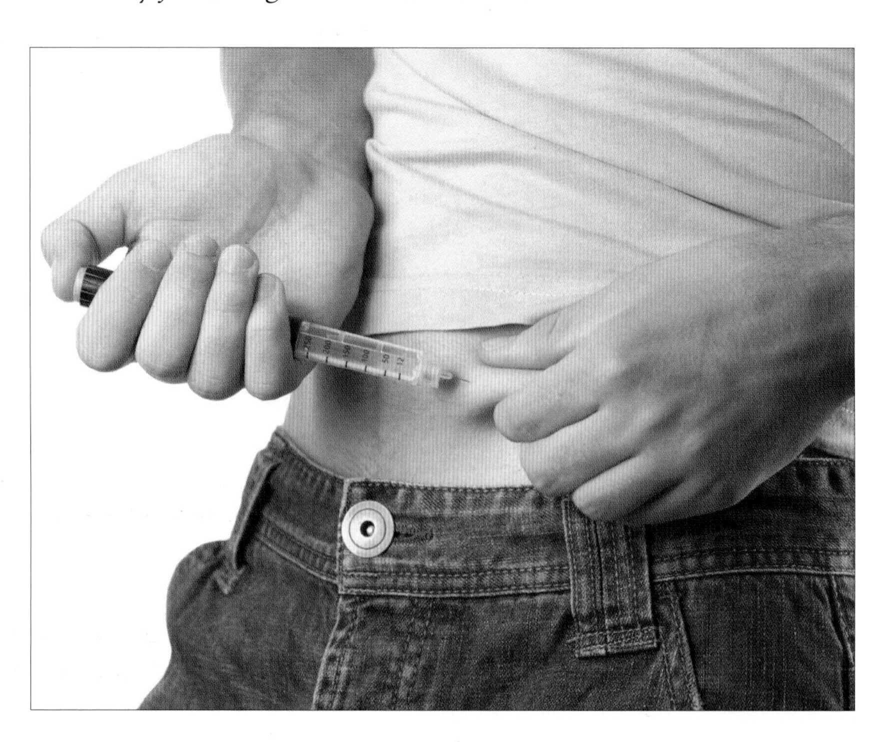

Figure 6.12 People with Type 1 diabetes have to inject themselves with insulin, sometimes several times a day.

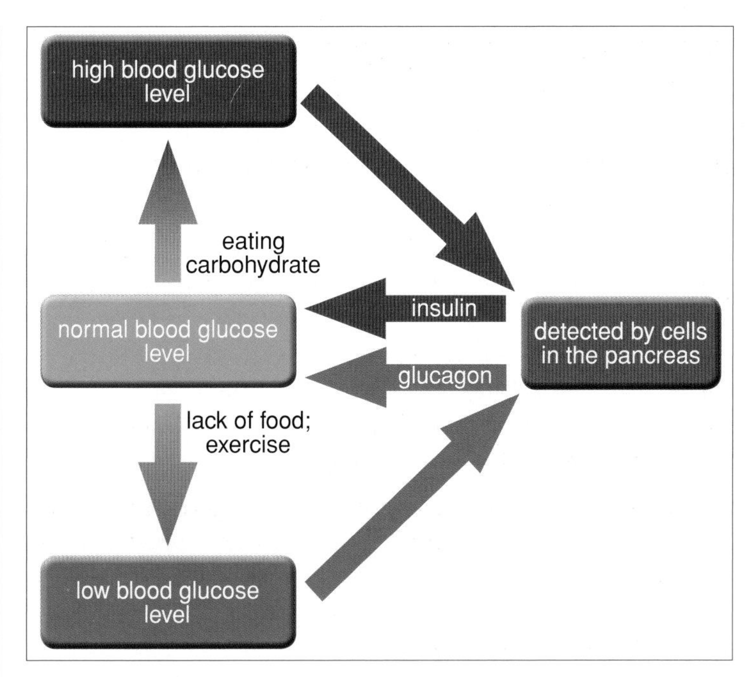

Figure 6.13 Negative feedback control of blood glucose levels.

Negative feedback

In healthy people, a rise in blood sugar sets in motion a series of events that results in the level being lowered again. In the same way, low blood sugar causes a process that raises the level. This mechanism is an example of **negative feedback**, and is summarised in Figure 6.13. It involves insulin and another hormone from the pancreas, glucagon, which raises blood sugar.

PRACTICAL HOW DO YOU KNOW IF THERE IS SUGAR IN URINE?

This activity helps you with:

★ experimental design
★ evaluation of experimental methods.

It is possible to do a complex and precise chemical analysis of urine to find the levels of sugar in it, but such tests are time consuming and expensive. A much simpler test can be done using testing sticks, which can be obtained from a chemist. A section of the stick changes colour when placed in urine and the colour change can be used to measure the amount of sugar in the urine.

You will test four samples of 'urine' to assess how well the diabetic patients supplying the samples are controlling their diabetes. Patients who are controlling their diabetes well will have no sugar in their urine. The more sugar present, the worse the control. The test used here is the Benedict's test for reducing sugars. Doctors would use special testing strips, which change colour without boiling.

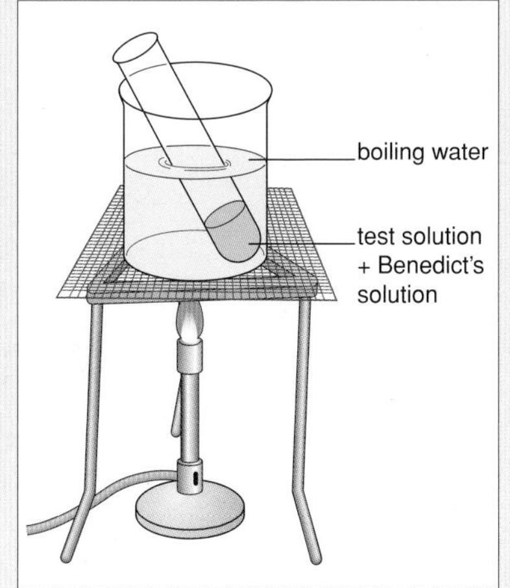

Figure 6.14 Apparatus for carrying out Benedict's test.

Apparatus

* Benedict's solution
* 4 'urine' samples labelled A–D
* 250 cm^3 measuring cylinder
* stopwatch
* 250 cm^3 beaker (to use as a water bath)
* 4 boiling tubes
* tripod
* gauze
* heating mat
* Bunsen burner
* eye protection

Risk assessment

- **Wear eye protection.**
- **You will be supplied with a risk assessment by your teacher.**

Procedure

1 Add 10 cm^3 of each urine sample to a boiling tube.
2 Test each urine sample by adding an equal amount of Benedict's solution, shaking gently to mix and placing in a boiling water bath.

continued...

PRACTICAL contd.

3 Leave in the water bath for one minute.
4 The solution will change from blue-green to orange to brick red, depending on how much sugar is in the urine.
5 Place patients A–D in order of how good their diabetic control appears to be.

Analysing your results

1 If this test is going to be useful, it must provide repeatable, reproducible and accurate results. How could you evaluate this method of testing urine?
2 How could you adjust the experimental method to give a *figure* for the amount of sugar in each sample (i.e. quantitative results)?

Extension

If left, the sugar naturally breaks down over time. Design an experiment to find how quickly the urine needs to be tested if accurate results are going to be obtained.

Controlling body temperature

Animals are kept alive by a series of chemical reactions that take place in their cells. These reactions are controlled by chemicals called **enzymes**, which are affected by temperature. If the body temperature is not kept constant, essential reactions could stop. Mammals and birds control their body temperature precisely using various mechanisms. Other animals have to use different means, for example moving into the Sun or shade to warm up or cool down. These mechanisms are not so precise.

The skin and temperature control

The skin is a complex structure that contains several different types of sense organ (Figure 6.15). It also performs various actions that help to control temperature.

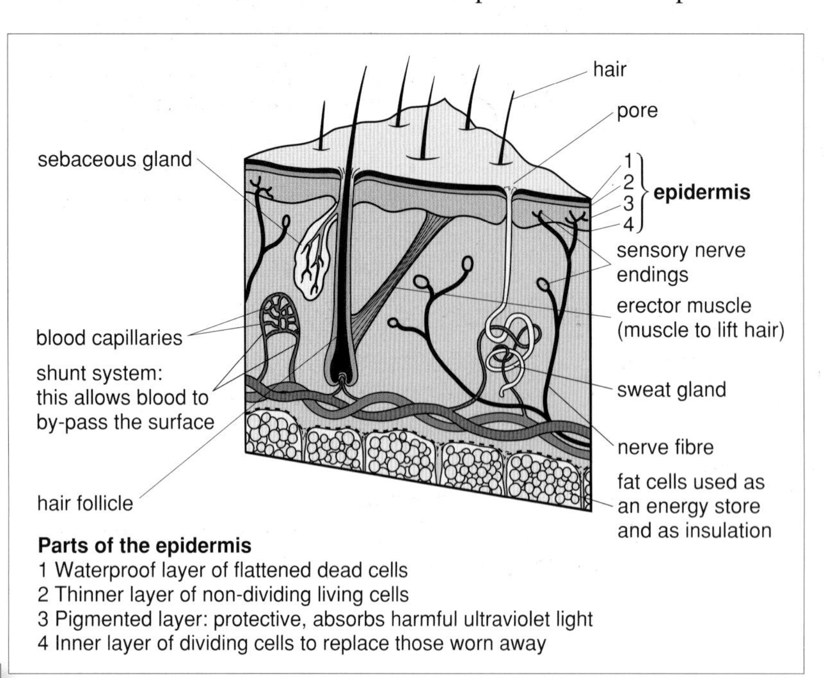

Figure 6.15 A section through human skin.

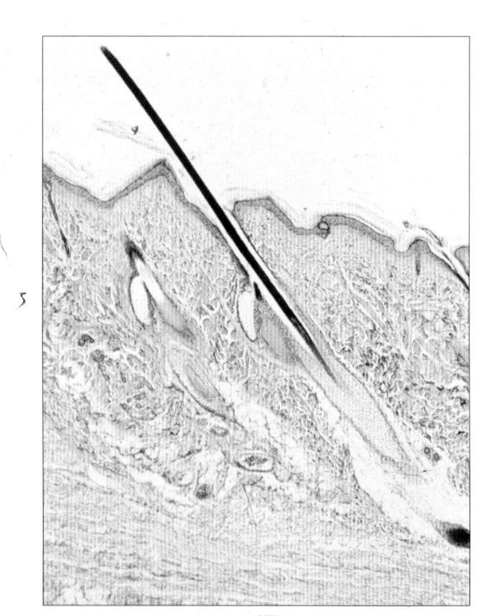

Figure 6.16 Photograph of a section through human skin. Can you identify any of the parts labelled in Figure 6.15?

- It produces sweat when the external temperature is high. The heat of the skin is used to evaporate the sweat and this in turn cools the skin.
- In hot weather, blood vessels near the surface of the skin dilate (get wider). This causes more blood to flow through them, so heat is lost to the atmosphere and the body cools down. In cold weather the vessels constrict (get narrower), so less blood goes to the surface to stop you getting colder. This is why you tend to go red when you're hot and pale when you're cold.
- The hairs in the skin stand on end in the cold to provide a thicker insulating layer to keep heat in. This isn't particularly effective in humans, because we don't have that much hair, but it's important in other mammals.
- When it gets cold, you shiver. The muscle contractions produce heat and so warm the blood a little.

These actions are illustrated in Figure 6.17.

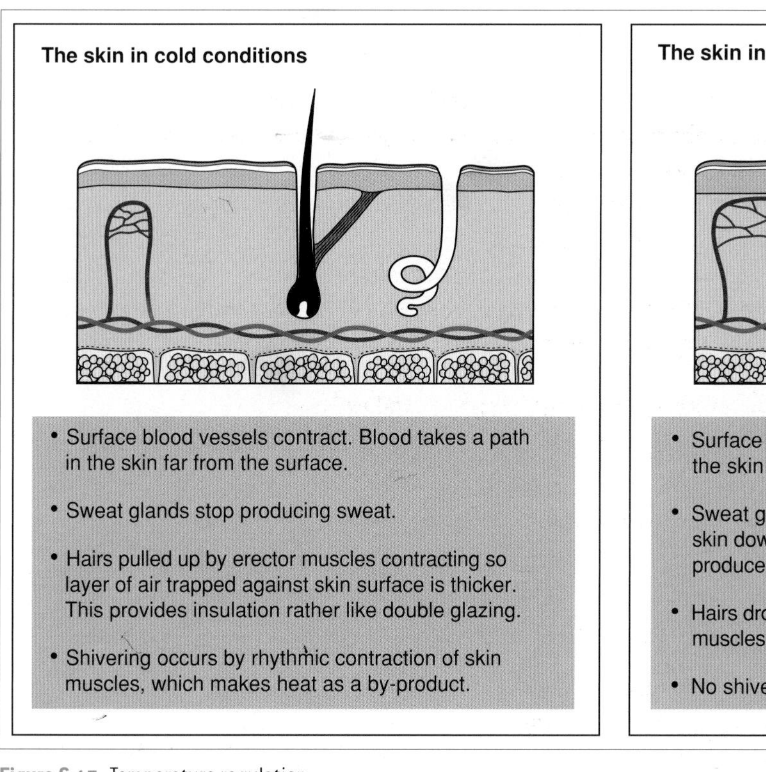

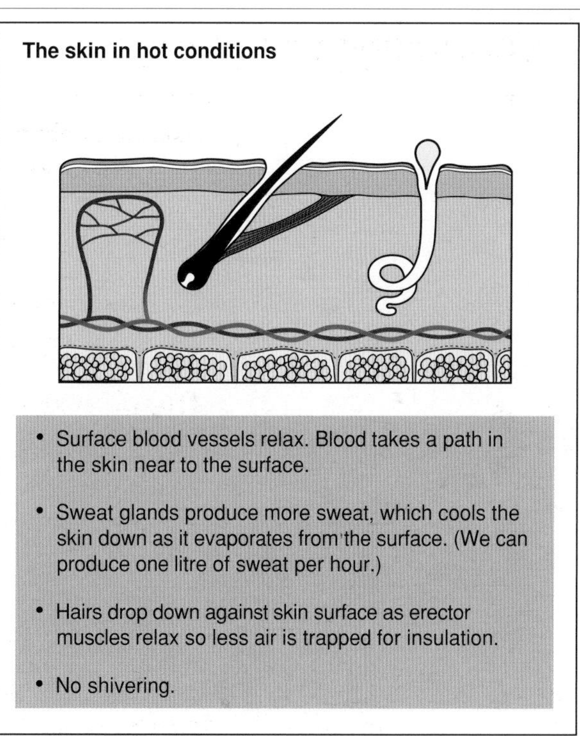

Figure 6.17 Temperature regulation.

QUESTION

5 Suggest something (other than moving into the Sun or shade as suggested above) that animals that cannot control their body temperature could do in order to:
  a warm up
  b cool down.

Why might antiperspirants be a bad idea?

People use antiperspirants to stop themselves sweating, but sweating is good – it is one of the ways that your skin helps keep the body's internal temperature constant, which is essential for good health.

However, antiperspirants are not really a bad thing. Not much air circulates under the armpits, so the sweat doesn't evaporate very well anyway. And they do stop you from smelling bad!

Chapter summary

- Sense organs are groups of receptor cells that respond to specific stimuli.
- The information from the sense organs is sent to the central nervous system (the brain and spinal cord) as electrical signals, along neurones.
- Plants respond to light (phototropism) and gravity (gravitropism). These responses are caused by hormones.
- Animals need to keep certain conditions (e.g. temperature, glucose levels, water levels) relatively constant so their bodies can work effectively.
- Diabetes is a condition caused by a lack of the hormone insulin. If it is untreated, blood sugar levels can rise to a fatally high level.
- Blood sugar control is an example of negative feedback.
- The skin has an important role in temperature control.

Health

Why do people get fat?

In order to survive and keep active, we need energy. Energy is needed for all living processes, and the more physical activity we do, the more energy we need. Different foods contain different amounts of energy, but the main type of food the body uses for energy is **glucose**, a sugar which we get from eating any **carbohydrate**. If we eat more carbohydrate than we need at the time, the body stores it in the liver for future use as a substance called **glycogen**. If we keep eating more food than we need, this store becomes full. The body then changes the carbohydrate into **fat**, which is stored under the skin and around the internal organs. In other words, we 'get fat'. These fat stores also increase directly if excess fat is eaten. If we eat less and exercise more, these stores get used up, but, because the body uses the glycogen store before the fat, it can take some time.

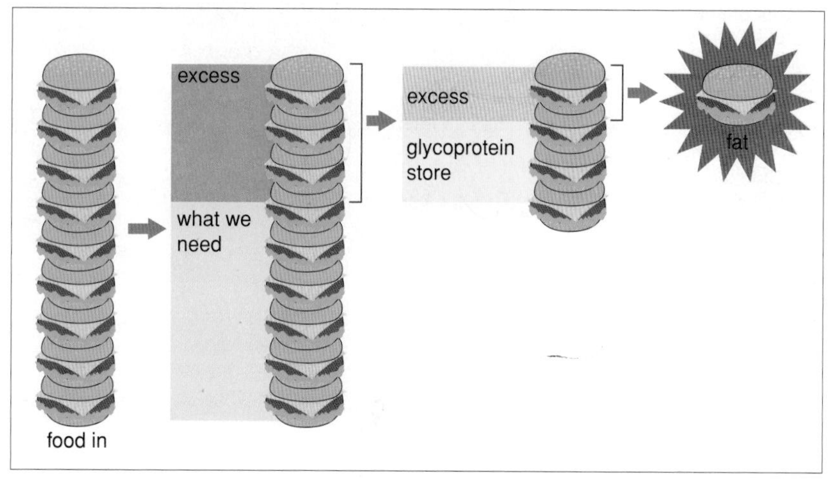

Figure 7.1 How carbohydrates are converted into fat.

If someone is overweight, it means that they have been taking in more energy than they need, probably for some time. To lose weight it is usually necessary to cut down on the energy taken in and also take more exercise to increase the energy used. To keep the weight down, a new, healthier food and exercise regime will have to be continued throughout life.

How can I tell if a food is unhealthy?

There is really no such thing as an 'unhealthy food'. All food provides useful nutrients, but it is important that certain types of food are not eaten to excess, otherwise health problems will arise. **Protein** is important in our diet for growth and repair of tissues.

It can be eaten in relatively large quantities without any ill effects because the body does not store it or convert it into fat. If you eat more protein than you need, the excess is usually broken down and/or excreted from the body. **Vitamins** and **minerals** are needed by the body to maintain health. They are only found in small quantities in foods, so we are very unlikely to take in too much. **Water** is an essential part of the diet. As we saw in Chapter 6, the amount of water in the body is controlled by hormones.
The foods we have to be careful not to eat too much of are **carbohydrates** and **fats** because they build up in the body and can lead to weight problems when stored as fat. On the other hand, both are a useful supply of energy and so cannot be cut down too much.

Most processed food has a table of nutrients on the packaging. This can be useful in deciding how much of it should be included in the diet. Figure 7.2 shows an example of a nutrient table from a packet of biscuits, and how the information could be used.

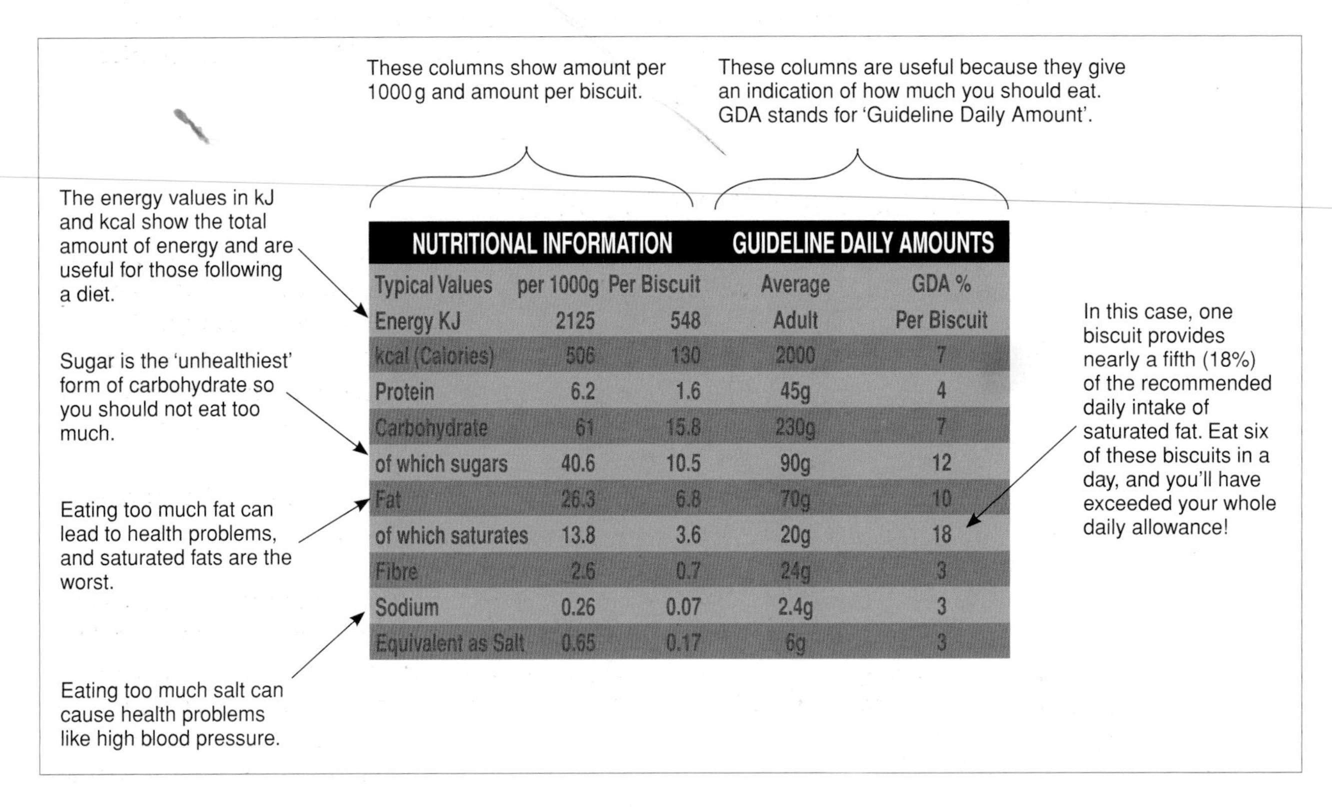

NUTRITIONAL INFORMATION			GUIDELINE DAILY AMOUNTS	
Typical Values	per 1000g	Per Biscuit	Average	GDA %
Energy KJ	2125	548	Adult	Per Biscuit
kcal (Calories)	506	130	2000	7
Protein	6.2	1.6	45g	4
Carbohydrate	61	15.8	230g	7
of which sugars	40.6	10.5	90g	12
Fat	26.3	6.8	70g	10
of which saturates	13.8	3.6	20g	18
Fibre	2.6	0.7	24g	3
Sodium	0.26	0.07	2.4g	3
Equivalent as Salt	0.65	0.17	6g	3

Figure 7.2 A nutrient table for a pack of biscuits. This is only an example and not all of these columns are always given. Some foods do not have a nutrient table at all.

Another cause for concern is the number of **food additives** in the food we eat. Additives are used to add flavour or colour, to increase shelf life, etc. They are identified by a code, which is always preceded by the letter 'E', so they are sometimes referred to as 'E numbers'. All E additives are basically safe, because in order to get an E number they are tested. However, some of them can have side effects if eaten in large quantities so it is best to limit the intake in the diet. The additives are not given in the nutrient table, but are included in the list of ingredients which is always found on packaged food.

SHOULD I EAT CRISPS?

This activity helps you with:
- ★ analysing data
- ★ communication skills.

Look at the information in Table 7.1.

Table 7.1 Nutrient table from a packet of crisps.

Average values	Per 100 g	Per bag	Guideline daily amount (%)
Energy (kcal)	514	257	13
Protein (g)	6.9	3.5	7
Carbohydrate (g)	53	26.5	12
of which sugars (g)	1.2	0.6	0.6
Fat (g)	30.5	15.3	22
of which saturates (g)	3.9	2.0	10
Fibre (g)	4.5	2.3	10
Sodium (g)	0.7	0.4	16
Equivalent as salt (g)	1.8	0.9	16

On the basis of these figures, if someone asked you whether they should eat crisps, what would you recommend, and why? In your answer, consider calorific value, sugars, fats and salt levels.

How much energy does food contain?

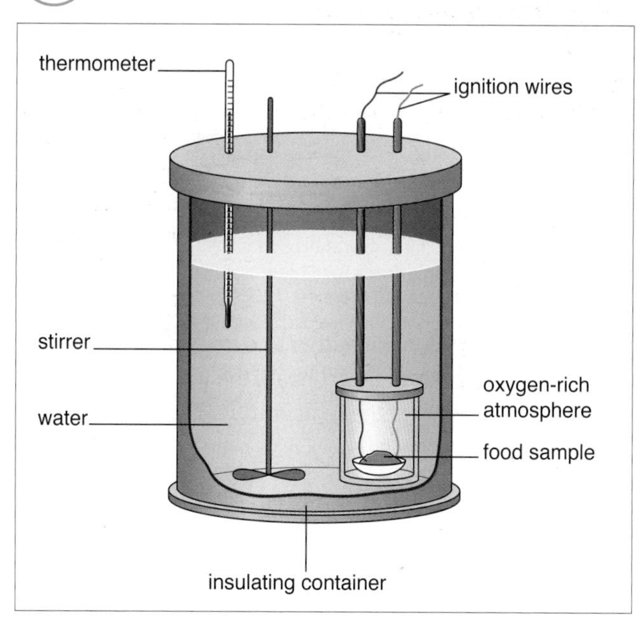

Figure 7.3 A bomb calorimeter.

People can tell how much energy is in packaged food by looking at a nutrient table, but how is this figure arrived at? Food scientists get the figures by burning the food so that the energy is released as heat. They measure the heat given off to see how much energy a known mass of food produces.

A special piece of apparatus called a food (or 'bomb') calorimeter is used for this. Figure 7.3 shows an example.

It is very important that the food is completely burnt (to get all the energy from it) so the sample is burned in an oxygen-rich atmosphere. The heat given off is measured by the increase in temperature of the water. The stirrer ensures that the heat is evenly spread and the container is insulated so that no heat is lost to the atmosphere.

PRACTICAL HOW DO PEOPLE KNOW HOW MUCH ENERGY IS IN FOOD?

This activity helps you with:
- ★ assessing experimental error
- ★ improving the design of apparatus.

Apparatus
* clamp stand
* boiling tube
* mounted needle
* thermometer
* measuring cylinder ($100\,cm^3$)
* snack food (e.g. some sort of potato-based snack that can be pierced by a mounted needle)
* Bunsen burner
* heatproof mat
* top pan balance
* eye protection

The energy from burning food can be measured in the laboratory using simpler apparatus than the bomb calorimeter in Figure 7.3.

The energy, in joules, in 1 g of the food tested can be calculated as follows:

$$\text{energy (J)} = \frac{\text{rise in temperature (°C)} \times \text{volume of water (cm}^3\text{)} \times 4.2}{\text{mass of food (g)}}$$

1000 joules = 1 kilojoule
So, the answer can be divided by 1000 to convert it into kilojoules.
One calorie is the amount of energy needed to raise the temperature of $1\,cm^3$ of water by 1 °C. 1 calorie = 4.2 joules.

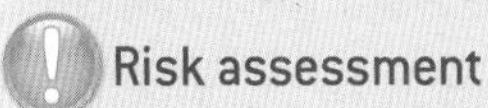

Risk assessment

- **Wear eye protection.**
- **You will be supplied with a risk assessment by your teacher.**

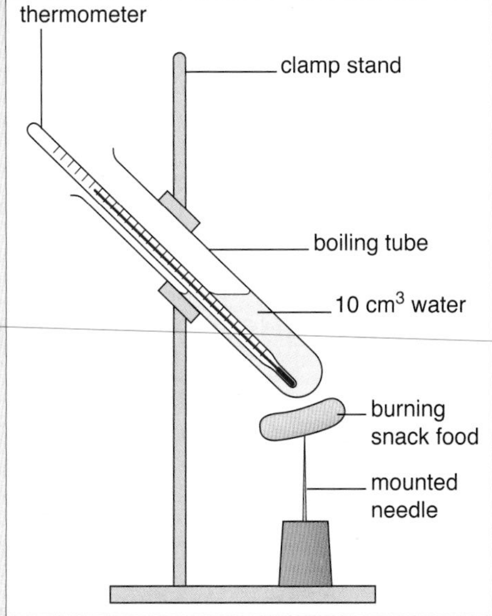

Figure 7.4 Apparatus used to measure the energy value of a snack food.

Procedure

1. Weigh a piece of the snack food and record the weight.
2. Measure $10\,cm^3$ of water using the measuring cylinder and pour it into the boiling tube.
3. Set up the apparatus as shown in the Figure 7.4.
4. Measure and record the temperature of the water in the boiling tube.
5. Carefully spear the snack food on the mounted needle, taking care not to lose any.
6. Set light to the snack food using the Bunsen burner and immediately hold the burning food under the boiling tube so that it heats the water.
7. When the food is completely burnt, record the temperature of water in the boiling tube again.
8. Calculate the amount of energy (in kilojoules) in 1 g of the snack food.

Discussion Points

The values obtained from this technique are usually inaccurate. These are possible sources of error compared with using the calorimeter:

- The food is burned in air rather than oxygen, which may mean that the food is not completely burned.
- The water is not stirred.
- The apparatus is not insulated.
- The thermometer bulb is very near to the heat.
- A lot of heat escapes into the surrounding air.

1. How serious an error do you think each of the above factors may have caused in your experiment?
2. Could the apparatus be modified in any way to reduce some of these errors?
3. How could you discover *how* inaccurate the calculated value might be?

How does my lifestyle affect my health?

Everyone makes lifestyle choices. Certain activities are known to either cause potential harm to your health or to have health benefits. Activities that benefit health may not appeal to some people, and harmful habits might be enjoyed, so everyone has to make choices about their own personal attitudes to a healthy lifestyle. However, such choices cannot really be made without reliable information about the effects of different activities.

The following are known to have health benefits:

- regular exercise
- balanced diet
- eating adequate amounts of a variety of fruit and vegetables.

These are known to be harmful to health:

- smoking
- excessive consumption of alcohol
- taking recreational drugs
- overeating
- having a diet high in saturated fat or salt
- sedentary lifestyle (lack of exercise)
- stress.

There are also some activities that have both beneficial and harmful effects, and others where there is not yet enough scientific evidence to be sure if there is an effect on health. For example, taking aspirin stops pain and may benefit people with heart conditions, but it also can cause the lining of the stomach to bleed. Some people believe that using a mobile phone may cause some damage to the brain, but the evidence is inconclusive.

TASK DOES ALCOHOL DAMAGE YOUR BRAIN?

This activity helps you with:
- ★ analysing data
- ★ judging the strength of evidence
- ★ evaluating experiments.

It is well known that alcohol is a drug that damages the liver, and also has temporary effects on the brain. There is also some evidence that drinking alcohol can cause long-term brain damage, though this has not been established.

A study in 2001 looked at one aspect of brain damage – shrinkage of the brain's frontal lobe. The study looked at 1432 people who had different levels of alcohol consumption, although none was alcoholic. The level of drinking was as follows:

- 667 were 'abstainers' (they didn't drink alcohol at all)
- 157 were light drinkers
- 362 were moderate drinkers
- 256 were heavy drinkers.

The graph in Figure 7.5 compares abstainers and heavy drinkers of different ages.

continued...

TASK contd.

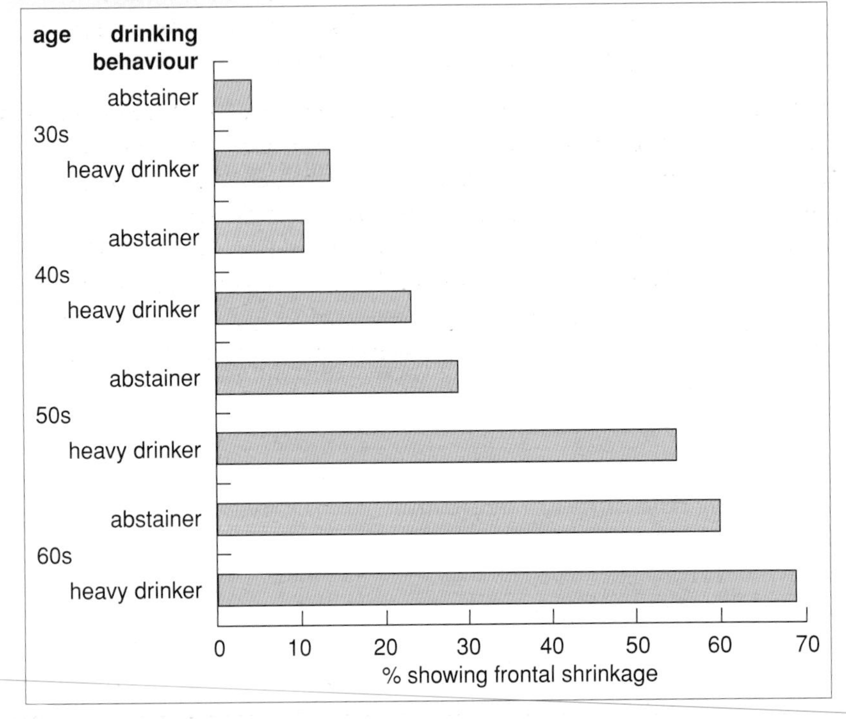

Figure 7.5 Graph showing percentage of frontal lobe shrinkage in heavy drinkers and non-drinkers of different ages.

1. Why do you think the researchers divided the data into different age groups?
2. The data provide evidence that frontal shrinkage is greater in heavy drinkers than in abstainers. State two features of the data that add strength to the evidence.
3. Can you say that heavy drinking *causes* brain shrinkage?

Extension

Use the internet to produce a report on the effects that alcohol and other drugs have on the chemical processes in the body.

Discussion Point

Do you think the sample size is sufficient to draw valid conclusions?

What strategies can we use to prevent, treat and cure disease?

There is a saying, 'prevention is better than cure', and that is certainly true of diseases. Once someone gets a disease they will suffer from certain symptoms specific to that disease, which will generally be unpleasant. If the body's immune system cannot defeat the disease on its own, the patient will require drugs or other treatments, either to cure the disease or to make it more bearable. This is inconvenient for the person concerned and expensive, either for the person or the taxpayer, or both. The drugs can also have unpleasant side effects.

The medical profession uses three important words to describe disease:

- **Chronic** conditions are long lasting and often (though not always) incurable. Medical treatment usually focuses on managing the disease to remove or reduce the symptoms and maintain a good quality of life. Examples include diabetes, epilepsy, most cancers, heart disease and arthritis.
- **Acute** conditions have a sudden onset and don't usually last for very long. That may be because they can be quickly cured (naturally or with drugs) or because the patient dies soon after onset. Examples include the common cold, flu and acute liver disease.

QUESTIONS

1 Muscular dystrophy is an incurable condition that reduces life expectancy. The common cold is a disease that causes relatively mild symptoms. Although there is no medical 'cure' for the cold, the body's natural defences overcome the disease after several days. Which of the medical terms: chronic, acute and severe, should be applied to:
   a muscular dystrophy
   b the common cold?

2 Suggest possible causes of the steady decline shown in the graph in Figure 7.6.

- **Severe** means that the disease is serious and in some cases life threatening. People sometimes think that the words chronic and acute imply that the disease is really serious, but they don't. You can have severe chronic diseases, and severe acute diseases, but not *all* chronic and acute diseases are severe.

Chronic diseases need managing. This usually involves taking certain drugs to slow down the disease, counteract it or reduce symptoms. Sometimes other technology is used to help the condition (e.g. diabetics test their blood using a glucose meter, which allows them to know how much insulin to inject and to manage their diet).

In acute disease the focus is usually on drug treatments, but surgery may also be necessary and technology plays a big part in devising surgical techniques that are effective and involve minimal risk to the patient.

TASK

HOW IS HEART DISEASE TREATED?

This activity helps you with:
★ analysing data.

Heart disease, which can take various forms, is a common condition throughout the world. Over the years, a huge amount of research has taken place to develop drugs and medical technologies to treat the disease. A lot of effort has also gone into prevention measures.

Figure 7.6 shows the death rates from heart disease in Europe between 1995 and 2010.

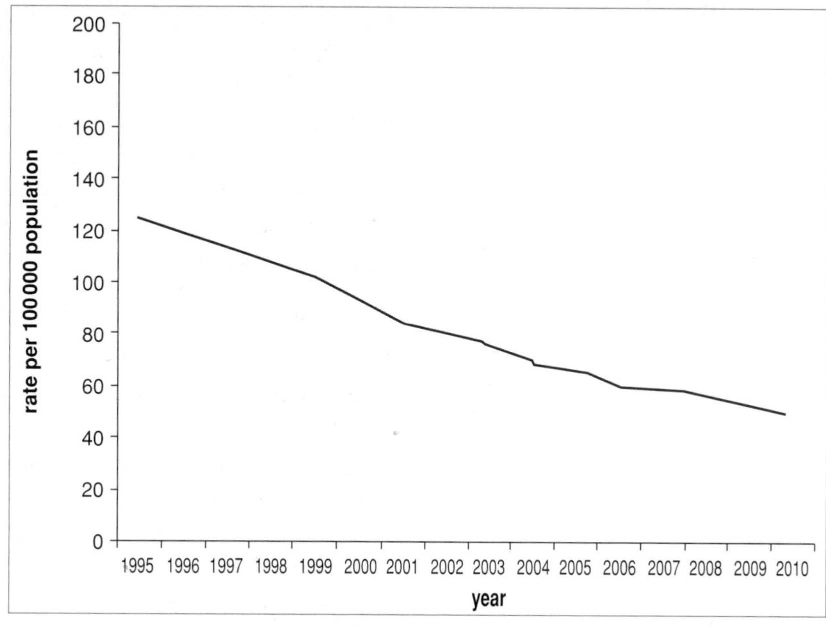

Figure 7.6 European death rates from heart disease, 1995–2010.

Some measures taken to treat heart disease are listed in Table 7.2. Current research includes the development of artificial hearts and growing heart cells from stem cells to replace damaged heart tissue.

Table 7.2 Some of the measures taken to treat heart disease.

Prevention	Advice on healthy diet; promotion of regular exercise regime; medical checks of at-risk patients
Drugs	Drugs to reduce blood cholesterol, reduce blood pressure, treat angina, prevent rejection of heart transplants
Technology	Advances in surgical techniques for heart operations including transplants, pacemakers, stents (for keeping blood vessels open)

Do all drugs have side effects?

A drug is a chemical that alters the way the body works in a certain way. The main effect is usually the one that is needed and the reason that the drug is taken. However, drugs often have other effects that are not needed, called side effects. These may be minor or serious. If there are serious side effects from taking a drug, a decision has to be made as to whether the benefit from the drug is worth the side effect. If not, the drug will not be released for use. Aspirin is a commonly taken drug that has a number of side effects. These are minor provided the tablets are taken in small quantities and with water, although some people with pre-existing medical conditions may need to avoid using aspirin.

PRACTICAL THE EFFECT OF ASPIRIN ON AMYLASE ENZYME

This activity helps you with:
- ★ presenting data
- ★ analysing data.

Apparatus
- * paper cup, containing 40 cm^3 tap water
- * 40 cm^3 starch suspension
- * 5 cm^3 ethanol
- * 5 cm^3 salicylic acid (aspirin) solution
- * iodine solution
- * 4 test tubes
- * test tube rack
- * 3 × 5 cm^3 plastic syringes
- * 1 cm^3 plastic syringe
- * stopwatch
- * glass-marking pen
- * eye protection

One common side effect of drugs is to inhibit the activity of enzymes that may be important in the body. This experiment looks at whether aspirin affects the activity of a given enzyme.

The enzyme used is amylase. In the body amylase catalyses the conversion of starch into maltose. One of the places it is found is in saliva, which starts the breakdown of starch in food. The activity of the enzyme can be tested with iodine. Iodine turns blue-black in the presence of starch, but remains orange-brown with maltose.

$$\text{starch} \xrightarrow{\text{amylase}} \text{maltose}$$

Hypothesis

Aspirin inhibits the action of amylase enzyme.

Risk assessment

- **Wear eye protection.**
- **You will be supplied with a risk assessment by your teacher.**

Procedure

1 Number the test tubes from 1–4.

2 Using a 5 cm^3 syringe, put 5 cm^3 starch suspension into each tube.

3 Fill a clean 5 cm^3 syringe with salicylic acid solution and add 0.5 cm^3 to tube 2, 1 cm^3 to tube 3 and 2 cm^3 to tube 4.

4 Take water from the paper cup into your mouth. Swill it around thoroughly to get a sample of saliva rich in amylase, then spit it out into the cup.

continued...

PRACTICAL *contd.*

Table 7.3 Colour chart for iodine solution

Colour	Unit scale
Brown, pale	4
Brown, dark	3
Brown-purple, pale	2
Brown-purple, dark	1
Blue	0

5 Using a clean 5 cm^3 syringe, add 2 cm^3 saliva solution to each tube and gently rotate the tubes to mix their contents. Start the stopwatch.
6 After 10 minutes, use the 1 cm^3 syringe to add 0.5 cm^3 iodine solution to each tube (see Figure 7.7). If the colours are faint add a further 0.5 cm^3 iodine solution to each tube. Estimate the colour intensities of each tube using the scale in Table 7.3. Zero on the scale means the starch has been unaffected by the amylase; 4 on the scale means the starch has been completely converted to maltose by the amylase.
7 Record your results in a suitable table.
8 Display your results in a graph.

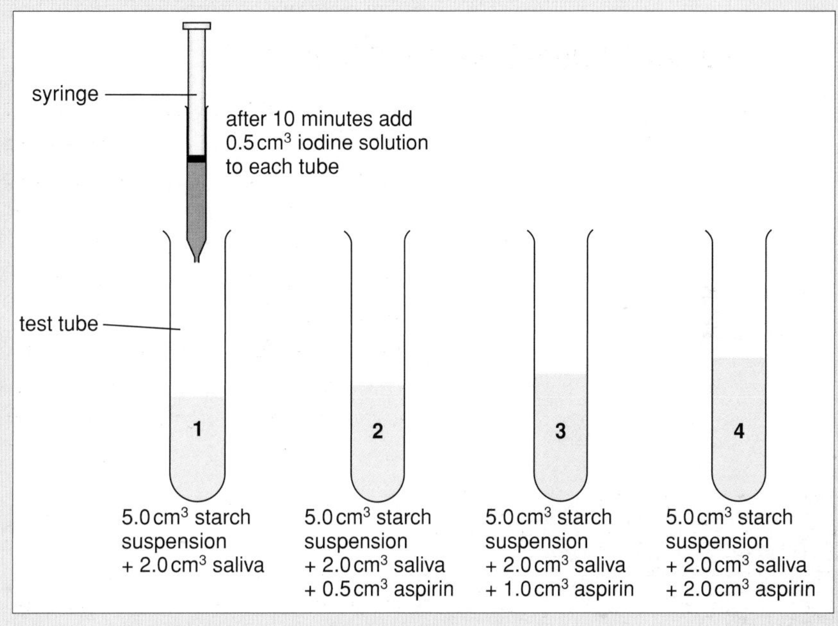

Figure 7.7 Apparatus set-up for the experiment.

Analysing your results

To what extent do your results support the hypothesis 'Aspirin inhibits the action of amylase enzyme'?

TASK

SHOULD ANIMAL TESTING BE USED TO TEST NEW DRUGS FOR SIDE EFFECTS?

This activity helps you with:
★ researching information
★ judging secondary information for validity and bias.

Suppose a new drug is discovered that has the potential to cure or treat a human disease. It is likely that the drug will have some side effects, but how serious those might be will not be known. The scientists, at this stage, cannot test the drug on human volunteers (the law will not allow it) because in extreme cases it might kill them. The alternative is to test the drugs on animals, usually mammals, because humans are mammals. In Britain the law says that any new medical drug must be tested on at least two different types of live mammal, and that one must be a large non-rodent type animal.

Many people are against animal testing but the arguments are complex and there is no simple solution. Consider the following points:

- In the UK animal testers have to be licensed. If they mistreat the animals or use them for unnecessary research they may lose their licence. However, 'not mistreating' means being as humane *as possible*. Sometimes the tests are unpleasant (for example, taking blood samples) but necessary – such tests are not classed as 'mistreating' the animal.

continued...

- There are alternatives to animal testing [for example, testing on human cells grown in the laboratory] but these are not suitable in all cases and a body is more complex than isolated cells, so the results may not be relevant.
- Humans and rats are different animals, and protestors say that this means that results using experimental animals are not necessarily relevant to humans. Testers say such differences can be taken account of in their conclusions. It is likely that animal test results are not entirely applicable to humans but provide some useful information.
- Some people believe that it is unethical to use animals because they cannot 'volunteer' for tests and that animal life should be regarded as being just as valuable as human life. If you believe that, then animal testing is clearly wrong, even if it saves human lives.
- In the past, drugs that would not have been licensed for use without animal testing *have* saved human lives.
- It is likely that animals do not have emotions like humans. They may not feel fear in the way we do, but it is very difficult to be certain of this.

Do some further research into the issue of animal testing and come to a personal conclusion about its value and whether/how it should be continued. Use evidence to justify your opinions, taking care to avoid bias.

Figure 7.8 What are your views on animal testing?

Chapter summary

- Different foods have different energy content and nutritional make-up.
- Excess energy from food is stored as fat by the body.
- Care has to be taken with the amounts of sugar, fat, salt and additives in the diet.
- Personal lifestyle choices can significantly affect health.
- Taking alcohol and other drugs affects chemical processes in the body and can be harmful to health.
- Science and technology provide answers to, and treatments for, many health problems.
- Some conditions can be prevented if suitable information is available.
- Drugs can be used to treat disease but nearly always have side effects.
- Animal testing has proved useful in developing helpful drugs, but there are ethical issues about its use.

Elements and the Periodic Table

What is an element?

Elements are the basic building blocks of matter. They cannot be broken into anything simpler by chemical means. Each element has its own symbol. Elements are made up of **atoms**, and all the atoms of an element are of the same type.

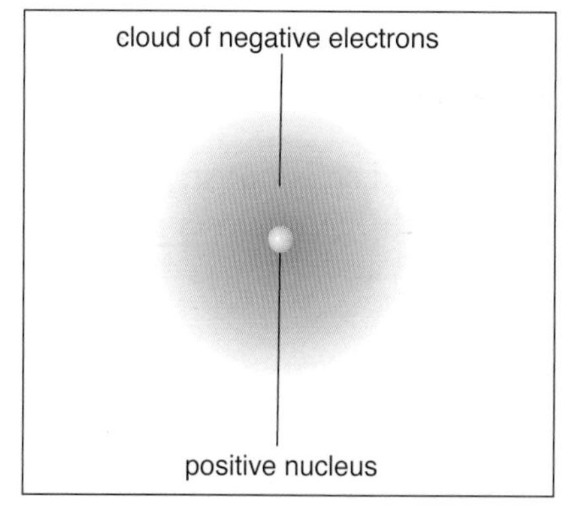

Figure 8.1 Model of the structure of an atom.

- Each atom contains a small positively charged central region called the **nucleus**.
- The nucleus is made up of two types of particle – **protons** (which have a positive charge) and **neutrons** (which have no charge).
- Light, negatively charged **electrons** surround the nucleus (see Figure 8.1) and are attracted to it. (Positive attracts negative.)
- Every atom of a particular element has the same number of protons, known as the **atomic number**. Each element has its own atomic number. For example, hydrogen has the atomic number 1; lithium has the atomic number 3; chlorine has the atomic number 17.
- The mass of an atom of an element is called its **relative atomic mass**. The units for this are atomic mass units (amu).
- The nucleus contains nearly all the mass of the atom.

What is the Periodic Table of the elements?

Figure 8.2 Dmitri Mendeléev (1834–1907).

The Periodic Table is a chart showing all the known elements, arranged in a logical way that allows chemists to predict the properties of individual elements. The first successful Periodic Table, which has developed into the one we still use today, was created by the Russian scientist **Dmitri Mendeléev** in 1869.

Mendeléev was interested in trying to organise the elements in a useful way. At the time, they were only classified by their relative atomic mass (known at that time as the atomic weight). Mendeléev noticed that if he arranged the elements in order of increasing relative atomic masses, the elements showed similar properties at regular intervals. To make this pattern fit the observed facts, though, he had to leave gaps in his table for elements that had not been discovered at the time, such as germanium, gallium and scandium (see Figure 8.3). Mendeléev predicted the properties of these elements. For example, for the element we now know as germanium he predicted the following:

- It should have a relative atomic mass of 72 (actually 72.6).
- Its density would be 5.5 (actually 5.47).
- It would form a liquid chloride, XCl_4, which would boil below 100 °C (actually germanium forms $GeCl_4$, which boils at 84 °C).

QUESTIONS

1 What group are the following elements in?
   a carbon (C)
   b calcium (Ca)
   c bromine (Br)
2 What is the atomic number of the following elements?
   a gold (Au)
   b iron (Fe)
   c radium (Ra)
3 Find the names (not just the formulae) of the following elements:
   a an element in the same period as sulfur (S)
   b the element in Group 5 that is in the same period as lithium
   c the element that has the atomic number 41.

Series	Group I	Group II	Group III	Group IV	Group V	Group VI	Group VII	Group VIII
1	1 H							
2	Li 2	Be	B	C	N	O	F	
3	3 Na	Mg	Al	Si	P	S	Cl	Fe Co Ni
4	K 4	Ca	?	Ti	V	Cr	Mn	
5	5 Cu	Zn	?	?	As	Se	Br	Ru Rh Pd
6	Rb 6	Sr	?	Zr	Nb	Mo	?	

Figure 8.3 Part of an early form of Mendeléev 's Periodic Table, showing how he left gaps for undiscovered elements.

It was later found that when the elements were arranged in order of their atomic numbers, inconsistencies found in the earlier periodic tables based on atomic weights/relative atomic masses were corrected. Nowadays the elements are arranged in order of their atomic numbers. The modern Periodic Table is shown in Figure 8.4. Each column is called a **group**. Each horizontal row of elements is called a **period**.

Hydrogen has some unusual properties and so is put in a group of its own. Helium is usually placed with hydrogen in a separate period of two elements. In between Groups 2 and 3 are a collection of metals known as the **transition metals**.

1 (I)	2 (II)											3 (III)	4 (IV)	5 (V)	6 (VI)	7 (VII)	0
								H 1									He 2
Li 3	Be 4											B 5	C 6	N 7	O 8	F 9	Ne 10
Na 11	Mg 12											Al 13	Si 14	P 15	S 16	Cl 17	Ar 18
K 19	Ca 20	Sc 21	Ti 22	V 23	Cr 24	Mn 25	Fe 26	Co 27	Ni 28	Cu 29	Zn 30	Ga 31	Ge 32	As 33	Se 34	Br 35	Kr 36
Rb 37	Sr 38	Y 39	Zr 40	Nb 41	Mo 42	Tc 43	Ru 44	Rh 45	Pd 46	Ag 47	Cd 48	In 49	Sn 50	Sb 51	Te 52	I 53	Xe 54
Cs 55	Ba 56	La 57	Hf 72	Ta 73	W 74	Re 75	Os 76	Ir 77	Pt 78	Au 79	Hg 80	Tl 81	Pb 82	Bi 83	Po 84	At 85	Rn 86
Fr 87	Ra 88	Ac 89															

Figure 8.4 The modern Periodic Table, showing symbols and atomic numbers (elements 58– 71, the lanthanides or rare earth elements, and elements with atomic number greater than 89 have been omitted for simplicity). Metals are shaded green and non-metals purple.

What's the difference between a metal and a non-metal?

Most of the elements in the Periodic Table are metals. As the Periodic Table can be used as an indicator of chemical properties, it is no surprise to find that metals and non-metals are each grouped together in it (see Figure 8.4).

Metals have the following properties:

- good conductors of electricity (**high electrical conductivity**)
- good conductors of heat (**high thermal conductivity**)
- **malleable** (can be beaten into sheets)
- **ductile** (can be drawn in wires)
- **lustrous** (freshly exposed surfaces are shiny)
- usually high densities
- high melting points
- high boiling points.

Non-metals are brittle, dull and of low density. They are generally **poor conductors** of heat and electricity, with **low melting points and boiling points**.

Some elements lie between the two, for example silicon and germanium are **semi-conductors** and are very important in electronics. The **graphite** form of carbon has all the characteristics of a non-metal except that it is a good conductor of electricity.

PRACTICAL HOW CAN YOU MEASURE THERMAL CONDUCTIVITY?

This activity helps you with:
- ★ fair testing
- ★ evaluating experiments
- ★ presentation of results.

Apparatus
- * 4 metal rods of the same size but of different metals
- * 4 pins
- * Vaseline
- * tripod
- * Bunsen burner
- * heatproof mat
- * stopwatch

Risk assessment

- **You will be supplied with a risk assessment by your teacher.**

All metals are good conductors of heat, but some are not as good as others. This experiment looks at one way of comparing thermal conductivity.

Procedure

1. Put the rods on a tripod so that the ends are close together.
2. Use Vaseline to stick pins on the other end of each rod.
3. Start the stopwatch and heat the ends without the pins equally with a Bunsen burner.
4. Time how long it takes for each pin to drop off, and record your results.

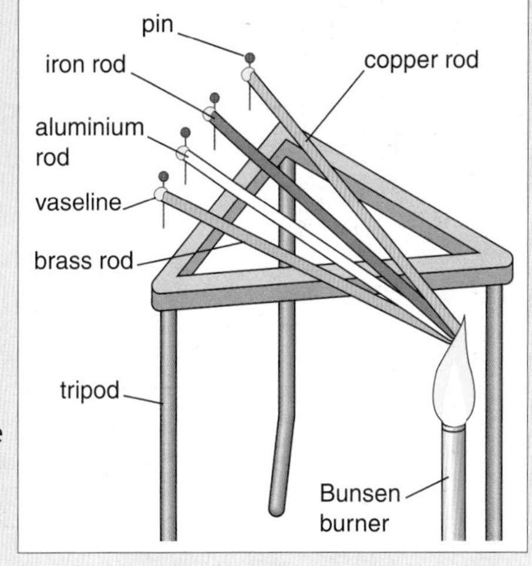

Figure 8.5 Set-up for testing thermal conductivity.

Analysing your results

1. Which metal bar lost its pin first and which was last?
2. Which metal was the best thermal conductor and which was the worst?
3. How was the test made fair?
4. Ideally, the amount of Vaseline used to attach each pin should be the same, as larger quantities would take longer to melt, but this is very hard to achieve. Look at your results and explain whether you think this lack of consistency had a significant effect in this experiment.
5. Do you think it would be useful to draw a graph of these results? If so, what sort of graph would it be?

TASK HOW CAN THE PERIODIC TABLE BE USED TO PREDICT PROPERTIES?

This activity helps you with:
- ★ understanding the Periodic Table
- ★ drawing graphs
- ★ improving your research skills
- ★ making predictions.

The groups in the Periodic Table show trends in various properties of the elements in them. These trends can be used to predict unknown properties of other elements in the group. As an example, look at Table 8.1, which has two values missing.

Table 8.1 Some properties of Group 1 of the Periodic Table.

Element	Melting point (°C)	Boiling point (°C)
Lithium	180	1 347
Sodium	98	883
Potassium	64	?
Rubidium	39	688
Caesium	?	678

These data can be used to predict the missing values (the melting point of caesium and the boiling point of potassium). The table shows clear trends in the melting and boiling points of these metals but a graph needs to be drawn to see the exact nature of each trend.

1 Draw graphs of the data to show the trends. Normally, because the elements form a discontinuous variable, a bar chart would be drawn. However, in this case it is easier to make predictions if you draw a line graph.
2 Use your graphs to predict the melting point of caesium and the boiling point of potassium.
3 Look up the actual values on the internet or in a book and see how close your prediction was. Record the true values.
4 Your prediction was probably quite accurate but not exactly correct. Suggest one reason why this method would not be expected to give an exact value.

Chemical properties

The Periodic Table can also be used to predict chemical properties of the elements. Many chemical properties show a certain trend as you go down the group. For example, the Group 1 metals all react with water to form hydroxides and hydrogen. The equation for this is:

$$2X + 2H_2O \rightarrow 2XOH + H_2$$

X represents the Group 1 element; you would just substitute the element's symbol into the equation. This reaction is **exothermic** – it produces heat. As you go down the group, the reaction becomes more violent and more heat is generated. The reactions are described in Table 8.2.

Table 8.2 Reaction of Group 1 elements with water.

Symbol	Element	Reaction with water
Li	Lithium	The metal fizzes and gradually reacts and disappears.
Na	Sodium	The metal reacts quickly and the heat given off melts the sodium, a molten ball of which rushes across the water surface. Often (but not always) enough heat is generated to set light to the hydrogen given off.
K	Potassium	The reaction is similar to that with sodium but faster, and the heat given off will always set the hydrogen alight.
Ru	Rubidium	Rubidium reacts so violently that the metal and the water shoot out of the container.
Cs	Caesium	Caesium explodes on contact with water, possibly shattering the container.
Fr	Francium	No measurable quantity of francium has ever been isolated. It is highly unstable and radioactive and there is probably no more that 400 g on the planet at any one time. It has never been reacted with water, but the reaction would certainly be incredibly dangerous!

Chemistry and compounds – explaining how things react

As early as 4000 years BC, the Ancient Egyptians were practising what we would recognise as aspects of modern chemistry. By 1000 BC they were making metals from their ores, perfecting fermentation, making all sorts of pigments for cosmetics and pottery and making medicines and perfumes from plants, amongst a whole host of other chemical processes. We know this because the Ancient Egyptians kept records in their hieroglyphics and temple art.

Figure 8.6 Egyptian 'chemists'.

The foundations of modern chemistry were laid by Greek and Arabic scientists who 'invented' scientific methods during the first millennium AD, but the first real 'chemist' was Robert Boyle who in 1661 published his book, '*The Sceptical Chymist*', in which he first outlined his ideas about the formation of new materials and the distinction between mixtures and compounds.

Figure 8.7 Robert Boyle (1627–1691).

It was Boyle who effectively defined what a compound is. Today, with our knowledge of modern science, we define compounds as being formed when two or more elements chemically combine to form a new substance. Although all compounds have names, the true nature of a compound is revealed by its formula. The chemical formula not only tells you what elements make up the compound, but also what ratio of atoms it contains.

There are two types of compound: **ionic** compounds (which are made up of electrically charged particles, like the sodium and chloride ions in sodium chloride or common salt) and **molecules**, which are electrically neutral.

One of the simplest molecular compounds is water, H_2O. The formula tells you that the molecule is made up of three atoms: two hydrogen atoms and one oxygen atom. A space-filler diagram can be drawn to illustrate the way that the atoms are joined together to make up the molecule (Figure 8.8).

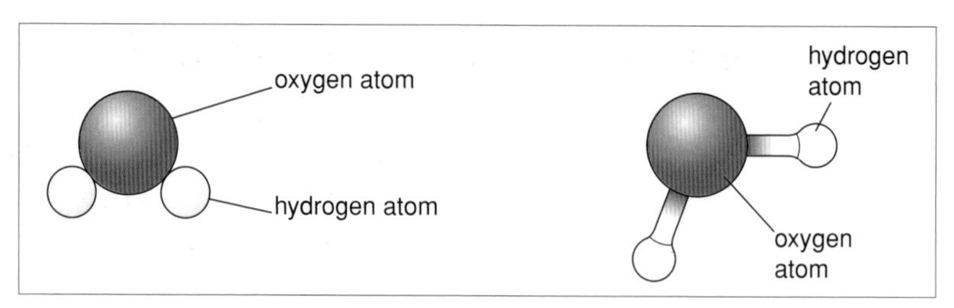

Figure 8.8 Two different models of a water molecule. The one on the left is called a space-filler diagram.

Space-filler diagrams are really useful as they mimic the plastic models that chemists use to represent atoms and molecules.

Chemical formulae are also really useful for the atom 'accountancy' that occurs when chemicals react together. In any chemical reaction, no new atoms are created and none are destroyed. Chemical reactions rearrange the way that the atoms are bonded together, with the formula telling you how the atoms are rearranged.

Water is formed when hydrogen gas burns in oxygen gas. Both hydrogen and oxygen molecules are diatomic, which means that each molecule contains two atoms. The formula of hydrogen gas is H_2, and oxygen is O_2. The equation for the combustion of hydrogen is therefore:

$$2H_2(g) + O_2(g) \rightarrow 2H_2O(l)$$

The letters in brackets tell you the state of the reactants and products (g = gas; l = liquid). The equation shows you that two molecules of hydrogen react with one molecule of oxygen forming two molecules of water. You can see that the formula of each molecule tells you how to work out the numbers of molecules needed to balance the equation. The chemical accountancy shows you that there are four atoms of hydrogen before the reaction, and four afterwards; and two atoms of oxygen before the reaction and two atoms afterwards – it balances!

PRACTICAL MAKING MOLECULES

This activity helps you with:
- ★ making chemical models
- ★ writing molecular formulae
- ★ drawing space-filler diagrams.

Apparatus
- * molecular modelling kit

Procedure

Your teacher will give you a molecular modelling kit and show you how to use it, including which coloured 'atoms' represent which elements. Molecular models are really useful because they show you a representation of the molecule in 3D.

1 Use the modelling kit to make models of the following molecules:
   - **a** water, H_2O
   - **b** carbon dioxide, CO_2
   - **c** methane, CH_4
   - **d** ammonia, NH_3

2 Draw the appropriate space-filler diagram for each one.

Your teacher will now show you some pre-made molecules.

3 Draw a space-filler diagram of each one and write the molecular formula.

How are ionic compounds formed?

When metals react with non-metals ionic compounds are usually formed. Ions are formed from electrically neutral particles like atoms when they transfer electrons. **Metal** atoms prefer to *lose* these electrons forming *positive* ions. **Non-metals** prefer to *gain* electrons forming *negative* ions. The salts that are formed when metals react with non-metals have names that reflect the metal ion and the non-metal ion. The metal ions have the same name as their atom, so sodium atoms become sodium ions. Non-metal ions have slightly different names. Oxygen forms oxides, fluorine forms fluorides, chlorine forms chlorides, bromine forms bromides and iodine forms iodides.

How do you write the formulae of simple ionic compounds?

The formulae of simple ionic compounds can be written using the formulae of the ions given in Tables 8.3 and 8.4. Whenever ionic compounds are formed from metals and non-metals, the resulting compounds are electrically neutral – the number of positive charges balances the number of negative charges. Sodium chloride is easy. The sodium ion is Na^+ and the chloride ion is Cl^-. The formula is $(Na^+)(Cl^-)$, which we usually write as NaCl.

The calcium ion is Ca^{2+}. This means that two chloride ions are needed to balance the one calcium ion. The formula for calcium chloride is $(Ca^{2+})(2Cl^-)$, which we usually write as $CaCl_2$.

Table 8.3 Formulae of some positive ions.

Ion	Formula
Group 1	
Lithium	Li^+
Sodium	Na^+
Potassium	K^+
Group 2	
Magnesium	Mg^{2+}
Calcium	Ca^{2+}
Strontium	Sr^{2+}

Table 8.4 Formulae of some negative ions.

Ion	Formula
Group 6	
Oxide	O^{2-}
Group 7	
Fluoride	F^-
Chloride	Cl^-
Bromide	Br^-
Iodide	I^-

QUESTIONS

4 Copy and complete this table:

Compound	Formula of positive ion	Formula of negative ion	Formula of compound
Calcium bromide	Ca^{2+}	Br^-	$CaBr_2$
Sodium oxide			
Magnesium bromide			
Potassium chloride			
Calcium oxide			
Sodium iodide			
Potassium iodide			

5 What are the names of the following compounds? For each one, write down the ions contained in the compound:

a LiCl **b** NaF **c** MgI_2
d MgO **e** $SrBr_2$ **f** Li_2O
g CaF_2 **h** K_2O

Writing the formulae of other ionic compounds

The formulae of other compounds can be written from the formulae of their ions. The formulae of some common ions will be given in the examinations (at the back of the exam paper) as shown in Tables 8.5 and 8.6.

Table 8.5 Formulae for positive ions.

Charge: +1		Charge: +2		Charge: +3	
Sodium	Na^+	Magnesium	Mg^{2+}	Aluminium	Al^{3+}
Potassium	K^+	Calcium	Ca^{2+}	Iron(III)	Fe^{3+}
Lithium	Li^+	Barium	Ba^{2+}	Chromium(III)	Cr^{3+}
Ammonium	NH_4^+	Copper(II)	Cu^{2+}		
Silver	Ag^+	Lead(II)	Pb^{2+}		
		Iron(II)	Fe^{2+}		

Table 8.6 Formulae for negative ions.

Charge: −1		Charge: −2		Charge: −3	
Chloride	Cl^-	Oxide	O^{2-}	Phosphate	PO_4^{3-}
Bromide	Br^-	Sulfate	SO_4^{2-}		
Iodide	I^-	Carbonate	CO_3^{2-}		
Hydroxide	OH^-				
Nitrate	NO_3^-				

QUESTION

6 Write down the formulae of the following. (*Hint*: Don't forget how to use brackets.)
- **a** sodium hydroxide
- **b** calcium hydroxide
- **c** iron(III) oxide
- **d** barium chloride
- **e** copper(III) sulfate
- **f** ammonium sulfate
- **g** magnesium sulfate
- **h** sodium carbonate
- **i** aluminium oxide
- **j** sodium phosphate

EXAMPLE

Q Find the formula of calcium nitrate.

A The calcium ion is Ca^{2+} and the nitrate ion is NO_3^-. The formula is $(Ca^{2+})(2NO_3^-)$, which we usually write as $Ca(NO_3)_2$. Notice how the brackets are used: $Ca(NO_3)_2$, not $CaNO_{32}$.

TASK

WHAT'S THE MAGIC FORMULA?

This activity helps you with:
- ★ working as part of a team
- ★ using scientific models
- ★ identifying information from models
- ★ writing formulae
- ★ naming compounds
- ★ identifying compounds as molecular or ionic
- ★ identifying the ions in a compound if it is ionic.

It was the Swedish chemist, Jöns Jacob Berzelius who in his paper of 1814, came up with the system of chemical element nomenclature that we use today. It was Berzelius who first used the system of single and double letters for element symbols and then showed how these could be used to describe compounds.

The chemical formula of a compound tells us everything about it – the magic formula!

Your teacher will give you (or put up on a screen) cards, each with the formula of a particular compound.

Figure 8.9 Jöns Jacob Berzelius (1779–1848).

1 Choose a partner and work as a team.
2 Identify as much information about each compound as possible from its formula.
3 Construct a suitable table to help you like this one. The details of two compounds have been filled in for you.

Formula	Name	Molecular or ionic compound	Atoms present (and number)	Ions present (if ionic)
CO_2	Carbon dioxide	Molecular	1 × C, carbon 2 × O, oxygen	–
$MgCl_2$	Magnesium chloride	Ionic compound	1 × Mg, magnesium 2 × Cl, chlorine	Mg^{2+} Cl^-

Chapter summary

- Elements are the basic building blocks of all substances and cannot be broken down into simpler substances by chemical means.
- Elements are made up of only one type of atom.
- Atoms contain a positively charged nucleus and orbiting negatively charged electrons.
- The modern Periodic Table was first developed by Dmitri Mendeléev.
- Mendeléev used creative thought to predict the properties of undiscovered elements.
- The rows in the Periodic Table are called periods and the columns are called groups.
- Metals have the following properties:
  - They are good conductors of heat and electricity.
  - They are malleable and ductile.
  - They are generally hard, dense and shiny.
  - They usually have high melting points and boiling points.
- Non-metals are generally poor conductors of heat and electricity and have low melting and boiling points.
- Metals are located on the left and middle of the Periodic Table. Non-metals are on the right-hand side.
- A few elements have properties that are intermediate between those of metals and non-metals (e.g. silicon).
- The groups in the Periodic Table show trends in the physical and chemical properties of the elements within them.
- In a chemical reaction, atoms are rearranged but none are created or destroyed.
- New substances called compounds are formed when atoms of two or more elements combine. Each compound has its own chemical formula.
- The chemical formula of a compound tells you the names of the elements, the number of atoms of each element and the total number of atoms present in the compound.
- Space-filler type diagrams and structural formulae can be drawn for simple molecules – they show which atoms are joined to which.
- When ionic compounds are formed, electrons are transferred from metal atoms to non-metal atoms, forming positively charged metal ions and negatively charged non-metal ions.
- You can write chemical formulae for ionic compounds given the formulae of the ions that they contain.

9 Metals

Welsh heavy metal?

There is something about Wales and metals! The metal mining, production and working industries have been part of the backbone of Welsh industry since prehistoric times. Metal tools first appeared in Wales around 2500BC – initially copper, closely followed by the alloy bronze. Gold jewellery started to appear at the same time, and one of the UK's best Bronze-Age treasures is The Mold Cape, a gold shoulder decoration made out of a single gold ingot, discovered in a burial tomb in Bryn yr Ellyllon, Flintshire in 1833. The cape is over 3500 years old. Human beings must have been mining and working metals for many hundreds of years before producing a thing of such beauty with so much skill.

Iron and steel have long been associated with South Wales. The abundant coal and iron ore seams in the valleys of Glamorgan and Monmouthshire fuelled the Industrial Revolution between 1730 and 1850. At the same time, the rise of the coal and iron industries led to a huge population explosion in South Wales, creating its own social problems. The population of Monmouthshire alone rose from 45 000 in 1801 to 450 000 in 1901. Metal (and coal) working quite literally created its own new geographic and human landscape.

Figure 9.1 The Mold Cape is made from a single gold ingot.

Figure 9.2 A South Wales iron works in the 1800s.

Figure 9.3 Anglesey Aluminium plant in Holyhead.

Figure 9.4 Port Talbot steelworks.

In September 2009, aluminium smelting from its ore ceased at the Anglesey Aluminium plant in Holyhead. Aluminium had been produced at the plant since 1971, but with the de-commissioning of the nearby Wyfla Nuclear Power Station, the company was unable to secure a viable electricity supply, and ore smelting ceased – the plant currently continues to smelt recycled aluminium. Aluminium is produced using electrolysis which, on the industrial scale, requires huge amounts of electrical power. There are only two remaining aluminium ore smelting plants in the UK.

The current 'jewel in the crown' of Welsh metal-working is the mighty steelworks in Port Talbot, producing 5 million tonnes of slab steel per year. The site boasts some of the most up-to-date steel-making facilities in the world and produces steel mainly for the car-making industry and domestic goods.

How do we get metals from the ground?

The largest gold nugget ever discovered was the 'Welcome Stranger' nugget, discovered in 1869 by two Cornishmen, Richard Oates and John Deason – both former tin-miners, in Victoria, Australia. It weighed an incredible 72 kg! They found it just below the surface of the ground at the base of a tree, and when they exchanged it at the bank, they were paid £9381, which would be nearly £2 million pounds today. Not bad for a day's work! Figure 9.5 shows a modern replica of the Welcome Stranger, together with the original photo.

Figure 9.5 The Welcome Stranger and a modern replica.

Only six metals are found naturally in their pure form – gold and silver being the obvious examples. The other metals found in their natural state are platinum, copper, iron (but only from meteorites) and mercury (but only with the mineral cinnabar). The largest silver nugget was found in 1894 in Colorado, USA and weighed 835 kg!

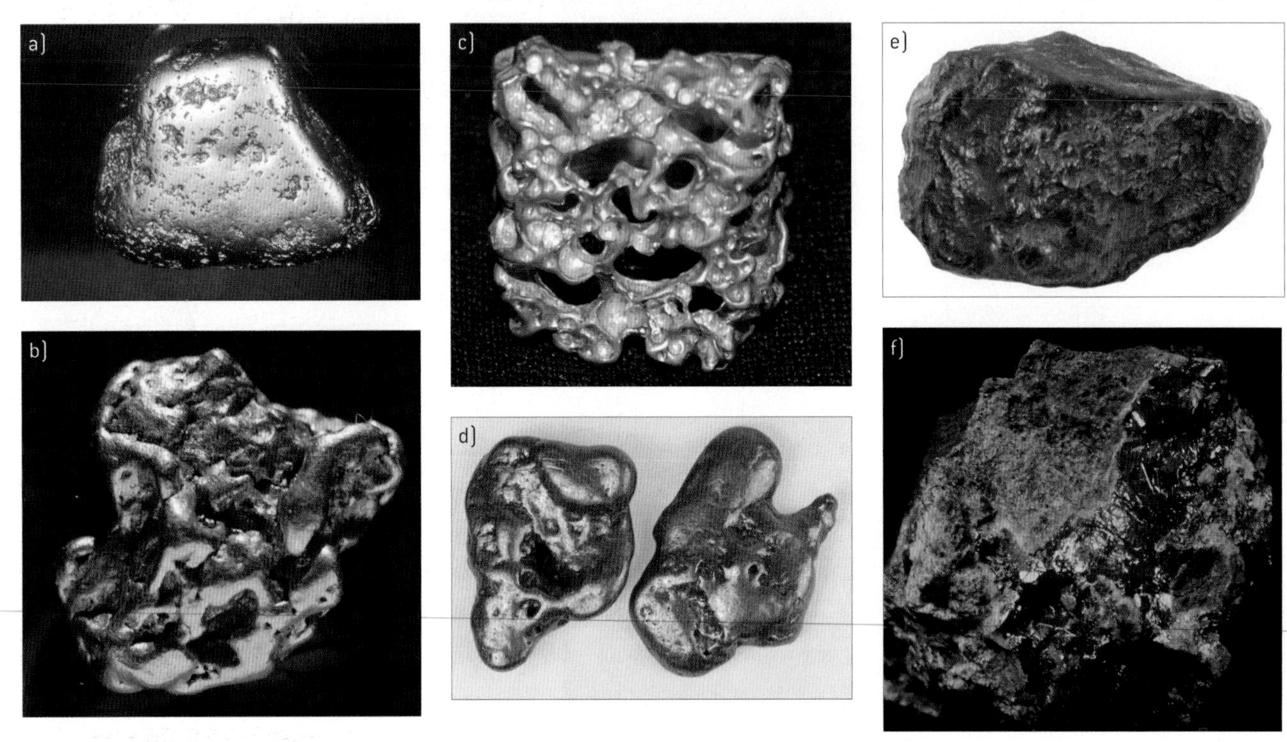

Figure 9.6 These metals are found in their pure form: a) platinum; b) gold; c) silver; d) copper; e) iron (meteorite); f) mercury (with cinnabar).

All of the metals found naturally are very unreactive. As metals become more reactive they are only found in combination with other elements, mostly non-metals, as **ores** – rocks containing metal compounds. The metals have to be extracted from their ore either by using chemical reactions involving heat, or by **electrolysis**.

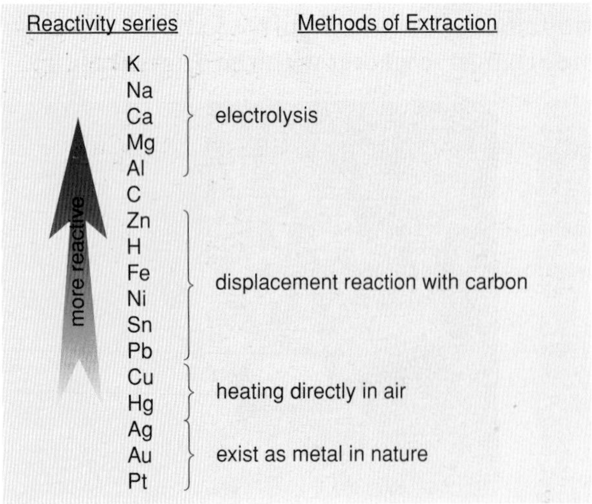

Reactivity series	Methods of Extraction
K, Na, Ca, Mg, Al	electrolysis
C	
Zn, H, Fe, Ni, Sn, Pb	displacement reaction with carbon
Cu, Hg	heating directly in air
Ag, Au, Pt	exist as metal in nature

more reactive (arrow pointing up)

Figure 9.7 The reactivity series, showing methods of extraction.

Some common ores are shown in Table 9.1.

The method of extraction of a metal from its ore depends on its place in the reactivity series (Figure 9.7). The metals at the bottom of the reactivity series can be found naturally, those in the middle are generally extracted by a displacement reaction involving carbon, and the most reactive metals can only be extracted by electrolysis.

Any metal in the reactivity series can displace any metal below it in the series from a compound containing that metal. You can see that carbon is included in the reactivity series, even though it is not a metal. Carbon can be used to extract those metals below it in the reactivity series from their ores.

Table 9.1 Examples of some common ores.

Element	Ore	Formula of ore	Example
Iron	Haematite, an oxide ore	Fe_2O_3	
Iron	Iron pyrites (fool's gold)	FeS_2	
Iron	Siderite, a carbonate ore	$FeCO_3$	
Aluminium	Bauxite, an oxide ore	$Al_2O_3.2H_2O$	
Lead	Galena, a sulfide ore	PbS	
Magnesium	Epsomite (Epsom salts), a sulfate ore	$MgSO_4.7H_2O$	
Titanium	Rutile	TiO_2	

PRACTICAL DISPLACING METALS

This activity helps you with:
★ working in a team
★ designing a methodical way to perform an experiment
★ designing a table to record experimental observations
★ recording observations
★ using observations to draw conclusions
★ writing word and balanced symbol equations.

In this practical task you will use metals higher up in the reactivity series to displace metals lower down in the series from their salts.

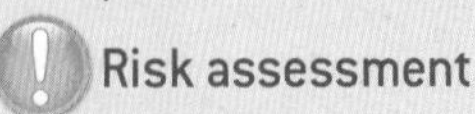

Risk assessment

- **Wear eye protection.**
- **Some of the chemicals that you will use are harmful. You will be supplied with a risk assessment by your teacher.**

Apparatus

* spotting tile
* small pieces of various metals
* metal salt solutions containing the available metals
* teat pipettes
* eye protection

Procedure

This experiment is best carried out in pairs. This is quite a difficult experiment to organise. You and your partner need to work systematically in order to work out which metals will displace one another. The reactivity series will help you but you may not be able to observe the reaction in the laboratory (the reaction may take some time, or require some heating).

Your teacher will tell you which metals are available to you and which salt solutions you can use. Commonly available metals are: magnesium, aluminium, zinc, iron, nickel, tin, lead and copper. You may be able to use chloride, nitrate and sulfate salts of each of these metals.

1 List the metals and metal salts that you have available.
2 Design a grid based on the spotting tiles' array, to test as many of the metals as you can with as many of the metal salt solutions – you may need more than one spotting tile to do this. Figure 9.8 shows you one of the ways of laying out your experiment.
3 Design a table to record your observations – these will include any colour changes of the salt solutions and the metals, and any solids that appear in the cavities of the spotting tiles.
4 Fill the spotting cavities with rows of each of the salt solutions using a clean teat pipette for each solution.
5 Add a small piece of one of each of the metals to a cavity, following your grid.
6 Perform your experiments methodically on the spotting tiles. Make and record any observations.
7 Follow your teacher's directions when clearing your experiment away. Do not pour the contents of the spotting tiles down the sink.

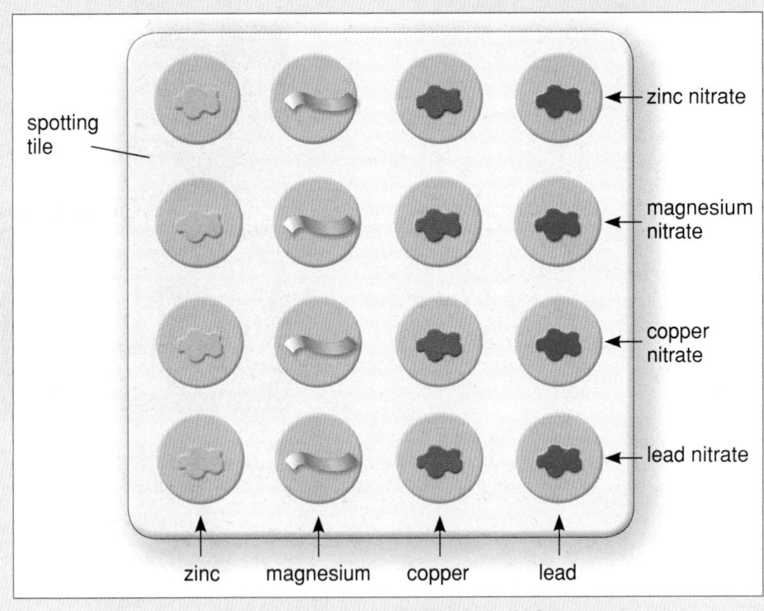

Figure 9.8 An example of how to arrange the metals and solutions.

continued...

PRACTICAL contd.

Example reaction

Silvery magnesium metal will displace copper from a blue solution of copper sulfate. The salmony pink copper that is formed immediately reacts with oxygen in the solution, forming black copper oxide. The displacement reaction can be described by the following equation:

$$\text{magnesium} + \text{copper sulfate} \rightarrow \text{magnesium sulfate} + \text{copper}$$

$$Mg(s) + CuSO_4(aq) \rightarrow MgSO_4(aq) + Cu(s)$$

Analysing your results

1 Use your observations to construct your 'experimental' reactivity series.
2 How does your reactivity series compare to the series in Figure 9.7?
3 Your teacher will give you the formula of each of the metal salts that you have used. For each successful reaction (one where you observed a reaction), write the displacement reaction as a word equation and a balanced symbol equation.

Discussion Point

Why is it not possible to do this experiment in the school laboratory using the metals lithium, sodium, potassium or calcium?

The thermit reaction

A more reactive metal will remove oxygen from the oxide of a less reactive metal when a mixture of the two is heated – this is called a **competition** reaction. One interesting and spectacular example of this is the thermit (or thermite reaction (see Figures 9.9 and 9.10). When a mixture of powdered aluminium and iron(III) oxide is ignited by a high-temperature fuse, molten iron is formed. This reaction is used in the rail industry to weld rails together on a track.

Figure 9.9 Rail welding using the thermit reaction.

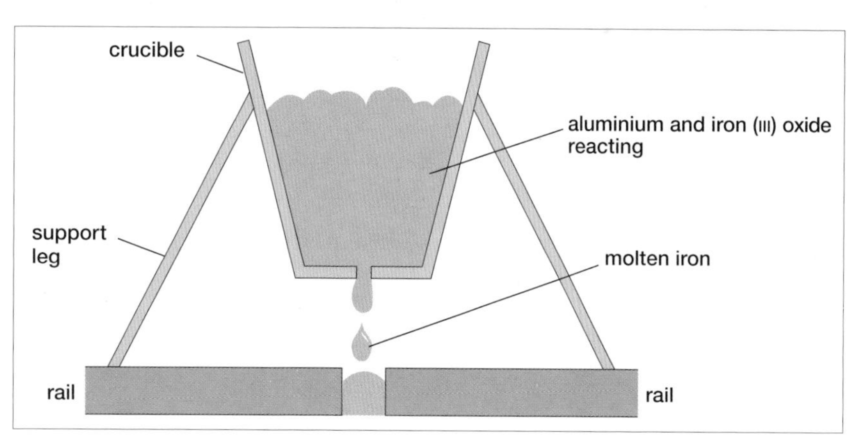

Figure 9.10 The thermit reaction.

PRACTICAL OBSERVING THE THERMIT REACTION

This activity helps you with:
★ carefully and safely observing a vigorous reaction
★ writing equations for a reaction
★ annotating a diagram.

Risk assessment

- **Wear eye protection (your teacher must wear a full face shield).**
- **You must be at least 4 m away from the reaction, which must be behind safety screens.**
- **Your teacher will provide you with a risk assessment.**

Procedure

Your teacher will demonstrate to you the highly energetic and exothermic thermit reaction. This reaction is very spectacular, and perfectly safe provided that you follow the risk assessment provided by your teacher.
You can download the experimental notes that your teacher will use from: www.practicalchemistry.org/experiments/the-thermite-reaction,172,EX.html

1 Your teacher will give you a diagram of the apparatus.
2 Observe the reaction and then annotate the diagram, showing the chemical reactions that occur.

Making metals – oxidation or reduction?

In industrial reactions like the thermit reaction where the reaction is used to weld rails together, or in a blast furnace where carbon is used to extract iron from its ore, the trick is to use a more reactive element to 'displace' the target metal from its oxide. In both of these examples the target metal is iron, as iron(III) oxide. During these reactions, the iron has oxygen removed from it (and is said to be reduced) and the more reactive element (aluminium or carbon in these cases) gains oxygen (and is said to be oxidised).

- reduction = loss of oxygen
- oxidation = gain of oxygen

The equation for the thermit reaction is:

aluminium + iron(III) oxide → aluminium oxide + iron

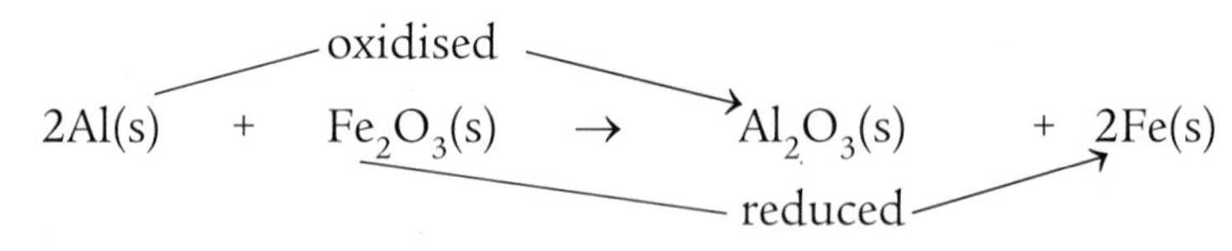

Making iron

Iron is a metal that is extracted from its ores by chemical reduction with carbon, which is more reactive than iron. Although the iron and steel industry has decreased in size in recent years, especially in Wales, it is still a very important industry worldwide.

The extraction is carried out in a blast furnace, as shown in Figure 9.11.

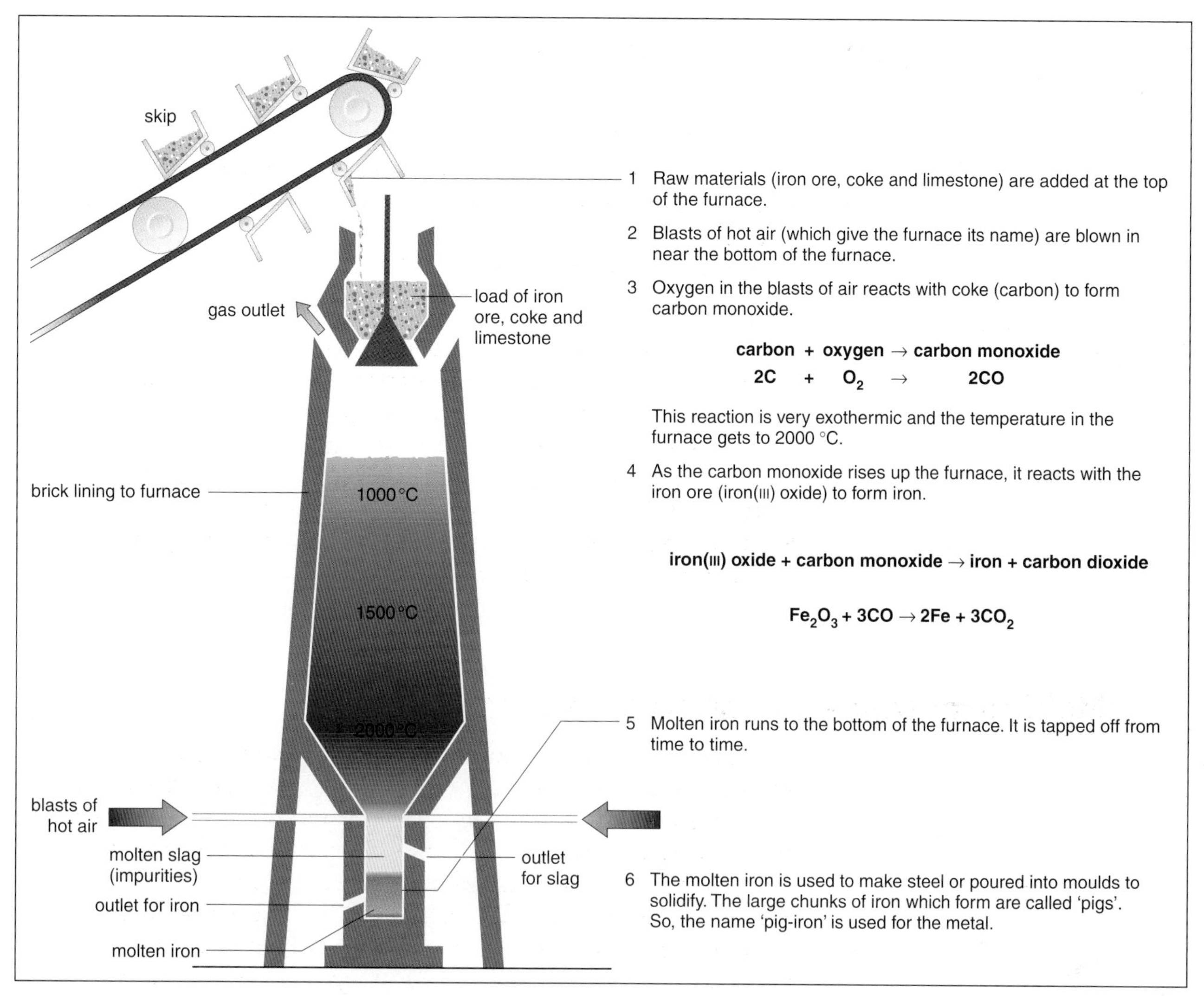

Figure 9.11 A blast furnace.

Iron ore, coke and limestone are heated in the blast furnace to make iron. The problem is to turn the iron(III) oxide in the iron ore into iron. This means removing oxygen. Hot air is blown into the furnace, where it combines with the coke (which is mostly carbon) to form carbon monoxide and liberate heat. The carbon monoxide reacts with the iron ore high in the furnace to form molten iron, which collects at the base of the furnace. The equation for the reaction is:

carbon monoxide + iron(III) oxide → carbon dioxide + iron

$$3CO(g) + Fe_2O_3(s) \rightarrow 3CO_2(g) + 2Fe(l)$$

In this reaction the iron is reduced and the carbon (in the carbon monoxide) is oxidised.

The coke itself may take part in the reduction of the iron:

carbon + iron(III) oxide → carbon monoxide + iron

$$3C(s) + Fe_2O_3(s) \rightarrow 3CO(g) + 2Fe(l)$$

The iron is reduced and the carbon is oxidised.

The limestone removes sandy materials from the iron ore to produce a molten slag of calcium silicate that floats on top of the molten iron. The slag is used as hardcore for roads and in building. The waste gases from the top of the furnace are used to preheat the blast of air at the bottom. In some furnaces this air is enriched with oxygen.

The iron that is produced by the blast furnace is often called pig iron (because the original moulds used to make iron ingots in an iron works resembled piglets suckling on a sow). It is a brittle metal as it contains a significant amount of carbon, up to 4.5%. To make the iron into the more useful steel, some of the carbon needs to be removed by blowing oxygen through it. Steel is harder and much less brittle than pig iron.

TASK GREEN STEEL?

This activity helps you with:

★ studying and evaluating the sustainability of scientific processes and their effects on society, the environment and the economy.

Steel is such an important material that it is recycled on a large scale. Recycling steel:

- saves up to 50% of the energy costs compared with producing new
- helps to conserve iron ore
- cuts down the emission of greenhouse gases from the furnaces.

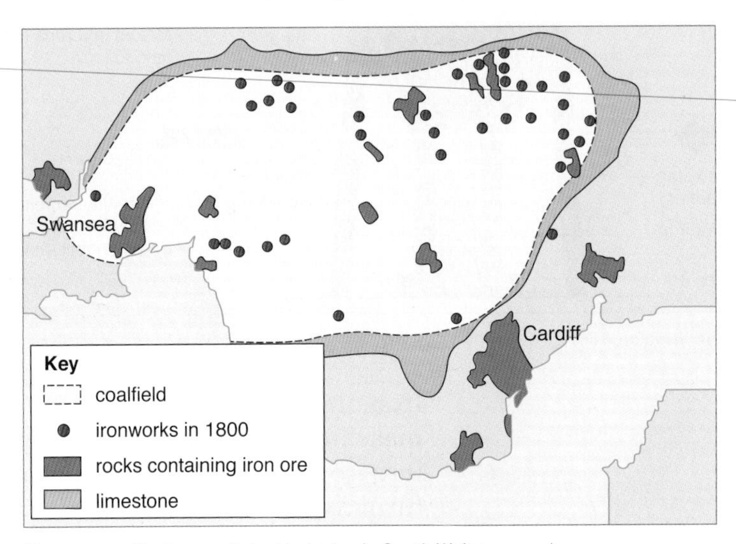

Figure 9.12 The iron and steel industry in South Wales.

The iron and steel industry in Wales initially grew out of the local availability of the raw materials coal, iron ore and limestone. However, its size has diminished in recent years, with the closure of a number of plants. The effect on former mining and iron-working communities has been devastating. Many thousands of people lost their jobs as the mines and plants closed, and it is only fairly recently that new investment and alternative jobs have returned in numbers to the area. The whole landscape has been shaped by the coal and iron industries. Regeneration and landscaping have been costly and lengthy, and the area still contains significant industrial sites that have yet to be cleared.

At Port Talbot steelworks, the vast majority of all the raw materials for the production of steel are now imported via Port Talbot docks, one of the largest set of docks in the UK, capable of handling the huge bulk container ships that bring the raw materials to the steelworks.

continued...

TASK contd.

Figure 9.13 Port Talbot docks.

1 Why is steel such an important metal?
2 What are the global advantages of recycling steel?
3 Why do you think that many of the mines and iron and steel works in South Wales closed?
4 What was the effect of the large-scale closure of the mines and steel works on the local population?

Discussion Points

1 Why is it so difficult to return former industrial sites to safe use today?
2 There is still plenty of coal and iron ore in South Wales. Why do you think that most of the raw materials for the Port Talbot iron and steel works are shipped in by bulk container from places like Brazil?

Electrolysis and the making of aluminium

Electrolysis is a chemical reaction that is brought about by an electric current passing through a conducting liquid. The conducting liquid is called the electrolyte and contains positively charged ions called **cations**, and negatively charged ions called **anions**. The current enters the electrolyte via two solid conductors called **electrodes**. The positive electrode is called the **anode**, and the negative electrode is called the **cathode**. The current is carried through the electrolyte by the movement of the ions. The negative anions move towards the anode. The positive cations are attracted to the cathode.

A typical set-up for an electrolysis experiment is shown in Figure 9.14.

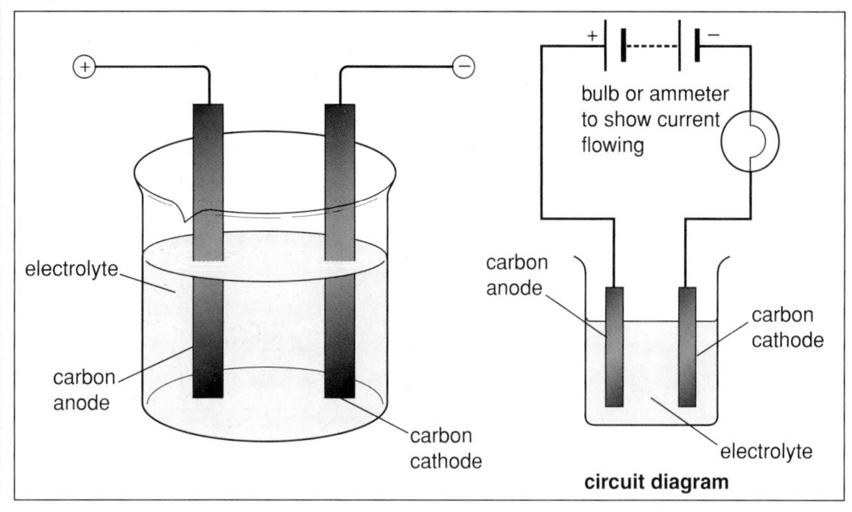

Figure 9.14 Electrolysis apparatus.

QUESTIONS

1 Explain what is meant by the following words: electrode, electrolyte, anion, cation, cathode, anode

2 Write ionic electrode equations for the electrolysis of the following molten salts:
   a sodium chloride (Na^+Cl^-)
   b magnesium oxide ($Mg^{2+}O^{2-}$)
   c lithium iodide (Li^+I^-)
   d calcium oxide ($Ca^{2+}O^{2-}$)
   e lead bromide ($Pb^{2+}Br^-$)
   f copper chloride ($Cu^{2+}Cl^-$)

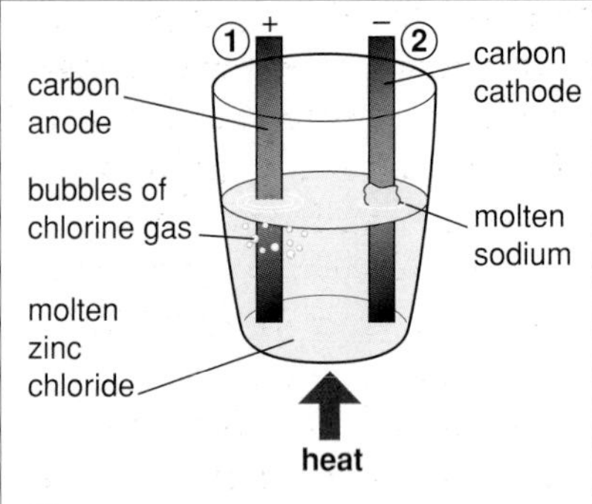

Figure 9.15 The electrolysis of molten zinc chloride.

The simplest case of electrolysis is one where the electrolyte contains only two ions. One example is an electrolyte of molten zinc chloride, $ZnCl_2$. Zinc chloride melts at quite a high temperature, so a crucible is required and care must be taken in setting up the electrolysis. Carbon rods are used as the electrodes. Molten zinc chloride contains only zinc ions (the cations) and chloride ions (the anions).

PRACTICAL THE ELECTROLYSIS OF ZINC CHLORIDE

This activity helps you with:

★ observing an electrolysis experiment of a molten salt
★ testing for chlorine gas
★ writing ionic electrode equations.

Procedure

Your teacher will show you a demonstration of the electrolysis of molten zinc chloride. You can download the experimental notes that your teacher will use from:
www.practicalchemistry.org/experiments/electrolysis-of-zinc-chloride,50,EX.html

In this demonstration experiment, an electric current is passed through the molten zinc chloride electrolyte and chlorine gas is formed at the anode, with zinc metal being formed at the cathode.

1 Draw a schematic diagram showing the electrolysis of zinc chloride.
2 Write ionic electrode equations to show how the ions in the molten salt form chlorine gas and zinc metal at each electrode.

The manufacture of aluminium by electrolysis

All the very reactive metallic elements are usually extracted from ores by electrolysis because they cannot easily be displaced by chemical reduction reactions. Aluminium is a very reactive metal although it can be used for window frames and for other construction purposes. It does not seem to be reactive because a very thin film of protective aluminium oxide forms on the surface of the metal. This film protects the metal from any further reaction.

The usual source of aluminium is the ore, bauxite. The bauxite is treated chemically to remove impurities, and is finally turned into the white solid aluminium oxide, Al_2O_3. Aluminium oxide is sometimes called alumina. It has a very high melting point. For aluminium oxide to be electrolysed (Figure 9.16), it has to be dissolved in molten cryolite (a mineral – sodium aluminium fluoride – with a high melting point that acts as a solvent for

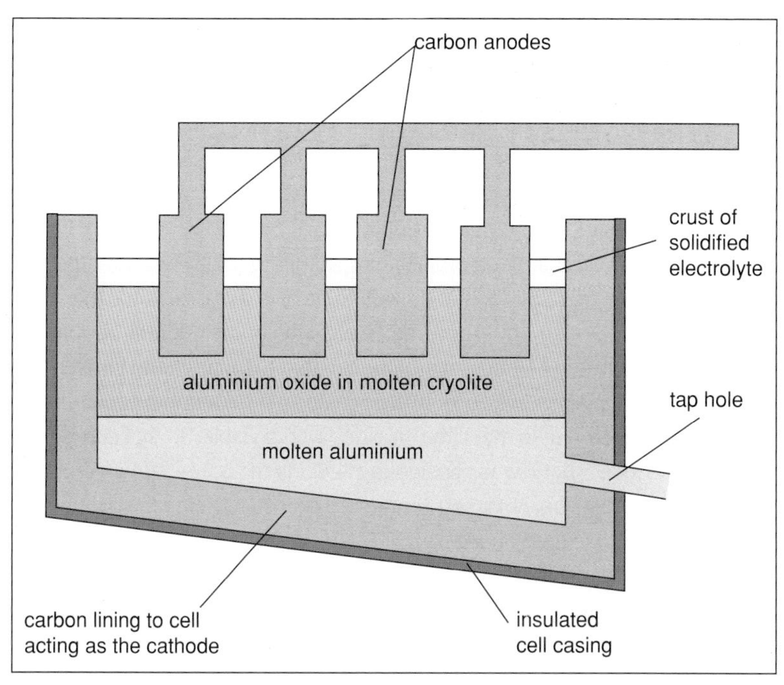

Figure 9.16 The electrolysis of aluminium oxide.

aluminium oxide). This brings the working temperature of the electrolyte to about 950 °C. Oxygen gas is formed at the carbon anodes, and at high temperature the anodes react with oxygen, burning away and having to be replaced from time to time. The carbon lining of the cell is also the cathode, and aluminium is formed here as the molten metal. Periodically, aluminium metal is removed, the crust is broken, and more aluminium oxide is added.

The aluminium ions are attracted to the cathode where they gain electrons and form aluminium metal:

$$Al^{3+} + 3e^- \rightarrow Al$$

aluminium ion → aluminium atom

The oxide ions are attracted to the anode where they lose electrons and form oxygen gas.

$$2O^{2-} - 4e^- \rightarrow O_2$$

oxide ion → oxygen molecule

TASK THE SAD TALE OF ANGLESEY ALUMINIUM

This activity helps you with:

★ investigating the links between science and economics, society and the environment.

The electrolytic production of aluminium requires very large quantities of electricity. The aluminium smelting plant at Holyhead on Anglesey (see Figure 9.17) consumed 250 MW, which once represented over 12% of all the electrical power consumed in Wales. Aluminium plants are often sited where there is cheap power, for example near hydroelectric power sources. In the case of Anglesey Aluminium, the plant was built close to Wylfa nuclear power station. The two nuclear reactors at Wylfa produced a combined electrical power output of 980 MW – and Anglesey Aluminium smelting works consumed about a quarter of the whole output from the power station.

Figure 9.17 Anglesey Aluminium smelting works (left) and Wylfa nuclear power station (right).

continued...

The factors that are considered when building an aluminium oxide smelting plant are very important. In Wales, the Anglesey site was chosen because it offered a deep-sea port and good road and rail links to customers in the UK and European Union. Aluminium and coke are imported by sea. However, the biggest reason was that the electrical power could be supplied by the National Grid from nearby Wylfa nuclear power station.

On 20 July 2006, the Nuclear Decommissioning Authority (NDA) who own the site, announced that the Wylfa power station was to close in 2010 because the operation to produce electricity would become 'totally uneconomic'. On 15 January 2009, the owners of Anglesey Aluminium announced that the aluminium smelting plant was to close, with the loss of 500 jobs; they could not find another economic source of 250 MW of electricity to make the smelting works viable. On 25 February 2009, the NDA announced that it was considering keeping the power station open until 2014. In September 2009, Anglesey Aluminium ceased aluminium smelting. The whole plant will shut in September 2011.

The high energy costs involved in the production of aluminium make recycling very economical – Anglesey Aluminium's current activity. The energy cost per tonne of recycled aluminium is only about 5% of the energy cost per tonne of aluminium produced from bauxite. In addition, the electrolytic process consumes the carbon anodes and produces oxides of carbon – greenhouse gases.

1 What is the main consideration when planning an aluminium smelting plant?
2 What other considerations need to be taken into account?
3 Why is the Lochaber Aluminium smelting plant in Scotland situated next to a hydroelectric power plant?
4 Why did aluminium smelting at Anglesey Aluminium have to cease in September 2009?
5 Explain why the closure of a large plant such as Anglesey Aluminium would cause huge social problems in the local area.
6 Anglesey Aluminium estimates that the plant will take 12 months to decommission. Do you think that, once cleaned up, the former plant would be a potential site for a housing estate? What sort of activities do you think would be best suited to this sort of site?

Discussion Point

Why is aluminium recycling so important, both economically and environmentally?

Uses of metals

Three very important commercial metals are aluminium, copper and titanium. These metals are used for a whole host of different applications which are dependent upon the properties of the metal. Table 9.2 summarises some of the properties, applications and uses of these three metals.

Table 9.2 Properties, uses and applications of three important metals.

Metal	Properties	Uses and applications
Aluminium	• Strong • Low density (about 2.7 g/cm^3, compared with iron, 7.9 g/cm^3) • Good conductor of heat • Good conductor of electricity • Resistant to corrosion (because of its thin surface oxide layer)	• Overhead high-voltage power cables for the National Grid; its lightness enables the pylons to be lightweight structures (although these are made from steel) • Saucepans, aluminium cooking foil (linked to its good conduction of heat and non-toxicity) • Its strength and lightness make it suitable for window-frames and greenhouse construction • Drinks cans, because of its lightness and non-toxicity • The manufacture of aeroplane and car bodies, since it is light and has high tensile strength
Copper	• Very good conductor of heat • Very good conductor of electricity • Malleable – can be easily formed into different shapes • Ductile – can be easily drawn into a long wire without breaking • Attractive colour • Lustrous – 'shiny'	• Making alloys such as bronze and brass • Jewellery and ornaments • Copper piping for plumbing • Connecting wire in electrical circuits, motors and other electrical goods • Many stainless steel saucepans incorporate a copper bottom for extra heat conduction
Titanium	• Hard • Strong • Low density (4.5 g/cm^3) • Resistant to corrosion • High melting point (1941 °C, compared with iron, 1536 °C)	• Jet engine parts • Spacecraft parts • Parts for industrial plants • Automotive parts • Strengthening steel • Medical implants • Jewellery • Sports equipment (tennis rackets, cycle frames)

QUESTIONS

3 Why is aluminium used for overhead electricity pylons?

4 Explain why many food tins are made out of aluminium. Why do you think that titanium is not used for food tins?

5 Why are only the bottoms of good quality saucepans made with copper?

6 Why are metal hip replacement joints made from titanium, rather than aluminium, when titanium is twice as dense?

7 Window frames are commonly made out of aluminium. Why are titanium and copper not used for this application?

8 Why are the fan blades of jet engines made from titanium?

9 You are put in charge of designing a new type of kettle to go aboard the International Space Station. Sketch your ideas for the design and, in particular, state the materials that you would use for the different parts of your design. For each material, state why you have used it.

Alloys

QUESTIONS

10 What is an alloy?

11 Why might brass be a good material out of which to make musical instruments?

12 Adding tin to brass makes it form a very thin protective oxide coating. Why is this sort of brass very useful for making parts for boats, such as cleat hooks and door handles?

13 Combinations of iron, aluminium, silicon and manganese make brass wear and tear resistant. What sort of applications do you think that these types of brass would be good for?

Mixtures of metals are called **alloys**. Sometimes designers and engineers need metal parts with very specific properties for very specific applications. In many cases, the natural metals do not have these properties, or the cost of natural metals for these applications would be uneconomic. In these cases alloys are used. Materials scientists have become very skilled at mixing molten metals together to form new alloys with modified properties. **Brass** is a very good example of this. It is an alloy made of copper and zinc.

Brass is a relatively hard, shiny, golden yellow metal. Its composition can be varied by varying the proportions of copper and zinc, and even adding other metals like aluminium, which makes it harder. By customising the composition, brasses can be made with a wide variety of different properties, allowing the alloy to be used for pipe couplings, for musical instruments and for fixing screws, as well as a whole host of other applications. In addition, because brass is made from copper and zinc, it is relatively cheap.

Stainless steel is another alloy, which is used to make saucepans, cutlery and sinks. Metallic spectacle frames are made from a variety of alloys. Nickel alloys and titanium alloys are commonly used, and some modern frames are made from smart alloys that allow the shape to be restored after bending or deformation.

Figure 9.18 Brass is typically made from 35% zinc and 65% copper.

Figure 9.19 Steel is typically made from 99% iron and 1% carbon.

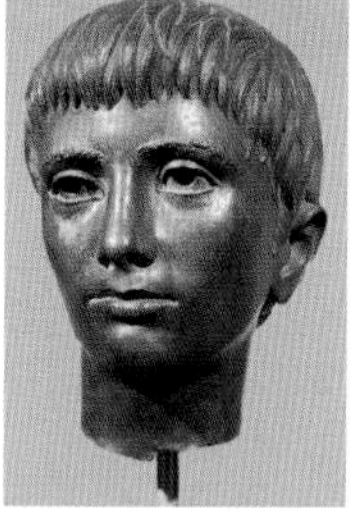

Figure 9.20 Bronze is typically made from 87.5% copper and 12.5% tin.

What are nano-particles?

Figure 9.21 This product contains metal ions.

When you stroll down the aisles of any supermarket, you will find products containing small amounts of metals such as silver, which are added to the products as antibacterial, antiviral or antifungal agents. The products cover a wide range, from hand-soaps to fridges.

So what's going on? Scientists have known for centuries that silver has an antibacterial effect. The Phoenicians, a race of people who lived about 1000BC in the eastern Mediterranean, used silver bottles to store water and vinegar to prevent it from spoiling, particularly on long sea voyages. The drinking water system on the International Space Station uses silver as a disinfectant.

Scientists now know that the silver ion (Ag^+) is bioactive and in sufficient concentration readily kills bacteria in water. The silver ions irreversibly damage key enzymes in the cell membranes of the micro-organisms. Silver also kills bacteria in external wounds in living tissue. Doctors and paramedics use wound dressings containing silver sulfadiazine or silver nano-materials to treat external infections.

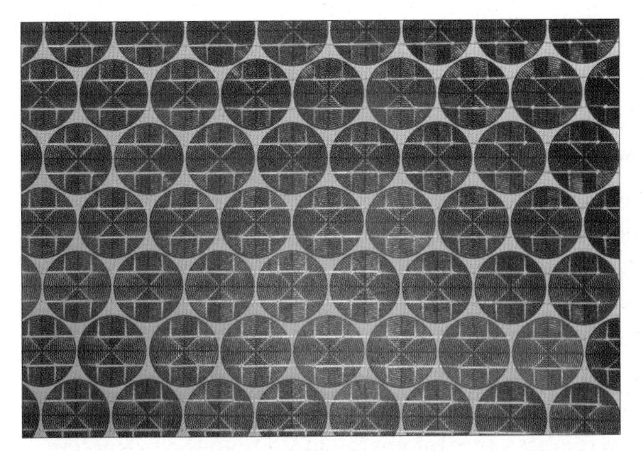

Figure 9.22 An array of photovoltaic solar cells.

Many products now have tiny particles of silver and other metals such as zinc added to them to increase their ability to kill micro-organisms. Many other metals, such as lead or mercury, have the same effect on micro-organisms, but silver is commonly used because it has the least toxic effect on humans.

The quantities of silver that are added to the products are quite small, and the silver is normally added as the compound silver nitrate. The addition of tiny particles like this is an example of 'nano-scale' technology, where the particles are between 1 and 100 nm (nanometres) in size. (1 nm = 1×10^{-9} m, or a billionth of a metre.)

Nano-particles sometimes have different properties from their bulk material. For example, nano-particles of gold or silicon absorb sunlight much more efficiently than thin, continuous layers of gold or silicon, which makes them much more useful in the design of photovoltaic solar cells.

Nano-technology is also being used to develop new batteries, particularly for electric vehicles. In this application, the electrodes of the battery are coated with nano-particles of silicon, titanium, carbon or manganese. The effects of these nano-scale materials is to make batteries that are less likely to catch fire, are higher powered, quicker to recharge and longer-lasting.

Nano-scale particle research is currently an area of intense scientific interest due to a wide variety of potential applications in biomedical, optical and electronic fields, but some people are worried about the wide-scale use of nano-particles, particularly silver, and particularly its use in antibacterial products such as soaps and disinfectants, and in clothing.

In 2008, the BBC reported that the Royal Commission on Environmental Pollution was calling for 'urgent regulatory action' on the nano-scale materials widely used in industry. The commission found that the materials have so far shown no evidence of harm to people or the environment, but it said there was a 'major gap' in research into the risks posed by the materials, which are found in over 600 products globally. The commission chair, Professor Sir John Lawton also said he would not recommend clothes with nano-scale silver. 'We are concerned about nano-scale silver in clothing getting into the environment because it could be potentially very damaging...their behaviour in the environment and in the body is hard to predict.' Nano-scale silver has been incorporated into fabrics in clothes to prevent the bacterial build-up that causes odours. But as it is worn away during

QUESTIONS

14 What is a nano-particle?

15 How does nano-scale silver kill micro-organisms?

16 Explain why nano-scale materials are used to make photovoltaic cells.

17 The Korean electronics firm, Samsung, makes a range of household appliances using nano-scale silver. What is the advantage of buying a fridge incorporating this technology?

18 Why do you think that major motor manufacturers are very interested in the use of nano-technology in the design of electric cars?

19 Explain why the Royal Commission on Environmental Pollution is worried about the wide-scale use of nano-scale silver.

20 What might be the effect of higher concentrations of silver in a river?

Discussion Point

Would you wear clothes containing fibres impregnated with nano-scale silver?

washing, nano-scale silver's bacteria-killing properties could wreak havoc on delicate ecosystems or municipal waste water systems that depend on bacteria. Sir John said: 'I wouldn't recommend nano-scale silver clothes and I wouldn't wear them myself'.

Chapter summary

- Ores found in the Earth's crust contain metals combined with other elements. These metals can be extracted using chemical reactions.
- Some unreactive metals (e.g. gold) can be found un-combined with other elements.
- The difficulty involved in extracting metals increases as their reactivity increases.
- The relative reactivities of metals can be investigated by performing displacement and competition reactions.
- Reduction is the removal of oxygen (from a metal compound); oxidation is the gain of oxygen (by a metal).
- Iron ore, coke and limestone are the raw materials used in the extraction of iron.
- The word and symbol equations for the reduction of iron(III) oxide by carbon monoxide are:

$$\text{carbon monoxide} + \text{iron(III) oxide} \rightarrow \text{carbon dioxide} + \text{iron}$$

$$3CO(g) + Fe_2O_3(s) \rightarrow 3CO_2(g) + 2Fe(l)$$

- The extraction of aluminium requires greater energy input than the extraction of iron. The method used to extract the most reactive metals (including aluminium) is electrolysis.
- The process of electrolysis of molten ionic compounds can be explained in terms of ion movement and electron gain/loss, using the terms electrode, anode, cathode and electrolyte.
- Aluminium is extracted on an industrial scale using large-scale electrolysis. This can be summarised by the following electrode equations in terms of charge and atoms:

$$Al^{3+} + 3e^- \rightarrow Al$$

$$2O^{2-} - 4e^- \rightarrow O_2$$

- There are considerable environmental, social and economic issues relating to the extraction and use of metals such as iron and aluminium. These include: siting of plants, fuel and energy costs, greenhouse emissions and recycling.
- The uses of aluminium, copper and titanium can be explained in terms of the following relevant properties:
  - aluminium – strong, low density, good conductor of heat and electricity, resistant to corrosion
  - copper – very good conductor of heat and electricity, malleable and ductile, attractive colour and lustre
  - titanium – hard, strong, low density, resistant to corrosion, high melting point.
- An alloy is a mixture made by mixing molten metals. Its properties can be modified by changing its composition.
- Nano-scale silver particles have antibacterial, antiviral and antifungal properties, and new uses are being developed in hygiene and medicine.
- The uses of nano-particles are related to their properties, which can be different from the bulk material.
- Although there are considerable potential benefits from using nano-scale materials, there are potential risks involved with current and future developments in nano-science.

Non-metals

Living with gases

Gases are everywhere! We live in a gas atmosphere – without the oxygen in the air our bodies would not function; without hydrogen, our Sun would not produce the sunlight that allows life to exist; without both of these gases we would not have water – our very existence on this planet is due to gases.

Figure 10.1 Gases are essential to life.

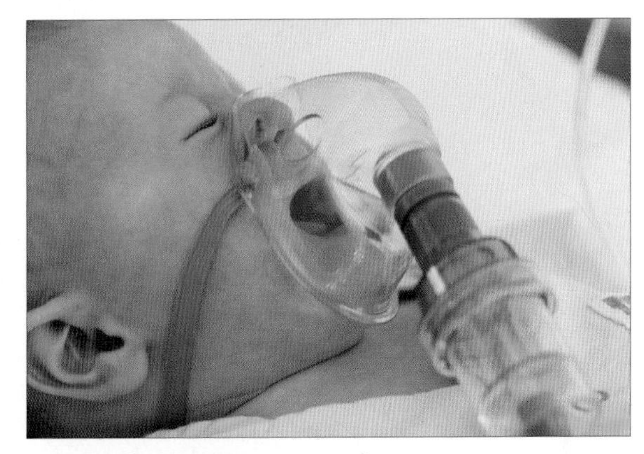

Figure 10.2 Oxygen life support.

Figure 10.3 This car is powered by a hydrogen fuel cell.

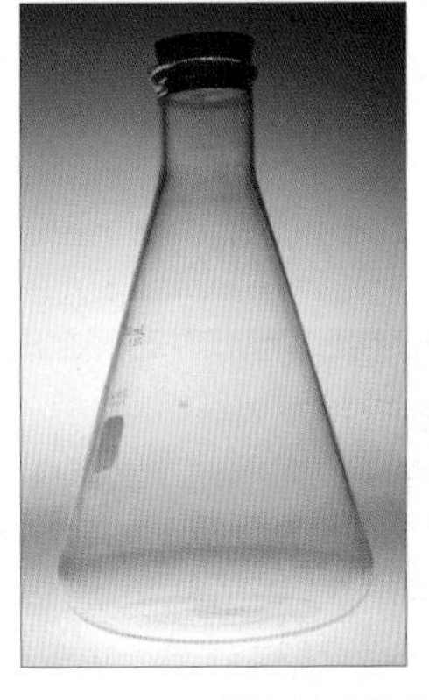

Figure 10.4 Chlorine kills bacteria.

Figure 10.5 Helium is lighter than air.

Figure 10.6 Neon and argon entertain.

What is air?

Air is a mixture of gases. The precise composition of air varies, mostly with altitude, but also due to the weather (clouds) and industrial and domestic chimney gases. The average composition of air at sea level is shown in Table 10.1.

Table 10.1 Average composition of air at sea level.

Gas	Chemical symbol	Percentage by volume
Nitrogen	N_2	78.08
Oxygen	O_2	20.95
Carbon dioxide	CO_2	0.036 (but varies)
Water	H_2O	Variable
Argon	Ar	0.93
Neon	Ne	0.0018
Helium	He	0.000 5
Krypton	Kr	0.000 11
Xenon	Xe	9×10^{-6}

Air is used as the raw material for extracting these different gases, using a process called fractional distillation. During the fractional distillation shown in Figure 10.8 the air is dried, carbon dioxide is removed and then the air is liquefied. The various gases are then separated according to their boiling points. The rarer noble gases, neon (boiling point: −246 °C) and argon (boiling point: −186 °C), are extracted by further fractional distillation. Neon and argon are both very unreactive gases, and can only be produced on an industrial scale by this process. Both gases are very important in the lighting industry. When an electric current is passed through them they both emit light with very particular and attractive colours – neon produces an orangey yellow colour whereas argon glows with an electric blue colour. The electric lightshow that is Las Vegas works because of neon and argon.

QUESTIONS

1 Calculate the combined percentage of the all the other gases in the atmosphere apart from nitrogen and oxygen.

2 Use Table 10.1 and your answer to Question 1 to plot a pie chart showing the composition of air. Include nitrogen, oxygen and 'other gases'.

3 How is oxygen extracted from air?

4 The boiling point of helium is −272 °C – just one degree above absolute zero. Explain why commercial helium is extracted from reserves deep underground (often associated with methane gas deposits), and not by liquefaction and fractional distillation of air.

Figure 10.7 Las Vegas – a spectacle of light.

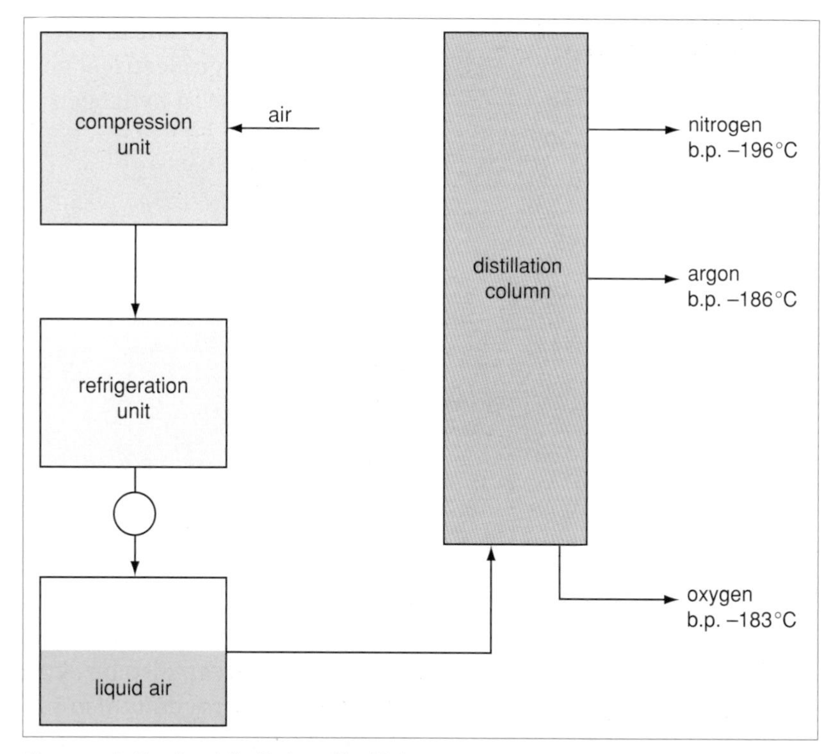

Figure 10.8 Fractional distillation of liquid air.

Hydrogen and oxygen

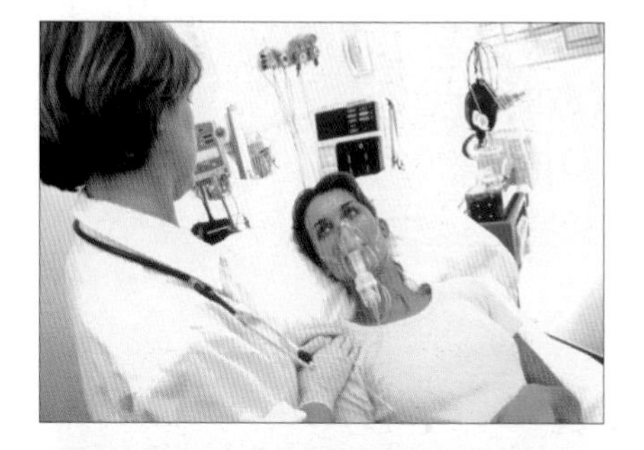

Hydrogen and oxygen are both **diatomic molecules**. This means that the atoms always join together with another atom to form a molecule with two atoms. The molecular formula of hydrogen gas is H_2 and that of oxygen gas is O_2.

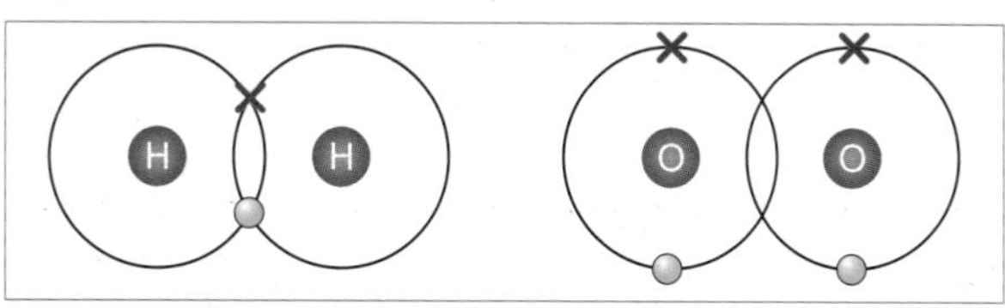

Figure 10.9 Molecular diagrams of hydrogen and oxygen.

Figure 10.10 Some uses of oxygen.

Hydrogen and oxygen are two of the most important gases that we use.

Oxygen is used:

- in medicine as an aid to breathing
- for producing high temperatures, as in oxyacetylene welding
- in liquid form as the oxidant in some rockets
- for enriching the air supply in some steel-making processes
- in high-altitude aircraft.

Hydrogen is used:

- in the petrochemicals industry for making certain types of hydrocarbon
- in the food industry for making margarine

- in welding to prevent impurities
- for cooling large electrical generators
- as a fuel for use in hydrogen fuel cells.

Figure 10.11 Some uses of hydrogen.

Oxygen can be extracted from air by the fractional distillation of liquid air, but it can also be extracted from water by the process of electrolysis. Hydrogen is also produced in the process, although on an industrial scale hydrogen is produced as a by-product of the petrochemicals industry.

In the electrolysis of water an electric current is passed through water. Oxygen gas forms at the anode (the positive electrode); hydrogen gas forms at the cathode (the negative electrode). Both electrodes are usually made of platinum metal strips. Overall, the equation summarising this process is:

$$\text{water} \rightarrow \text{hydrogen} + \text{oxygen}$$

$$2H_2O(l) \rightarrow 2H_2(g) + O_2(g)$$

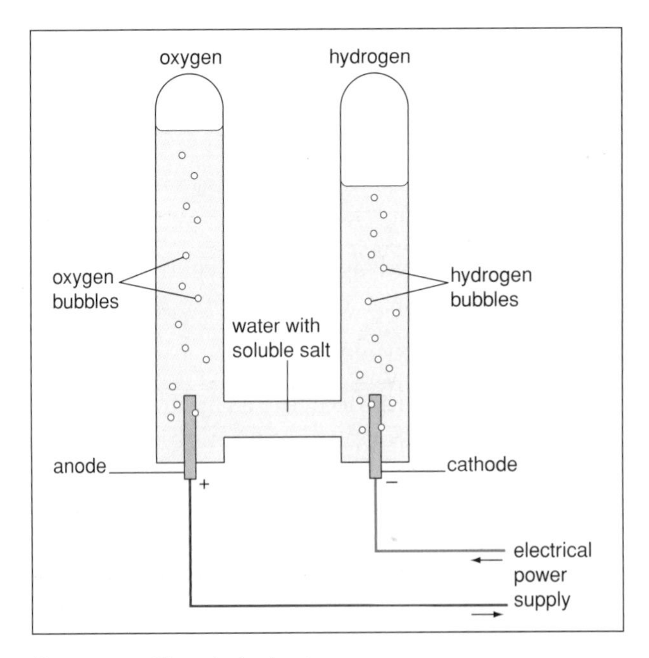

Figure 10.12 Electrolysis of water.

You can see from the equation and from the diagram, that twice as much hydrogen gas is produced than oxygen gas.

PRACTICAL MAKING AND TESTING HYDROGEN AND OXYGEN

This activity helps you with:
★ observing an electrolysis reaction
★ learning and carrying out the tests for hydrogen and oxygen gases.

Your teacher may show you the electrolysis reaction of water using a piece of equipment called a Hofmann voltameter (Figure 10.13).

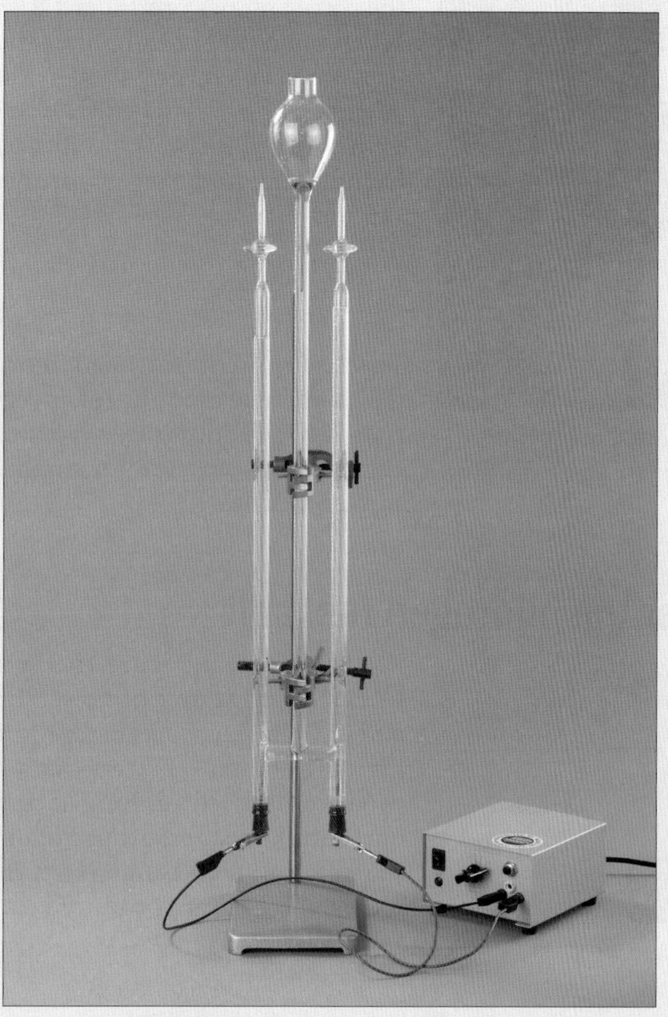

Figure 10.13 A Hofmann voltameter.

Apparatus
* splints
* test tubes of hydrogen and oxygen gas
* eye protection

Your teacher will produce test tubes of hydrogen and oxygen gas for you to test using the voltameter.

Risk assessment

- **Wear eye protection.**

Procedure

Test for hydrogen gas:

1 Apply a burning splint to the mouth of an inverted test tube containing hydrogen gas.
2 You should hear a 'squeaky pop'. This is a small,explosion as the hydrogen combines with oxygen in the air to form water.

Test for oxygen gas:

1 Apply a glowing splint to an inverted test tube containing oxygen gas.
2 The glowing splint should relight.

Hydrogen and oxygen – rocket fuel!

Figure 10.14 The Space Shuttle uses an explosive mixture of hydrogen and oxygen to propel it forwards.

The Space Shuttle uses liquid hydrogen and liquid oxygen stored in a huge orange external tank. The resulting controlled explosion of the two liquids burns at a temperature of 3300 °C and each of the three engines produces 1.8 MN (million newtons) of thrust. When operating at full power the three main engines are using up liquid fuel at a rate of 4000 litres per second. This is the same rate as emptying an averaged sized swimming pool in 25 seconds!

In the combustion chamber of the engines, liquid hydrogen is burning in liquid oxygen, forming water (which is immediately vaporised).

hydrogen + oxygen → water + energy

$$2H_2(l) + O_2(l) \rightarrow 2H_2O(g) + \text{energy}$$

The explosion of hydrogen in oxygen is extremely exothermic, releasing huge amounts of energy. In the Space Shuttle this energy is used to produce thrust from the rocket engines, but this sort of engine is not practical for smaller engines, such as those that power cars and lorries.

Enter the hydrogen fuel cell

The car shown in Figure 10.3 (page 103) uses a small hydrogen fuel cell as its source of power. In a hydrogen fuel cell, hydrogen and oxygen react together, producing an electric current. This is a bit like a standard battery but a hydrogen fuel cell will continue to produce electricity as long as it is fed with hydrogen and oxygen (air). Unlike a standard petrol or diesel engine, the combustion (waste) product is water. No carbon dioxide, sulfur dioxide or nitrogen oxides are produced – a car like this is a zero pollution

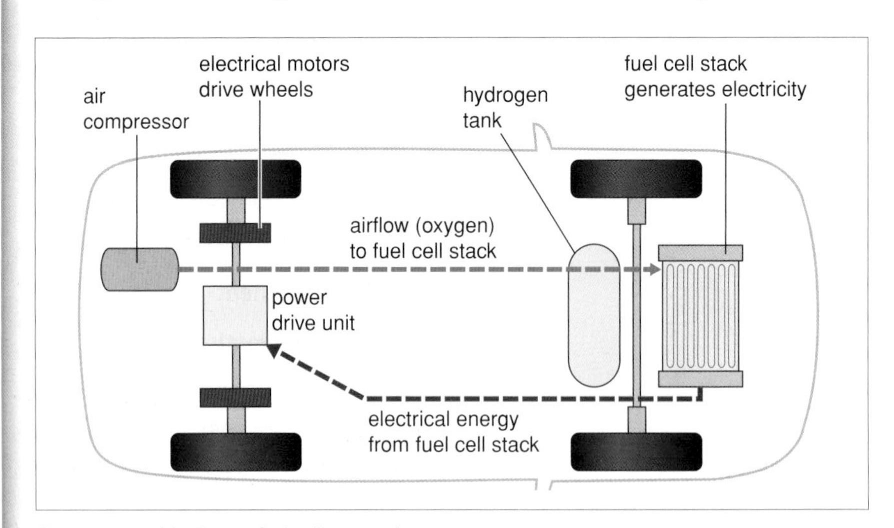

Figure 10.15 A hydrogen fuel cell system in a car.

QUESTIONS

5 How is hydrogen fuel stored in a car?

6 How might the very small number of hydrogen filling stations limit the usefulness of hydrogen fuel cell cars in the UK?

7 What is the chemical reaction that occurs inside the hydrogen fuel cell? Write:
  - a a word equation for this reaction
  - b a balanced symbol equation for this reaction.

8 How is the energy produced by the reaction of hydrogen and oxygen converted to kinetic energy of the car?

9 Why do you think that the car requires a large fuel cell stack?

QUESTION

10 The hydrogen storage cylinder in a hydrogen fuel cell car is very strong and has several fail-safe valves fitted to it. However, it is quite heavy – much heavier than the equivalent petrol tank. Why does the cylinder have to be so strong and why does this limit the range of the car?

car. Hydrogen is also the most abundant element in the Universe. We have a great deal of hydrogen on Earth, but most of it is bound up with other elements in other molecules and compounds, and this means that energy has to be put in to split the molecules and compounds up. In fact, more energy has to be put in to extract the hydrogen gas from hydrocarbons or water than you get out using the hydrogen as a fuel source. It is OK if this energy comes from renewable sources such as sunlight or wind power, but if the energy comes from burning fossil fuels – then what is the point? You might as well burn the fossil fuel!

A hydrogen fuel cell system is shown in Figure 10.15.

Currently the global network of hydrogen fuel stations is very limited. There are only five operational hydrogen fuel stations in the UK, although another eight are planned. A typical hydrogen fuel cell car has an operational range of approximately 240 miles – which makes refuelling a problem at present.

Fuel cells are very efficient devices, much more efficient than the equivalent combustion engine, but there are safety issues. The hydrogen fuel is stored in the car in a pressurised cylinder. In an accident, if the pressurised cylinder were to fracture then the resulting explosion could be very dangerous – hydrogen is much more flammable than petrol. However, is this any more dangerous than an exploding petrol tank?

Discussion Point

Why is it globally important that the hydrogen fuel produced for fuel cell cars be extracted using renewable energy?

Useful halogens

The halogens are a group of diatomic elements occupying Group 7 of the Periodic Table. They consist of: fluorine, F_2, chlorine, Cl_2, bromine, Br_2, iodine, I_2 and astatine, At_2.

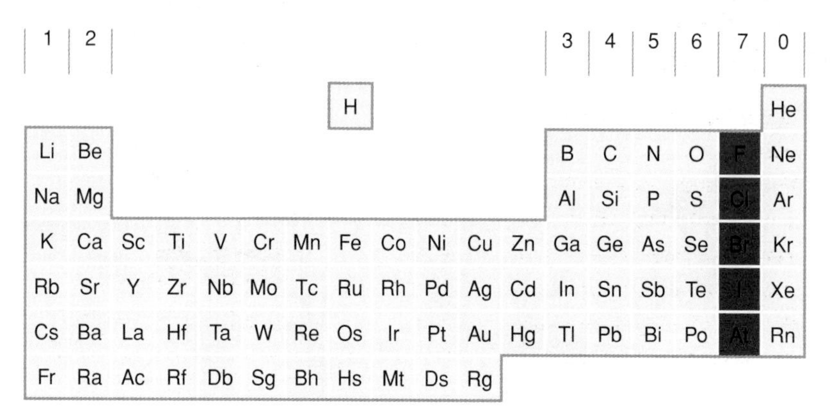

Figure 10.16 The position of the halogens in the Periodic Table.

Astatine is a rare radioactive material with few uses outside of the nuclear industry, but all the others have many uses in a wide range of different applications (Table 10.2).

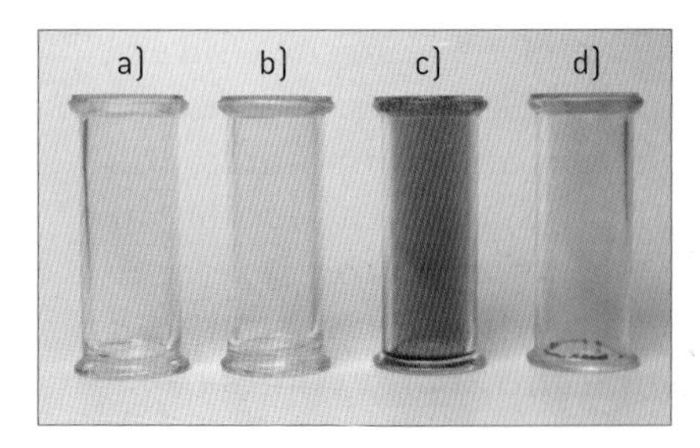

Figure 10.17 The halogens: a) fluorine b) chlorine c) bromine d) iodine

Table 10.2 Appearance, properties and uses of the halogens.

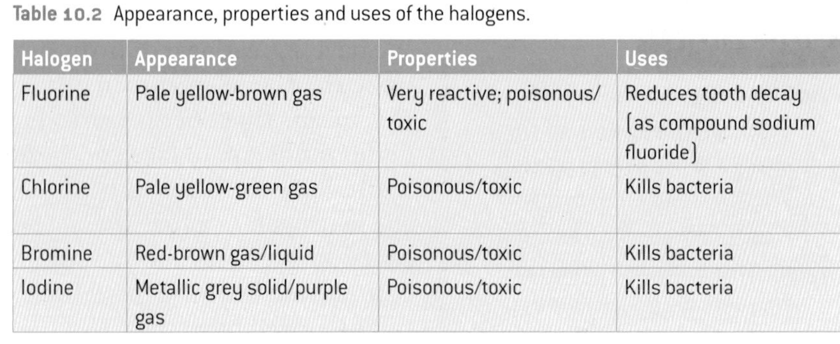

Halogen	Appearance	Properties	Uses
Fluorine	Pale yellow-brown gas	Very reactive; poisonous/ toxic	Reduces tooth decay (as compound sodium fluoride)
Chlorine	Pale yellow-green gas	Poisonous/toxic	Kills bacteria
Bromine	Red-brown gas/liquid	Poisonous/toxic	Kills bacteria
Iodine	Metallic grey solid/purple gas	Poisonous/toxic	Kills bacteria

Chlorine and iodine are found in seawater. Chlorine is one of the elements in sodium chloride, common table salt. About 1.9% of the mass of seawater is made up of chlorine. This means that there is approximately 19 g of chlorine in every litre of seawater. There is much less iodine than chlorine in seawater. Only about 0.05 p.p.m. (parts per million) of seawater is iodine. The raw material for making chlorine is sodium chloride, which is extracted from seawater. Iodine used to be extracted from seawater but this is no longer considered economically viable. Modern iodine production uses brine associated with deposits of oil or gas.

Chlorine is produced by the electrolysis of brine (concentrated sodium chloride dissolved in water). In this process, an electric current is passed through the brine. Chlorine gas is produced at the anode and hydrogen gas at the cathode.

Figure 10.18 Apparatus for the electrolysis of brine.

PRACTICAL MAKING CHLORINE GAS

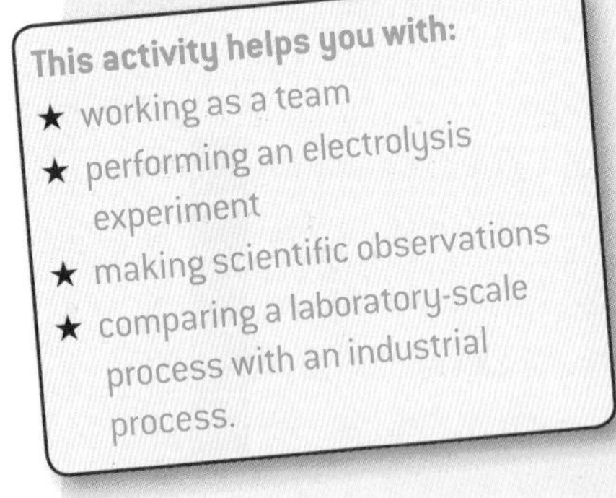

This activity helps you with:

- ★ working as a team
- ★ performing an electrolysis experiment
- ★ making scientific observations
- ★ comparing a laboratory-scale process with an industrial process.

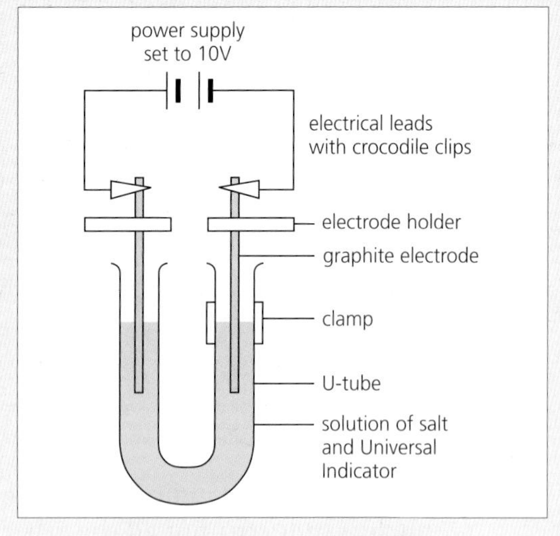

Figure 10.19 Apparatus for making chlorine gas.

Practically, this experiment is best carried out using a U-tube (Figure 10.19).

continued...

PRACTICAL contd.

Apparatus

* 0–12 V dc electrical power supply unit
* electrical leads
* crocodile clips
* graphite electrodes and holder
* U-tube electrolysis cell
* solution of brine and Universal Indicator solution
* eye protection

Risk assessment

- **Wear eye protection.**
- **You will be supplied with a risk assessment by your teacher.**

Procedure

1. Set-up the experiment as shown in the diagram using suitable electrode holders (not holed-bungs).
2. Make up a solution of brine using 2 spatulas of sodium chloride in 75 cm^3 of distilled water.
3. Add 4 drops of Universal Indicator solution.
4. Pour the green brine solution into the U-tube as shown in the diagram.
5. Connect up the external circuit and set the power supply to between 9 and 12 V, depending on the power supply that you are using.
6. Turn on the power supply and observe carefully what happens inside the U-tube.
7. Turn the power supply off as soon as you start to smell a 'bleachy swimming-pool' smell – this will probably be less than 5 minutes.
8. Disconnect all the electrical equipment from the electrolysis cell, but leave the cell to be emptied by your teacher or science technician.

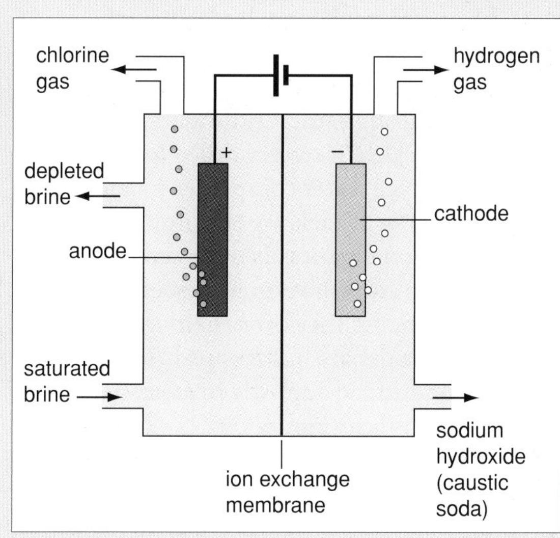

Figure 10.20 Industrial production of chlorine.

Figure 10.20 shows how chlorine is extracted from brine on an industrial scale. This is called the chloralkali process and involves the use of a non-permeable ion exchange membrane at the centre of the cell which allows the sodium ions (Na^+) to pass to the second chamber where they react with the hydroxide ions to produce sodium hydroxide, caustic soda ($NaOH$).

Analysing your results

1. Which element forms at the:
   a anode
   b cathode?
2. In the brine solution, the sodium chloride dissolves, forming an ionic solution of chloride ions (Cl^-), sodium ions (Na^+), hydrogen ions (H^+) (from the water) and hydroxide ions (OH^-) (from the water). What happens to the chlorine ions at the anode?
3. Why do you think that hydrogen gas is formed at the cathode, not sodium metal? (*Hint*: think about the reactivity series.)
4. Study the diagram of the chloralkali process. What are the similarities and differences between this technique and the technique that you used in the laboratory?

TASK THE CURIOUS CASE OF FLUORIDE, WATER AND TEETH

Read the article and then answer the questions that follow.

This activity helps you with:

- ★ identifying evidence
- ★ thinking about scientific ethical issues
- ★ studying a controversial scientific case study
- ★ examining how a scientific issue has changed over time
- ★ examining the link between science, society and government.

Fluoride ions (mostly from sodium fluoride) occur naturally in drinking water, with typical values below 0.5 mg/l. Some UK water companies, like those in the West Midlands add extra fluoride (up to about 1 mg/l) to drinking water to reduce tooth decay in the general population, even though it is known that very high concentrations can actually cause tooth decay, and fluoride has been linked to illnesses such as bone cancer. In 2002, doctors and dentists in Wales called on the Welsh Assembly to instruct Dŵr Cymru Welsh Water to add fluoride to the Welsh water supply. This request caused a great deal of controversy, since the addition of fluoride causes an ethical dilemma; we all have to use the public water supply – and individuals have no choice but to use it. Research published by National Statistics in 2003 has shown that Welsh children have a higher level of tooth decay than children in the West Midlands where fluoride has been added since 1964. In April 2005, the Western Mail newspaper reported that the Welsh Assembly had no plans to add fluoride to drinking water despite the call of Welsh doctors and dentists, backed up by the British Medical Association. The BMA published a summary paper in June 2004 stating that they could find no convincing evidence of any adverse risk to human health through water fluoridation. In 2008, Alan Johnson, the then UK Health Secretary called for all UK strategic health authorities to force water companies to add fluoride to drinking water as a key way of tackling the growing problem of tooth decay. He said 'I don't want this to be carried out in areas where there has been no consultation whatsoever, but every time the public hear the arguments they overwhelmingly go for fluoridation – the problem is, the debate has stopped.' Only 10% of England's water has fluoride added – mostly in areas such as the West Midlands and the North-East (Figure 10.22) – this is a relatively small proportion of the UK population and is targeted mostly at children in deprived areas who do not brush their teeth. As of 2011, the Welsh Assembly has still not instructed Dŵr Cymru Welsh Water to add fluoride to water supplies in Wales, despite it having the power to do so. Anti-fluoride campaigners are still calling for more research particularly into the long-term risks of fluoridation, as there are some (unsubstantiated) concerns about a possible increased risk of cancer, infertility, bone fractures and a condition known as fluorosis, which causes discoloration of the teeth. Alan Johnson pointed out that long-term fluoridation schemes have been in place since the 1940's in the United States with no ill effects. Over 70% of all US water supplies have fluoride added to them.

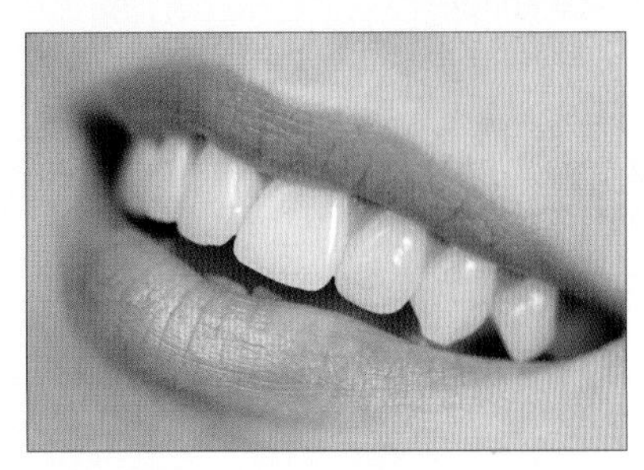

Figure 10.21 Do we need fluoride in our water?

continued...

TASK contd.

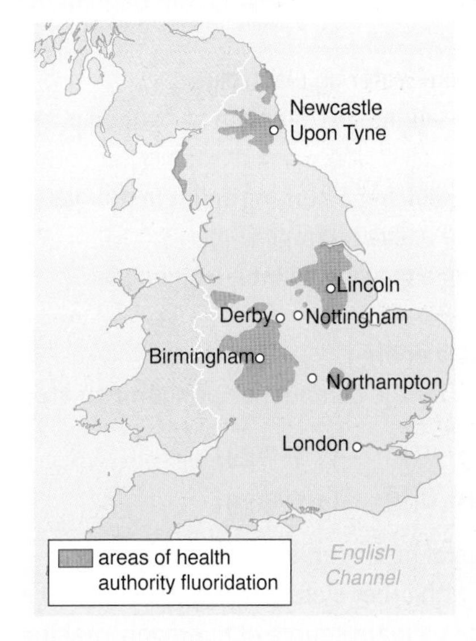

Figure 10.22 Areas in England with fluoridation.

Following the Health Secretary's statement, the British Medical Association (BMA) Cymru Wales, once again repeated their call for fluoride to be added to the Welsh water supply. Welsh Secretary of the BMA, Dr Richard Lewis, said: 'The BMA has been in favour of the fluoridation of mains water supplies for many years. We believe that the fluoridation of water is an effective public health strategy for reducing tooth decay in the population. The evidence shows that water fluoridation is one of the most effective ways of reducing tooth decay in the community. Different communities have different needs and it is essential that there is local debate and a democratic process before any final decisions are made.'

Questions

1 What is the evidence that fluoride in water improves dental health?
2 Why is adding fluoride to water controversial?
3 Why do some people think it is unethical to add fluoride to water supplies?
4 How do you think that scientists working with Government have carried out surveys to gather the evidence for supporting the addition of fluoride to water?
5 Although there have been suggestions of a heightened risk of cancer, infertility and bone fractures due to fluoridation of water, what long-term evidence appears to disprove this?
6 In Wales, who is ultimately responsible for the decision to add fluoride to water?
7 Why is the BMA Cymru Wales renewing its call to add fluoride to water? Why is it important for organisations like the BMA to support issues like this?

Discussion Point

Do you think that the Welsh Assembly should now vote in favour of adding fluoride to Welsh water supplies? Explain your reasoning.

Chapter summary

- Many non-metals, including nitrogen, oxygen, neon and argon, are found in the air.
- Hydrogen and oxygen can be produced from water by electrolysis.
- During the electrolysis of water, twice the volume of hydrogen as oxygen is produced.
- The test for hydrogen gas is performed by placing a burning splint in the gas – if it explodes with a 'squeaky pop' then the gas is hydrogen.
- The test for oxygen gas is performed using a glowing splint – oxygen gas relights the splint.
- Hydrogen gas burns in air, releasing usable energy.
- The word and balanced symbol equations for the combustion of hydrogen are:

$$\text{hydrogen} + \text{oxygen} \rightarrow \text{water} \quad (+\text{ energy})$$

$$2H_2(l) + O_2(l) \rightarrow 2H_2O(g) \quad (+\text{ energy})$$

- There are advantages and disadvantages of hydrogen as a fuel. Hydrogen is very abundant, but only in combination with other elements in compounds. It's combustion product is water, which also is a main source of hydrogen, making it a very renewable fuel, but there are storage and safety issues, and the costs of extraction at present are very high and only feasible if a suitable renewable energy source is used to produce it.
- Chlorine and iodine can be obtained from compounds in seawater but this is no longer considered to be an economically viable source of iodine.
- Chlorine, iodine, helium, neon and argon have many uses because of the following relevant properties:
  - chlorine – poisonous/toxic, kills bacteria
  - iodine – poisonous/toxic, kills bacteria
  - helium – very low density, very unreactive
  - neon – very unreactive and emits light when electric current passes through it
  - argon – very unreactive and emits light when electric current passes through it
- Sodium fluoride, taken in toothpaste or in the water supply, prevents tooth decay. Scientists have gathered evidence to establish this fact by a range of survey techniques.
- There are arguments for and against fluoridation of the water supply, including the ethical issue of removing freedom of choice for the individual.

Acids

A new mission to Venus?

In 1975 Venera 9, a Russian space probe, successfully landed on the planet Venus. Venera 9 was able to transmit pictures for 53 minutes before it corroded away! The highly concentrated sulfuric acid in the atmosphere of Venus literally ate away the space probe.

So, one of the many problems that humans face exploring Venus is that the acid in the atmosphere is so concentrated that most materials will react quickly with it. It destroyed the protective casing of the Venera 9 spacecraft. A human being walking on Venus seems a very unlikely event!

Figure 11.1 The space probe Venera 9 and the inhospitable surface of Venus.

How do we classify the acidity of materials?

Acids (and alkalis) are classified using the **pH scale**. The scale runs from 0 to 14 and actually measures the concentration of hydrogen ions in the substance. All acids contain H^+ ions in water – the higher the concentration of H^+ ions, the lower the pH and the stronger the acid. Hydrochloric acid is made when the gas hydrogen chloride (HCl) reacts with water, forming hydrogen ions and chloride ions:

$$HCl(g)\ (+\ water) \rightarrow H^+(aq) + Cl^-(aq)$$

Substances are alkaline if they contain hydroxide ions, OH^-. When the compound sodium hydroxide dissolves in water, hydroxide ions and sodium ions are formed:

$$NaOH(s)\ (+\ water) \rightarrow Na^+(aq) + OH^-(aq)$$

If the concentration of OH^- ions is high, then the pH is high and the solution is a strong alkali.

The pH scale is used to classify substances as acidic, alkaline or neutral. Substances with a low pH (less than 7) are classified as acidic, substances with a pH of 7 are classified as neutral, and substances with a pH higher than 7 are classified as alkaline. The pH scale can then be used to subdivide acids and alkalis into **strong** or **weak**. Figure 11.2 shows this and indicates the pH of some common household chemicals.

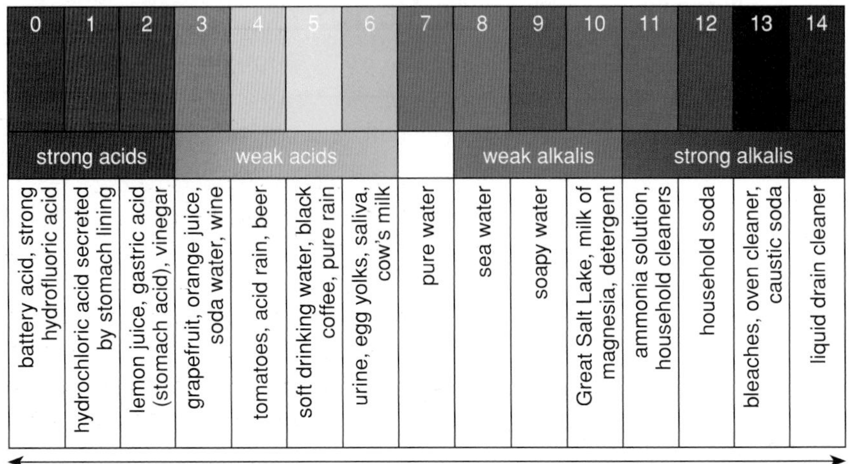

Figure 11.2 The pH scale.

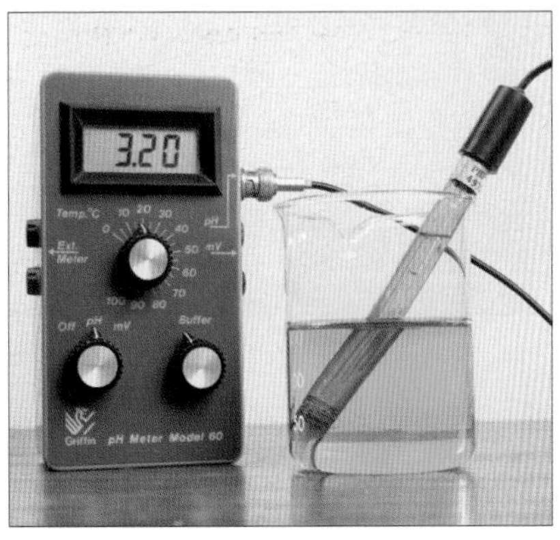

Figure 11.3 Universal Indicator paper and a pH probe and data logger.

pH can be measured in a variety of different ways. One of the best ways is to use a chemical indicator. These are chemicals that are one colour in acid conditions and another colour in alkaline conditions. Chemical indicators like this are very useful for investigating the reactions between acids and alkalis – these investigations use a chemical technique called a **titration**, where the indicator is used to show the end point of the reaction – when all of one chemical has reacted with the other. **Universal Indicator** is a clever mixture of many different chemical indicators. It comes as either a solution or a test paper (where a piece of filter paper has had the solution soaked into it and dried). Universal Indicator turns different colours depending on the pH (see Figure 11.3). It turns red in strong acids, yellow in weak acids, green in neutral solutions, blue in weak alkalis and purple in strong alkalis. Another, more accurate way to measure pH is to use an electronic pH sensor. These sensors are either stand-alone probes/meters or they come as part of a data-logger system.

PRACTICAL INVESTIGATING WAYS TO MEASURE pH

This activity helps you with:
★ working as part of a team in an organised and methodical way
★ planning a standard procedure
★ designing a table to record scientific measurements and observations
★ carrying out chemical experiments safely
★ making and recording measurements and observations
★ identifying substances as the result of experimental tests
★ evaluating experimental methods

Apparatus
* pH paper
* Universal Indicator solution
* data logger pH probe system
* electronic pH meter
* solutions labelled A–E
* test tubes
* test tube rack
* teat pipettes
* eye protection

Risk assessment
- **Wear eye protection.**
- **You will be supplied with a risk assessment by your teacher.**

Procedure
Your teacher will give you a variety of different methods of measuring pH. These will include chemical indicators (some as solutions and some as papers) and electronic pH meters. You will also be given a collection of five different colourless solutions. The solutions are labelled from A–E and will contain strong/weak/neutral solutions of acids, alkalis or water. Your task is to use each pH measuring technique to measure the pH of each solution and then use your measurements to classify each solution as strong acid, weak acid, neutral, weak alkali or strong alkali. You will then use your results, observations and experience of using the different methods to decide which is the best method for measuring the pH of a solution.

1 Work with a partner. You will need to be organised and methodical in order to complete the task accurately. Your teacher will show you a range of different apparatus that you can use as part of your investigation.
2 Plan a standard procedure with your partner, including which apparatus you will use and the method that you will use to do the experiments.
3 To complete the task you will need to design and complete a table to record your observations. You will also need a written analysis of the different techniques, discussing the strengths and weaknesses of each method, and your conclusion.

What happened to Venera 9?

The Russian scientists and engineers who designed and built Venera 9 knew that the atmosphere of Venus contained a large amount of sulfuric acid. So they knew that the space probe would only last a few hours on the surface before the acid reacted with the metal casing and destroyed the probe. They also knew that the probe had to be quite light in order to fit on the rocket launcher and be lifted into space and make the flight to Venus. If the probe was too heavy, the rocket and the probe would not be able to carry the correct amount of fuel to make the long journey to Venus. Therefore, the design was a compromise. The casing had to be made of a strong metal that could withstand the landing; the metal had to be as un-reactive as possible with concentrated sulfuric acid; and the metal panels had to be thin enough so that they would be light enough for the rocket and the probe to carry enough fuel to get them to Venus. A difficult dilemma!

How do metals react with acids?

The reaction of acids with metals is pretty fundamental to chemistry. Some metals react explosively with even weak acids, yet others hardly react at all – only with highly concentrated acids and at high temperatures (just the conditions that are found on Venus). There are three common acids – hydrochloric acid (HCl), sulfuric acid (H_2SO_4) and nitric acid (HNO_3). All three acids are found naturally. When hydrochloric acid reacts with metals it forms compounds called chlorides; sulfuric acid forms sulfates; and nitric acid forms nitrates.

Table 11.1 The reactivity series.

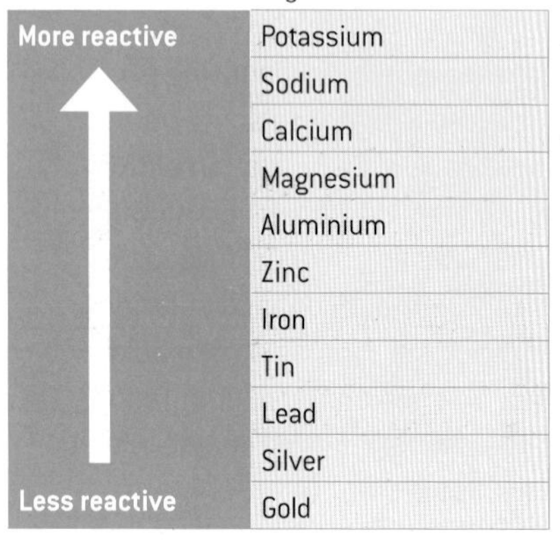

More reactive ↑	Potassium
	Sodium
	Calcium
	Magnesium
	Aluminium
	Zinc
	Iron
	Tin
	Lead
	Silver
Less reactive	Gold

Metals can be arranged in order of their reactivity with common substances like water and acids. A metal reactivity series, including some of the most common metals, is shown in Table 11.1.

The reactions of potassium, sodium and calcium with acid are very energetic. Each reaction produces a great deal of heat, and the hydrogen gas produced during the reaction explodes with oxygen in the air. These reactions need to be carried out under very controlled conditions and are not possible in the student laboratory.

potassium + hydrochloric acid → potassium chloride + hydrogen

$$2K(s) + 2HCl(aq) \rightarrow 2KCl(aq) + H_2(g)$$

QUESTIONS

1 Sodium and lithium react with acid in a similar way to potassium. Write word equations for the reactions of:

a sodium and sulfuric acid

b lithium and nitric acid.

2 When calcium reacts with acids it forms: calcium chloride, $CaCl_2$, with hydrochloric acid; calcium sulfate, $CaSO_4$, with sulfuric acid; and calcium nitrate, $Ca(NO_3)_2$, with nitric acid. Write word equations and balanced symbol equations for the reactions of calcium with:

a hydrochloric acid

b sulfuric acid

c nitric acid.

3 Rubidium (Rb) is an alkali metal that is even more reactive than potassium.

a Write balanced symbol equations for the explosive reactions of rubidium with hydrochloric acid, sulfuric acid and nitric acid.

b If the reaction of potassium with water is explosive, what special conditions do you think you would need to have in order to observe the reaction of rubidium with a concentrated acid?

PRACTICAL INVESTIGATING THE REACTIONS OF METALS WITH ACIDS

This activity helps you with:
- ★ working as part of a team
- ★ carrying out chemical reactions safely
- ★ making and recording scientific observations and measurements
- ★ looking for patterns in scientific observations and measurements
- ★ drawing conclusions based on scientific observations and measurements.

Apparatus
- * test tubes
- * test tube rack
- * thermometer
- * selection of common metals
- * dilute acid: hydrochloric; sulfuric; nitric
- * splints
- * eye protection

Risk assessment
- **Wear eye protection.**
- **You will be supplied with a risk assessment by your teacher.**

Procedure
Your teacher will give you a selection of different common metals and bottles of dilute hydrochloric, sulfuric and nitric acid.

1. Investigate the reactions of all the metals with all the acids. Put approximately 1 cm^3 of dilute acid in a test tube and add a small piece of metal. Observe the reaction. Some of the reactions are quite easy to see, and hydrogen gas is produced.
2. Test for hydrogen gas using a lighted splint and listening for the squeaky pop. Some of the reactions are very difficult to observe, or do not happen at all.
3. Observe each reaction and note any colour changes and/or amount of effervescence (bubbling).
4. Measure temperature changes using a thermometer.
5. Wash each test tube out with plenty of cold water before reusing it for another reaction.

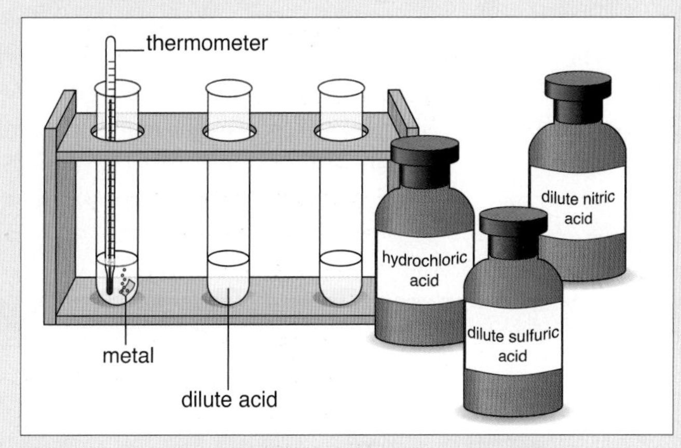

Figure 11.4 Apparatus for the investigation.

6. Copy and complete the following table. You need to comment on:
   - any colour changes
   - any effervescence (bubbling)
   - results of the hydrogen gas test (if relevant)
   - any temperature changes.

If there are no visible reactions, record 'NVR' ('no visible reaction').

Metal, symbol	Reaction with hydrochloric acid	Reaction with sulfuric acid	Reaction with nitric acid
Magnesium, Mg			
Aluminium, Al			
Zinc, Zn			
Iron, Fe			
Tin, Sn			
Lead, Pb			
Copper, Cu			

continued...

PRACTICAL contd.

Analysing your results

1. For each visible reaction:
   - **a** write a word equation
   - **b** find out the chemical formula for each metal salt produced
   - **c** write a balanced symbol equation.
2. Are there any variations in the reactions of individual metals with individual acids?
3. Using your table of observations as a whole, construct a reactivity series based on your observations. How does your series compare with the one in Table 11.1?
4. Why is it difficult to put metals like copper and lead into a reactivity series, based on this experiment?
5. What further experiments could you do to determine the reactivity series of the more un-reactive metals like copper and lead?

TASK

WHICH METAL WOULD YOU CHOOSE FOR A VENUS SPACE PROBE?

This activity helps you with:

★ looking for patterns in scientific data

★ drawing conclusions based on scientific data.

There are many other metals that are not included on the simple reactivity series in Table 11.1. When designing a space probe that will land on Venus, three different properties of the metal need to be considered. At the end of the day, space engineering is always a compromise. In this task you will study a more detailed reactivity series (Table 11.2) containing not only many more metals, but also information about the strength of the metals and their densities (a measure of how closely packed the matter in them is). A high density means an equivalent amount of a substance will weigh more. Study the table – metals become more reactive the closer they get to the top of the table.

Table 11.2 Properties of metals in order of reactivity.

More reactive	Metal, symbol	Reactivity	Density (kg/m^3)	Strength (GPa)
↑	Potassium, K	Reacts with water	890	3.53
	Sodium, Na	Reacts with water	968	10
	Lithium, Li	Reacts with water	534	4.9
	Strontium, Sr	Reacts with water	2 640	15.7
	Calcium, Ca	Reacts with water	1 550	20
	Magnesium, Mg	Reacts with acid	1 738	45
	Aluminium, Al	Reacts with acid	2 700	70
	Zinc, Zn	Reacts with acid	7 140	108
	Chromium, Cr	Reacts with acid	7 190	279
	Iron, Fe	Reacts with acid	7 874	211
	Cadmium, Cd	Reacts with acid	8 650	50
	Cobalt, Co	Reacts with acid	8 900	209
	Nickel, Ni	Reacts with acid	8 908	200
	Tin, Sn	Reacts with acid	7 365	50
	Lead, Pb	Reacts with acid	10 660	16
	Copper, Cu	Reacts with strong acids when heated and pressurised	8 940	128
	Silver, Ag	Reacts with strong acids when heated and pressurised	10 490	83
	Mercury, Hg (liquid – melting point = −38 °C	Reacts with strong acids when heated and pressurised	13 534 (liquid)	N/A
	Gold, Au	Reacts with strong acids when heated and pressurised	19 300	79
Less reactive	Platinum, Pt	Reacts with strong acids when heated and pressurised	21 450	168

continued...

TASK contd.

Discussion Point

Some metals can be mixed together, forming alloys. Alloys tend to have properties that are composites of the metals that make them up. If you were designing a new metal alloy for the Venus space probe, which metals would you try to alloy and why?

Questions

1 Which metal in the table has:
 a the lowest reactivity
 b the lowest density
 c the highest strength?
2 How do the three metals from your answer to question 1 rate as materials for building a Venus space probe?
3 Which metal do you think would be the best compromise choice for building a Venus space probe?

What is the atmosphere of Venus doing to the rocks?

The highly concentrated sulfuric acid in the atmosphere of Venus is gradually eating away some of the rocks that make up Venus's lithosphere (the rocky surface). Figure 11.5 shows a false-colour image of part of the surface of Venus constructed from radar images taken from the Magellan Space probe between 1990 and 1994. You can see from the images the effect of the chemical weathering due to the sulfuric acid 'rain'. The surface features are tending to blend together in much the same way as acid rain on Earth is weathering away the rocks used to construct buildings and statues.

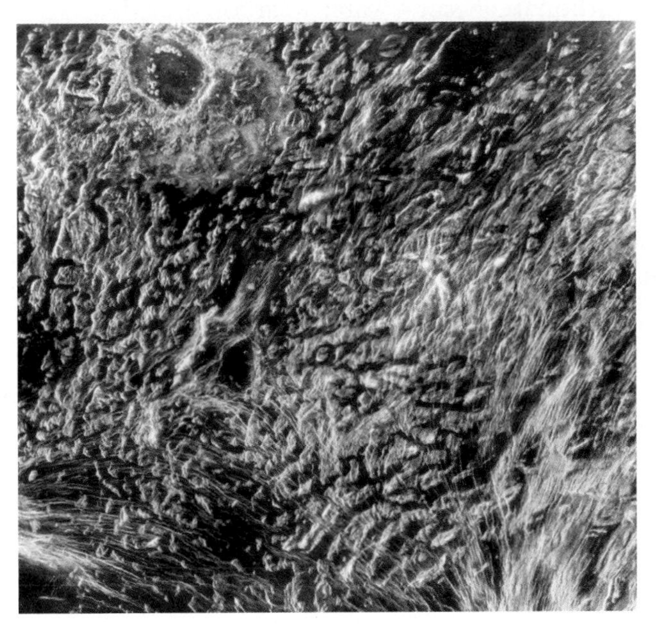

Figure 11.5 The surface of Venus is being eroded by the acidic atmosphere in much the same way as acid rain erodes limestone on Earth.

Acids react with metal oxides, hydroxides (both called **bases**) and carbonates. Most rocks found in the Solar System contain metal compounds like these, and the effect on Venus is to gradually remove these compounds from the lithosphere, replacing them with sulfates, formed from their reaction with the sulfuric acid.

The reaction of an acid with a base or an alkali (a base dissolved in water) is called **neutralisation**. In a neutralisation reaction of this kind, an acid reacts with a base or an alkali, forming a metal salt and water. On Venus, the water immediately reacts with more sulfuric acid.

acid + base → salt + water

For example:

sulfuric acid + magnesium oxide → magnesium sulfate + water

$H_2SO_4(aq) + MgO(s) \rightarrow MgSO_4(aq) + H_2O(l)$

acid + alkali → salt + water

For example:

hydrochloric acid + sodium hydroxide → sodium chloride + water

$HCl(aq) + NaOH(aq) \rightarrow NaCl(aq) + H_2O(l)$

Carbonates react with acids in a different way. The gas carbon dioxide is formed, together with the metal salt and the water:

acid + carbonate → salt + water + carbon dioxide

For example:

hydrochloric acid + calcium carbonate → calcium chloride + water + carbon dioxide

$2HCl(aq) + CaCO_3(s) \rightarrow CaCl_2(aq) + H_2O(l) + CO_2(g)$

Acid reactions with bases, alkalis and carbonates are all **exothermic** – this means that they give out energy in the form of heat.

PRACTICAL MEASURING THE HEAT OF NEUTRALISATION OF ACID/BASE REACTIONS

This activity helps you with:

★ working as part of a team
★ carrying out chemical reactions safely
★ making and recording scientific observations and measurements
★ looking for patterns in scientific observations and measurements
★ drawing conclusions based on scientific observations and measurements.

Apparatus

* test tubes
* test tube rack
* thermometer
* balance
* selection of metal bases
* dilute acids: hydrochloric, sulfuric, nitric
* 100 cm^3 beaker
* teat pipette
* eye protection

The reaction of sulfuric acid and the compounds in the rocks on Venus increases the surface temperature of the planet. Venus is already the hottest planet in the Solar System, and this is in part due to its extreme geochemistry. The amount of heat generated by a neutralisation reaction can be measured by the temperature change of the acid and water during the reaction.

Risk assessment

- **Wear eye protection.**
- **You will be supplied with a risk assessment by your teacher.**

Procedure

In this experiment you will be adding an excess of acid to a known fixed mass of base, so you will be measuring a temperature change per unit mass.

You will only need about 2 cm^3 of acid for each reaction. Wash each test tube out with plenty of cold water before reusing it for another reaction.

1. Place a piece of clean paper on top of the balance and tare the balance.
2. Measure out 0.5 g of one of the metal bases.
3. Carefully pour the metal base into a test tube.
4. Pour about 20 cm^3 of dilute hydrochloric acid into a small beaker.
5. Use the thermometer to measure the temperature of the acid and record this temperature.
6. Place the thermometer into the test tube
7. Use a teat pipette to transfer 2 cm^3 of acid into the test tube.
8. Measure and record the highest temperature of the excess acid during the reaction.
9. Calculate the temperature change of the reaction.
10. Repeat the procedure using the other metal bases.
11. Repeat the whole procedure using the other two acids.
12. Record your measurements in a table like this. You will need a separate table for each acid.

Metal base	Reaction with hydrochloric acid		
	Start temperature (°C)	Highest temperature (°C)	Temperature change (°C)

Analysing your results

1. Which reaction produced the highest temperature change?
2. Are there any patterns in the temperature changes produced by the three different acids?
3. Which metal base produced the highest average temperature change across all three acids?
4. For each reaction write:
   a. a word equation
   b. a balanced symbol equation.
5. How could a heat measurement be used to identify a particular metal base?

PRACTICAL MEASURING THE HEAT OF NEUTRALISATION OF ACID/BASE REACTIONS

This activity helps you with:

★ working as part of a team
★ following a complex standard procedure
★ carrying out chemical reactions safely
★ carrying out a titration safely
★ making and recording scientific observations and measurements
★ looking for patterns in scientific observations and measurements
★ drawing conclusions based on scientific observations and measurements.

Apparatus

* burette
* funnel
* volumetric pipette and filler
* thermometer (digital thermometers work best)
* conical flask
* several 100 cm^3 beakers
* indicator solution
* dilute acids: hydrochloric; sulfuric; nitric
* dilute alkalis: sodium hydroxide; potassium hydroxide

Risk assessment

- **Wear eye protection.**
- **You will be supplied with a risk assessment by your teacher.**

Procedure

In this experiment you will be measuring the heat of neutralisation of acid/alkali reactions. This experiment is in two parts. First you will need to perform a titration between the acid and the alkali to determine the end point of the reaction using a suitable indicator. Then you will measure the temperature change of the same reaction.

Work with a partner. Each pair in the class will do a different reaction, and then you will pool your results. All the acids and alkalis are the same concentration.

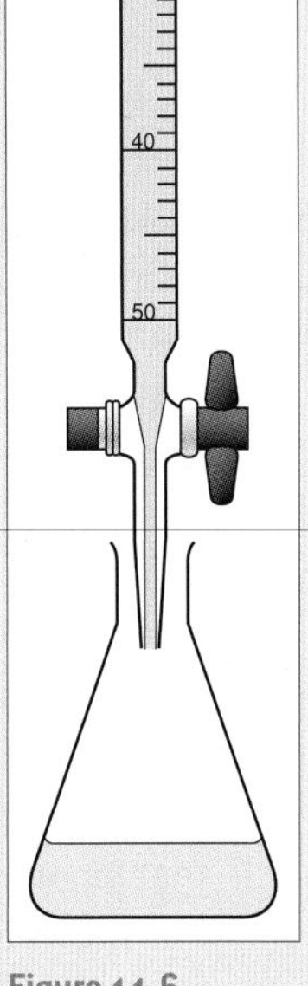

Figure 11.6 Titration apparatus.

Part A – Determining the end point of the reaction

Standard procedure

1. Use a volumetric pipette to pour 25 cm^3 of your chosen acid into a conical flask.
2. Add two or three drops of indicator (your teacher will tell you which is the most suitable).
3. Pour 100 cm^3 of your chosen alkali into a beaker.
4. Use a funnel to safely pour 50 cm^3 of the alkali into a burette mounted on a suitable stand.
5. Adjust the alkali in the burette until it reads 0 cm^3.
6. Slowly add the alkali to the acid in the flask, approximately 1 cm^3 at a time.
7. Record the volume of alkali needed to fully neutralise the acid in the flask – this is when the indicator (just) changes colour.

Part B – Measuring the heat of neutralisation

Standard procedure

1. Use the method in Part A to set up the reaction but ***do not add the indicator*** to the acid.
2. Place a thermometer in the acid and measure and record the temperature.
3. Quickly add the required amount of alkali (determined in Part A).
4. Measure and record the highest temperature.

Analysing your results

1. Write a word equation and balanced symbol equation for your reaction.
2. Pool your results with the rest of the class. Order your results from most exothermic to least exothermic.
3. How could you use this information to identify an unknown alkali?

Acids and carbonates

When acids react with carbonates they effervesce (produce bubbles of gas). The gas produced is carbon dioxide. On Venus, carbon dioxide gas is produced when the sulfuric acid in the atmosphere reacts with carbonates in the rocks. It passes into the atmosphere, increasing the concentration of carbon dioxide and adding to Venus's huge greenhouse effect.

If you are given a substance and you suspect it is a carbonate then you can test the gas given off when it reacts with an acid by passing the gas through limewater (calcium hydroxide solution).

When you bubble carbon dioxide into limewater, a white precipitate of calcium hydroxide is formed. If you pass the carbon dioxide through for a long time, the white precipitate disappears to leave a colourless solution.

PRACTICAL TESTING FOR CARBONATES

This activity helps you with:

- ★ working in a team
- ★ making scientific observations
- ★ testing for gases.

Apparatus

- * 2 test tubes (or boiling tubes)
- * delivery tube
- * dilute hydrochloric acid
- * calcium carbonate powder
- * spatula
- * limewater
- * teat pipette
- * eye protection

Risk assessment

- **Wear eye protection.**
- **Your teacher will supply you with a suitable risk assessment.**

There are two ways of doing this test:

Procedure 1

1. Add dilute hydrochloric acid to the suspected carbonate. Place the stopper in the tube (see Figure 11.7).
2. Pass the gas through a small volume of limewater.
3. A white precipitate, indicated by cloudiness, should confirm the presence of a carbonate. Some people say 'the limewater goes milky'.

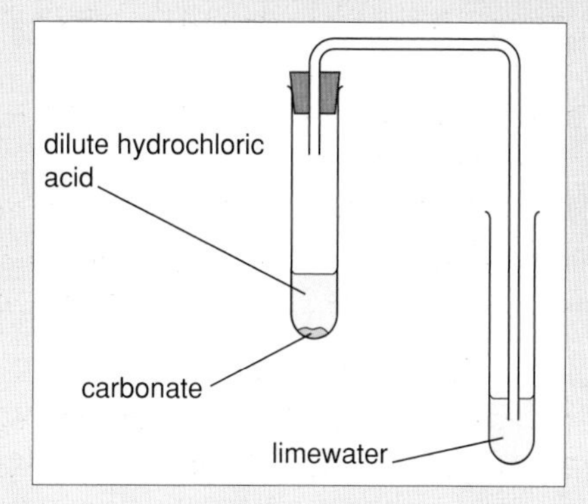

Figure 11.7 Apparatus for CO_2 test.

Procedure 2

Alternatively, suck up the suspected carbon dioxide in a clean teat pipette and then force the gas through no more than 1 cm^3 of limewater in a clean test tube. Cleanliness is essential.

PRACTICAL WHICH MAGNESIUM?

This activity helps you with:

- ★ working as part of a team in an organised and methodical way
- ★ producing a risk assessment for an experiment
- ★ planning a standard procedure
- ★ designing a table to record scientific measurements and observations
- ★ carrying out chemical experiments safely
- ★ making and recording measurements and observations
- ★ identifying substances as the result of experimental tests
- ★ evaluating experimental methods.

You will be supplied with Petri dishes containing four powders, labelled A, B, C and D. One is magnesium metal, one is magnesium carbonate, one is magnesium oxide and one is magnesium chloride. You do not know which letter corresponds to which powder. Your task is to perform chemical tests to identify which powder is which.

Apparatus

Read the standard procedure and then identify the apparatus that you need. This will need to be ordered via your teacher and your science technician.

Risk assessment

- **Wear eye protection.**
- **You will be supplied with a blank risk assessment and hazcards by your teacher.**

Standard procedure

1. Work with a partner.
2. Construct a suitable risk assessment using the hazcards supplied.
3. Identify the tests for each substance.
4. Collect suitable apparatus to perform each test.
5. Perform the test and record your observations and/or measurements in a suitable format.

Analysing your results

1. Use your observations and measurements to identify substances A, B, C and D.
2. Write suitable word and balanced symbol equations for your reactions.

The crystals of Venus?

Figure 11.8 The surface of Venus is covered in crystals.

The products of the chemical reactions occurring between the sulfuric acid and the compounds in the lithosphere of Venus result in some magnificent crystalline structures. The pictures taken by Venera 9 suggest that the space probe landed on a surface covered in crystals.

The combination of high concentrations of acid together with high temperatures and pressures make the surface of Venus a potential ‘crystal garden’. Crystals form when substances come out of solution (usually water) and start to form regular solid arrays of the particles of the substance. If the conditions are right, then the small crystals become ‘seeds’ for much larger crystals and sometimes huge crystalline structures can be formed.

In the laboratory, small crystals can be made by evaporating the solutions produced during neutralisation reactions of metal bases, alkalis and acids. If the metal salt produced as a result of the neutralisation reaction is

Figure 11.9 Large, regular crystals of copper sulfate.

soluble, then by gently heating the solution of the metal salt in an evaporating basin the water can be evaporated off, leaving small solid crystals of the metal salt in the bottom of the evaporating basin. These small crystals can then be used as 'seeds' to 'grow' larger crystals of the salt, by suspending them in a very concentrated solution of the metal salt. As the seed crystal sits in the concentrated solution of its salt, the water in the concentrated solution gradually evaporates, increasing the concentration of the salt even more. Eventually the salt starts to come out of the solution, adding to the small seed crystal as it forms and 'growing' the crystal.

PRACTICAL MAKING CRYSTALS OF COPPER SULFATE AND COPPER CHLORIDE

This activity helps you with:

- ★ working as part of a team
- ★ following a complex standard procedure
- ★ carrying out chemical reactions safely
- ★ making and recording scientific observations made through a microscope.

Copper chloride crystals can be made from copper oxide and dilute hydrochloric acid:

copper oxide + hydrochloric acid → copper chloride + water

$$CuO(s) + 2HCl(aq) \rightarrow CuCl_2(aq) + H_2O(l)$$

To make copper sulfate crystals, use copper oxide or copper carbonate and dilute sulfuric acid:

copper oxide + sulfuric acid → copper sulfate + water

$$CuO(s) + H_2SO_4(aq) \rightarrow CuSO_4(aq) + H_2O(l)$$

copper carbonate + sulfuric acid → copper sulfate + water + carbon dioxide

$$CuCO_3(s) + H_2SO_4(aq) \rightarrow CuSO_4(aq) + H_2O(l) + CO_2(g)$$

In each case the resulting copper salt solution needs to be evaporated in an evaporating basin over a low roaring Bunsen flame.

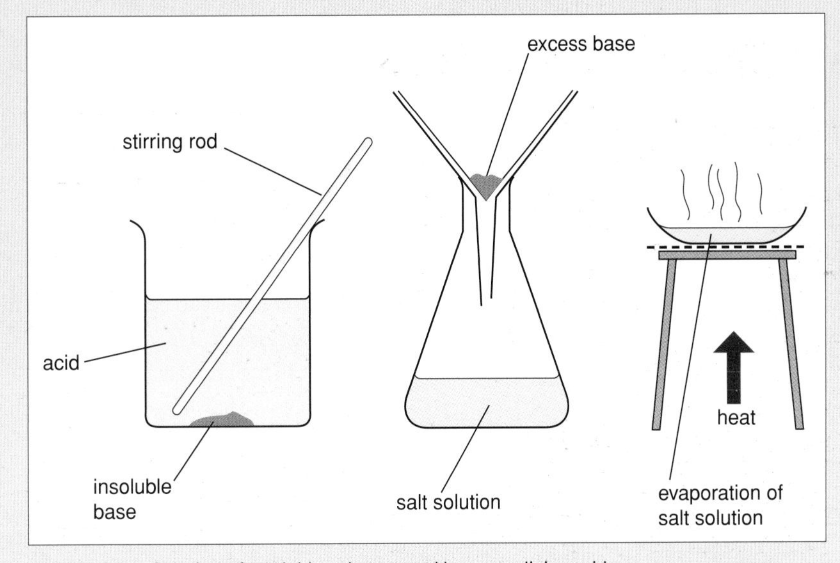

Figure 11.10 Solution of a soluble salt prepared by neutralising acid.

continued...

PRACTICAL contd.

Apparatus
* dilute hydrochloric acid
* dilute sulfuric acid
* copper carbonate powder
* copper oxide powder
* spatula
* beaker
* stirring rod
* funnel
* filter paper
* conical flask
* evaporating basin
* Bunsen burner; tripod; gauze; heatproof mat
* seed crystals
* concentrated copper salt solution
* cotton thread
* small beaker
* microscope
* sticky tape
* clear microscope slide
* eye protection

Risk assessment
- **Wear eye protection.**
- **You will be supplied with a risk assessment by your teacher.**
- **Take great care to turn off the Bunsen burner when there is still a small amount of liquid left in the evaporating basin, otherwise it will spit hot crystals at you.**

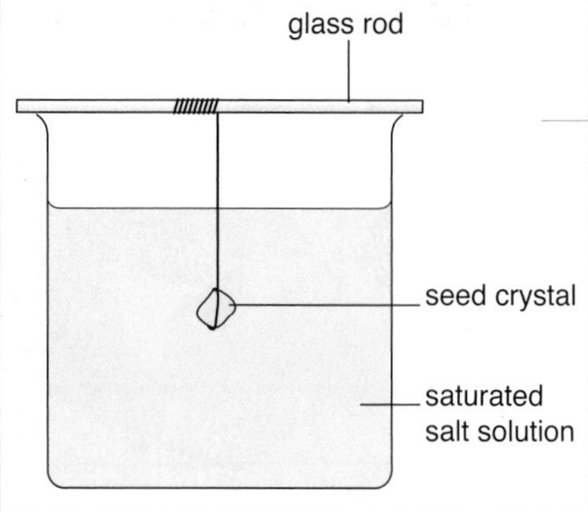

Figure 11.11 Who can grow the biggest crystal?

Standard procedure
1. Work with a partner.
2. Select one of the reactions shown above.
3. Carry out the experiment as shown in the diagram. Remember, copper carbonate will effervesce with the sulfuric acid.
4. Take care when the liquid has almost all evaporated from the evaporating basin (see risk assessment above).
5. Wait for the evaporating basin to cool down, then scrape a small amount of the crystals onto a microscope slide, cover them with a strip of clear sticky tape and observe the crystals under a microscope at low power.
6. Draw a sketch of some of your crystals. Your teacher may give you slides of other crystals to observe with the microscope. Sketch these crystals too – make sure you label each sketch with the name of the crystal and the magnification of the microscope.
7. The crystals that you have made are probably too small to act as seeds in a crystal growing garden. Your teacher will give you some larger seed crystals to grow larger crystals. Tie a short length of cotton around one of the seed crystals and attach the other end of the cotton to a glass rod. Suspend the seed crystal in a concentrated solution of the salt in a small beaker (Figure 11.11). Leave the beaker in a warm place, or on a window sill to 'grow'. The crystal will continue to grow until the level of the solution drops below the level of the suspended crystal.

QUESTIONS

4 Strontium is a reactive metal in the same Periodic Table group as calcium and magnesium.
   - **a** Which crystals can be formed from the reaction of strontium oxide (SrO) with sulfuric acid?
   - **b** How can the crystals be formed?

5 Write word and balanced symbol equations for the reactions of strontium carbonate with:
   - **a** hydrochloric acid
   - **b** nitric acid.

6 How would you grow large crystals of strontium sulfate?

For sale! Desirable residence on Venus

Figure 11.12 Venus – an attractive place to live?

Venus is a hellish place. The daytime temperature is hot enough to melt lead, the atmospheric pressure is crushing and there are huge billowing clouds of concentrated sulfuric acid everywhere. There is arguably no more challenging place for humans to send space probes, let alone live. Venus is approximately the same size and mass as the Earth, with approximately the same force of gravity, and, crucially, it is a 'reflection' of the Earth as it could be if we do not get to grips with our global warming. If humans do not control the release of carbon dioxide into the atmosphere then generations of people in the future will need to leave Earth because it will have become almost as hellish as Venus. Where can we go?

Discussion Point

If we don't want to end up with a planet like Venus, what can we do as individuals?

Chapter summary

- Substances can be classified as acidic, alkaline or neutral in terms of the pH scale.
- Acids and alkalis can be classified as either weak or strong depending on their pH.
- Acids react with some metals. The extent of the reaction depends on the metal's position in the reactivity series.
- The reaction of dilute acids with bases (and alkalis) is called neutralisation, and these reactions are exothermic (give out heat).
- The reaction of dilute acids and carbonates is also exothermic; carbonates effervesce in acid, giving out carbon dioxide gas.
- The test to identify carbon dioxide gas is to pass the gas through limewater. If it turns milky then the gas is carbon dioxide.
- Soluble salts, such as copper sulfate, can be made from the reaction of insoluble bases and carbonates with acids. These soluble salts can form crystals.
- Word equations and balanced symbol equations can be used to describe the reactions of metals, bases (including alkalis) and carbonates with hydrochloric acid, nitric acid and sulfuric acid.

Fuels and plastics

What's 'crude' about crude oil?

Crude oil is the name given to oil extracted from the Earth. It is called that because the word 'crude' is the opposite of 'refined', and refining is the name given to the processes that change the oil into usable substances.

Crude oil is not a simple chemical, and when you know how it has been formed you will understand why it isn't, and why crude oils from different places differ from each other in a number of ways.

Crude oil was formed from the remains of simple marine organisms (mainly plants) that lived millions of years ago. When they died their remains accumulated on the bottom of the oceans. Over a very long time, their bodies were covered by layers of mud, silt and sand that eventually formed rock. This created tremendous pressure and heat that, together with an absence of oxygen, turned the remains into the liquid we call crude oil.

The continuing pressure forced the oil into areas of porous rock, which we now call **reservoirs**. The process of oil formation also results in the formation of natural gas, so often gas and oil supplies are found in the same place, as in the North Sea.

The bodies of living things contain many different chemicals, so oil is a complex mix of substances called **hydrocarbons** (because they contain only carbon and hydrogen). The organisms whose remains formed the oil were very varied, so the exact composition of the oil in different reservoirs will not be the same.

The process of oil formation is a type of fossilisation. For this reason oil and gas (and coal) are called **fossil fuels**.

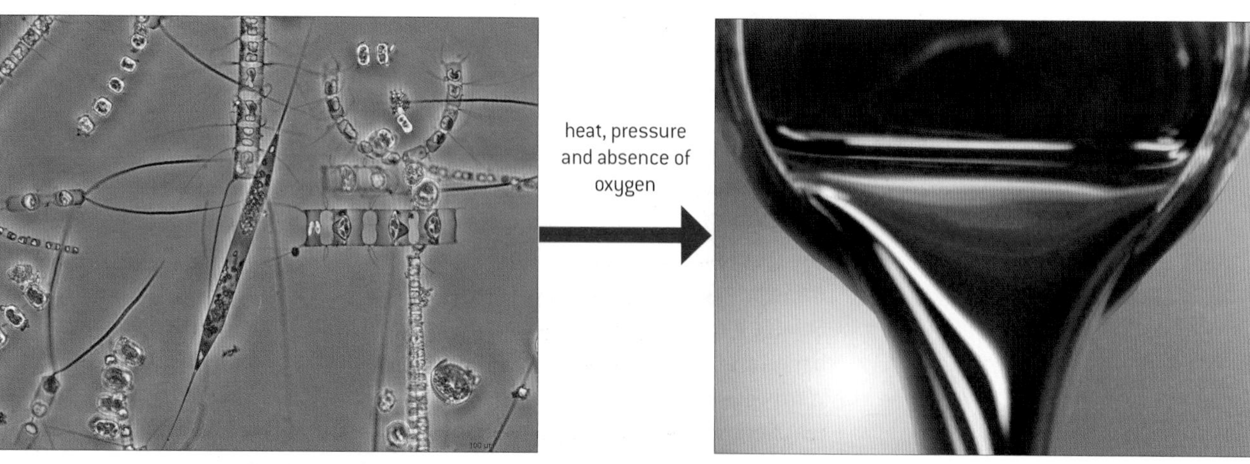

Figure 12.1 Oil formation.

TASK

WILL WE RUN OUT OF OIL?

This activity helps you with:
★ analysing graphs
★ identifying sources of error.

Oil is classed as a **non-renewable** source of energy because it takes millions of years to form and so effectively it cannot be replaced. Over the last century the human race has been using oil at an ever-increasing rate. There are different claims about when the first commercial oil well was dug, but it was around 1850. Before that, the oil reserves of the Earth had remained untouched for millions of years. Eventually, the oil reserves of the planet will run out. This will happen when the oil that remains is either too difficult to get out, or in such small quantities that it is not commercial to do so.

Figure 12.2 shows a prediction of future oil use and how the reserves will decrease.

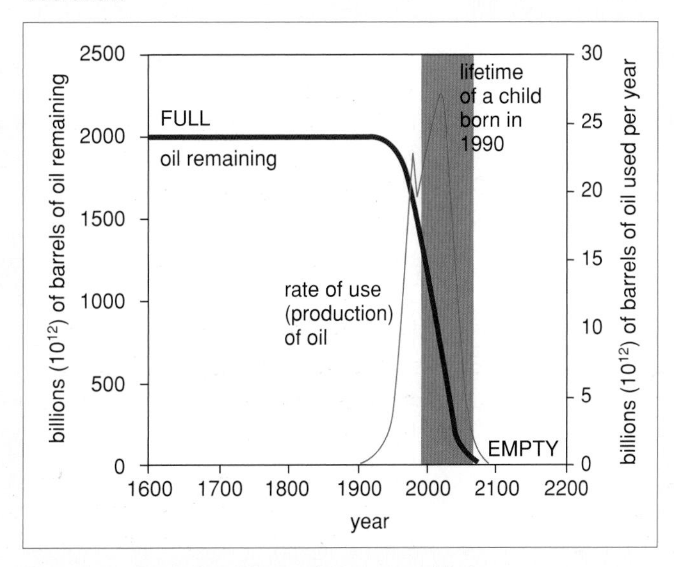

Figure 12.2 Prediction of future oil reserves and their use.

1 According to these predictions, when will the oil reserves run out?
2 At one point, the rate of use line goes above the 'oil remaining' line. Explain why this does *not* mean that the amount of oil being used is actually more that the total oil remaining (which is clearly impossible).
3 Predictions can never be absolutely accurate. Explain why there may be sources of error:
   a in the prediction for oil use
   b in the prediction for the oil remaining.

TASK

SHOULD WE USE LESS OIL?

This activity helps you with:
★ understanding the social, economic and environmental impact of reducing the use of oil.

Many people argue that we should use less oil, for a variety of reasons. As with most things in science, though, the situation is not straightforward. Using or not using oil has social, economic and environmental consequences. Oil is used to make the following products (but this is not a complete list):

- various types of fuel, including petrol, diesel and aeroplane fuel
- lubricant oils for use in all sorts of machinery
- plastics
- paint (some paints are water-based, but many are made with oil)
- clothing and material products like nylon and polyester
- packaging materials like polystyrene
- asphalt, which is used to make roads.

continued...

Tables 12.1, 12.2 and 12.3 list the social, economic and environmental impacts of reducing use of oil.

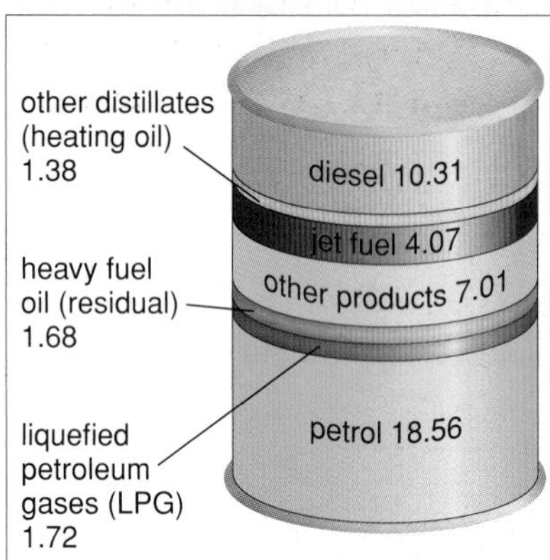

Figure 12.3 How a barrel of oil is used (gallons).

Table 12.1 Social impact of reduction in oil use.

Bad	Good
The oil industry provides jobs for huge numbers of people – estimated to be about 400 million worldwide. Reduction in use could mean many people would lose their jobs. The oil industry makes cheap plastics that allow people to purchase goods they would not be able to afford if made from more expensive materials. Cutting down on oil use might make things more expensive in the shops. Oil is easy to transport to power stations anywhere. Renewable energy power stations (using sunlight, water or wind) could only be built in certain places, and people would not like to have such power stations on their doorsteps. If airlines had to reduce their oil use, it would mean air tickets would go up in cost considerably. Many people might no longer be able to afford holidays abroad.	Oil spills can ruin the livelihoods of people affected by pollution, particularly fishermen and those working in the tourist industry. Such people would be more secure.

Table 12.2 Economic impact of reduction in oil use.

Bad	Good
The government gets huge amounts of money from taxes on petrol. These taxes go towards funding the building of roads, the NHS, education, looking after the poor, etc. If people lose their jobs because oil use is reduced, the UK taxpayers would have to pay more to provide unemployment benefit for those out of work.	Although the UK is an oil producer, it does not produce enough for its needs, and has to buy in oil from other countries. This bill would be reduced if people in the UK used less oil. Reduction in oil use would mean a growth in alternative energy supplies, and this would allow entrepreneurs to develop successful businesses to boost the economy and provide jobs.

Table 12.3 Environmental impact of reduction in oil use.

Bad	Good
An increase in the number of wind farms would mean large numbers of wind turbines across the country, and many people find these unsightly.	If less oil was being used, the incidences of oil pollution from tankers and oil wells would be reduced. The burning of oil releases lots of carbon dioxide into the atmosphere, which contributes to global warming. Incomplete combustion can produce toxic carbon monoxide. Less oil use would mean less of this pollution.

The issues listed in Tables 12.1, 12.2 and 12.3 are just some of those involved with the impact of reducing the amount of oil used. Of course, sooner or later we will have to reduce the use of oil, otherwise it will run out more quickly, possibly before the human race has fully developed alternative energy sources. The argument really involves decisions about how quickly and by how much we should reduce the use of oil.

Question

Look at the possible good and bad effects of reducing oil use in Table 12.1, 12.2 and 12.3. Which of the arguments do you think are the *weakest*, either in favour of or against cutting down on the use of oil? Explain how you reached your opinion.

How is oil made into petrol?

Crude oil has already been described as a complex mixture of different chemicals. Many of these chemicals can make useful products, but they have to be extracted from the oil first. Petrol is just one of many products that results from this **refining** process.

Figure 12.4 The Texaco oil refinery in Pembrokeshire, South Wales.

To refine it, the oil must be subjected to several processes. The first of these is **fractional distillation**. This involves separating the complex mixture of hydrocarbons in crude oil into simpler mixtures of hydrocarbons (fractions), depending on their boiling points (see Figure 12.5). Remember that these fractions are still mixtures, not simple compounds.

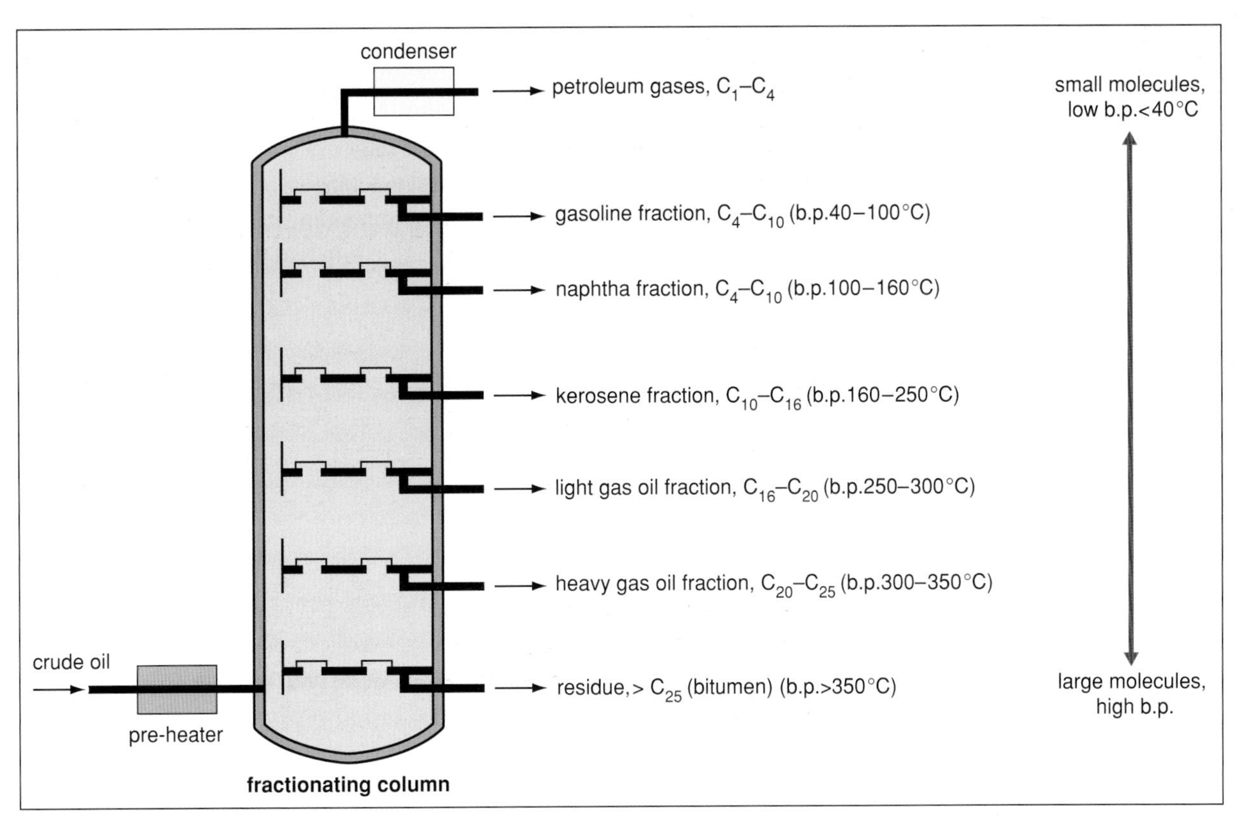

Figure 12.5 The fractional distillation of crude oil.

Crude oil is vaporised before it enters the base of the fractionating column. The temperature of the column decreases with height. As the vaporised crude oil rises up the fractionating column, it passes through bubble cap plates that collect condensed liquid at that temperature and allow vapours of liquids with lower boiling points to move higher up the column. Each plate contains many bubble caps like the one shown in Figure 12.6.

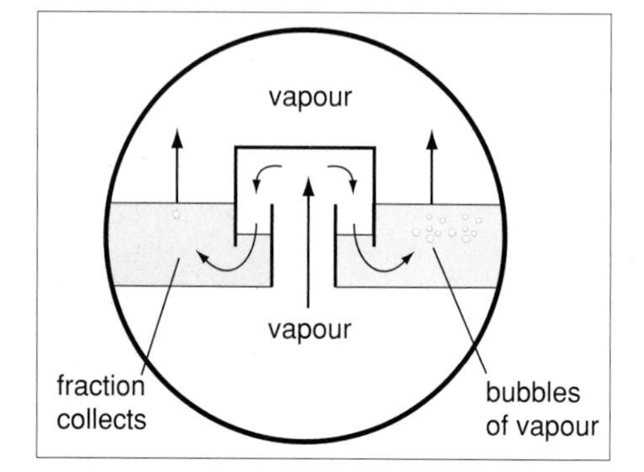

Figure 12.6 Bubble cap plates collect liquid and let vapour pass.

QUESTIONS

1. Why will the temperature reduce as you go up the fractionating column?
2. Explain clearly *exactly* how a bubble cap separates the liquid and the vapour.
3. Do research to find out some uses of the kerosene fraction.

Some of the products from the fractional distillation of oil are fuels. These include the lowest boiling point fractions, gases and gasoline (petrol).

PRACTICAL DOING YOUR OWN FRACTIONAL DISTILLATION

This activity helps you with:
★ designing experiments.

Apparatus

* Bunsen burner
* heatproof mat
* clamp stand
* side-arm 'hard glass' (borosilicate) test tube
* bent delivery tube and rubber connection tubing
* 4 small sample tubes (20 mm × 5 mm) minimum size
* thermometer 0–360 °C with cork to fit side-arm test tube
* teat pipette
* beaker (100 cm^3)
* 'hard glass' watch glass
* mineral or ceramic fibre
* wooden splints
* crude oil substitute
* eye protection

 Risk assessment

- **Wear eye protection.**
- **You will be supplied with a risk assessment by your teacher.**

continued...

PRACTICAL *contd.*

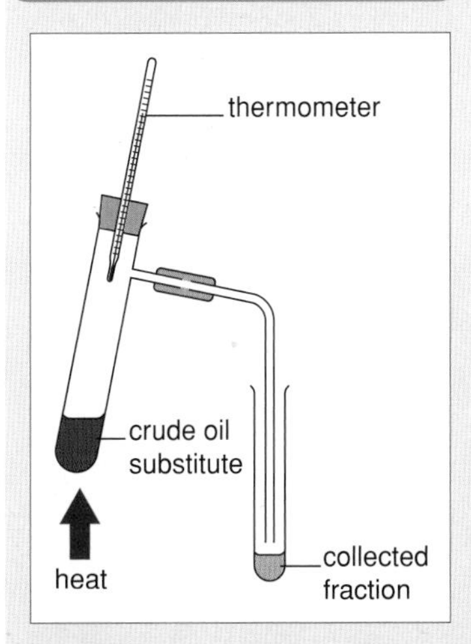

Figure 12.7

Procedure

1 Place about a 2 cm^3 depth of ceramic fibre in the bottom of the side-arm test tube. Add about 2 cm^3 of crude oil alternative to this, using the teat-pipette.
2 Set up the apparatus as shown in Figure 12.7 with one addition – a beaker of cold water around the collecting tube. The bulb of the thermometer should be level with, or just below the side-arm. Heat the bottom of the side-arm test tube gently, with the lowest Bunsen flame. Watch the thermometer.
3 When the temperature reaches 100 °C, replace the collection tube with another empty one. The beaker of cold water is no longer necessary.
4 Collect three further fractions, to give the fractions as follows:
   a room temperature to 100 °C
   b 100–150 °C
   c 150–200 °C
   d 200–250 °C
5 A black residue remains in the side-arm test tube. Test the four fractions for viscosity (how easily do they pour?), colour, smell and flammability. To test the smell, *gently* waft the smell towards you with your hand. To test for flammability, pour onto a hard glass watch glass and light the fraction with a burning splint.
6 Keep one set of fractions and mix them together. See that they combine to form a mixture very like the original sample.

Questions

1 Why was the beaker of cold water needed for the first sample but none of the others?
2 Why is the position of the thermometer bulb important?

Discussion Point

The fractions collected in this experiment still contain a mixture of chemicals. What evidence is there when doing the experiment that the liquid collected cannot be a single chemical?

What is a plastic?

There are many different types of plastic but in order to be called a plastic, a material has to have certain properties. The word 'plastic' comes from the Greek word *plastikos* which means *capable of being moulded or shaped*, and it is this property that makes a plastic a plastic. Plastic can be formed into a variety of shapes – sheets, fibres, tubes, bottles, boxes, utensils, etc.

Although they are all capable of being moulded or shaped, different plastics have different properties, and these different properties make them useful in different situations. **Thermoplastics**, for instance, soften when heated, whereas **thermosets** are resistant to heat.

Table 12.4 Properties and uses of different plastics.

Plastic	Properties	Uses
Polystyrene	A thermoplastic; resists attack by acids, alkalis, and many solvents; does not absorb water; excellent electrical insulator.	Drinking cups, meat trays, DVD and CD cases, refrigerator insulation, plastic cutlery, takeaway food trays, packaging materials
Polypropene	A thermoplastic; melts at a high temperature; light and strong; does not absorb water; excellent insulator; resistant to many corrosive chemicals	Dishwasher-safe food containers, fibres used to make durable carpets, medical equipment, car bumpers, pipes
High-density polyethene	A thermoplastic; rigid and strong; heat resistant; resistant to acids and alkalis	Freezer bags, milk bottles, toys, margarine tubs, detergent bottles, dustbins
Low-density polyethene	A thermoplastic; not as strong, and more flexible than the high-density form	Plastic carriers, food film wrap, bin bags
Polyvinyl chloride (PVC)	A thermoplastic; resistant to corrosion; can be made in rigid and flexible forms; an excellent insulator	Inflatables, waterbeds, flexible hoses, roofing membrane, window and door frames, vinyl records, clothing, wiring insulation
Polytetrafluoroethene (PTFE)	A thermoplastic; very resistant to heat and cold; not attacked by any common chemical; very 'slippery'; known by the brand name 'teflon'	Non-stick cookware; wet-weather gear and sports clothing
Epoxy resin	A thermoset; hard, rigid and sometimes brittle; resistant to chemicals; excellent adhesive properties	Epoxy glues; used in composite materials, e.g. fibreglass
Bakelite	A thermoset; hard, resistant to heat and chemicals; a good electrical insulator	In the past was extensively used in household appliances and electrical plugs; now used for billiard balls and board game (e.g. chess) pieces

QUESTIONS

4 Pick two of the thermoplastics from Table 12.4 and one use of each. Explain how the plastics' properties suit them for those uses.

5 Research the name of another thermosetting plastic and what it is used for.

How are plastics made?

Plastics are made from long-chain molecules known as **polymers**, made up of many smaller units called **monomers**. The process of joining the monomers together to form polymers is known as **polymerisation**.

Many of the monomers used to make plastic come indirectly from the fractional distillation of oil, from the naphtha or gas oil fractions. The chemicals used belong to a group called alkanes, but before they can be used they have to be processed further to form reactive monomers. They are heated either under pressure or with a catalyst (which speeds up the chemical reactions needed) in a process called '**cracking**' to make shorter alkane molecules and alkenes.

An example of an alkane used for cracking is decane, which forms octane and ethene.

decane → octane + ethene

Octane can be used in petrol manufacture, and ethene can be used for making a large range of chemicals, including polymers. Monomers such as ethene form **addition polymers**. Addition polymers are formed from one type of monomer. The monomer ethene forms the polymer poly(ethene) or polythene (see Figure 12.8).

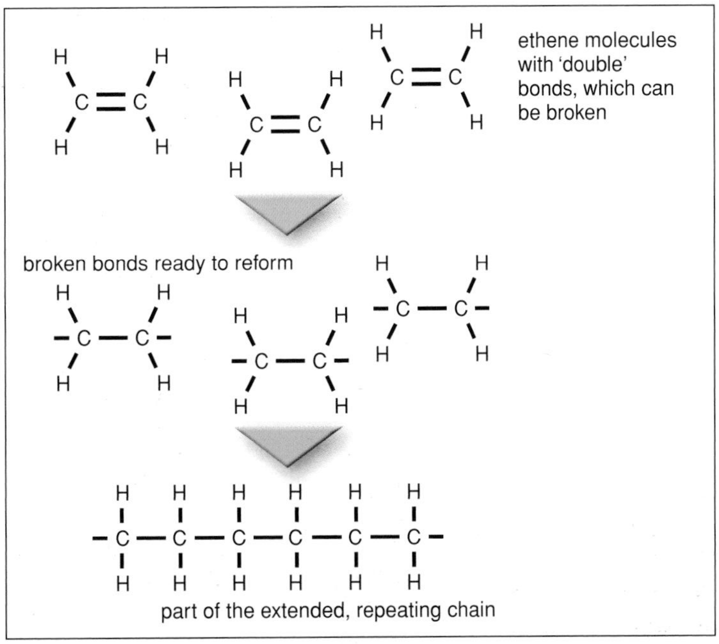

ethene molecules with 'double' bonds, which can be broken

broken bonds ready to reform

part of the extended, repeating chain

Figure 12.8 Formation of polythene from ethene.

There are other addition polymers made from compounds that are derived from ethene, all of which are useful everyday plastics, e.g. polytetrafluoroethene (PTFE) and poly(chloroethene) or polyvinylchloride (PVC). You can see in Figure 12.9 that in each case the monomer is slightly different and these monomers give the plastics different properties.

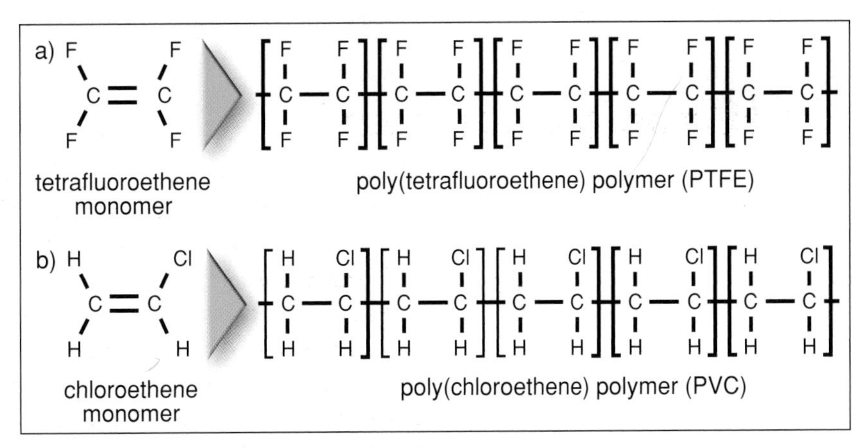

Figure 12.9 Formation of (a) PTFE and (b) PVC.

The process of plastic formation is summarised in Figure 12.10.

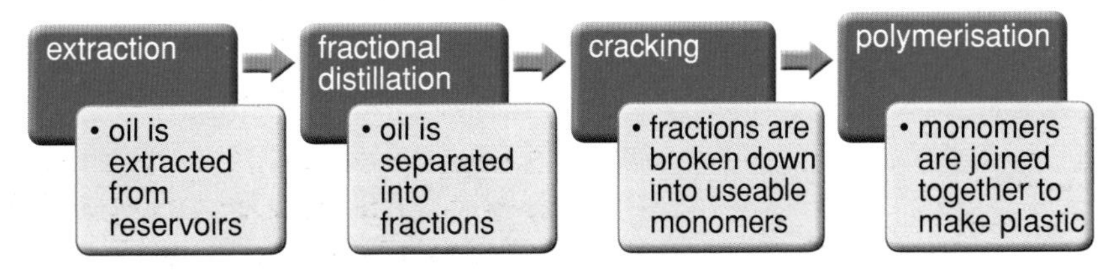

Figure 12.10 Stages in plastic formation.

Why is it important to recycle plastics?

Many plastics are very resistant to chemical attack, and nearly all of them are **non-biodegradable**, in other words, they do not rot. If they are put into normal household waste, the plastics will be buried underground where they will stay for hundreds of years. Because they do not rot away, plastics tend to fill up landfill sites so that new sites have to be found. The UK uses over 5 million tonnes of plastic each year, so this is a big problem.

It is far better if waste plastic is recycled. This has several advantages:

- less used plastic going to landfill
- less oil used for plastic production
- less energy consumed.

Although it is possible to recycle all plastics, some are more difficult and expensive to recycle than others. Some plastics have to be separated from others for recycling, so to assist this most plastic packaging has a symbol on it to indicate the type of plastic it is made from (see Table 12.5).

Table 12.5 Plastic recycling symbols. Types 1–3 are recycled in many areas of the UK, but types 4–7 are difficult to recycle and facilities may not exist in your area.

Symbol	Polymer type	Examples
1 PETE	PET Polyethene terepthalate	Fizzy drinks bottles Mineral water bottles Squash bottles Cooking oil bottles
2 HDPE	HDPE High-density polyethene	Milk bottles Juice bottles Washing up liquid bottles Bubble bath and shower gel bottles
3 V	Polyvinyl chloride	Usually in bottle form, however, not that common these days
4 LDPE	LDPE Low-density polyethene	Many types of packaging are made from these materials, for example, plastic formed around fresh meat and vegetables
5 PP	PP Polypropene	
6 PS	PS Polystyrene	
7 OTHER	Other All other resins and multi-materials	

TASK HOW GOOD IS THE UK AT RECYCLING PLASTICS?

This activity helps you with:

★ reading data from graphs
★ analysing graphs
★ developing your mathematical skills.

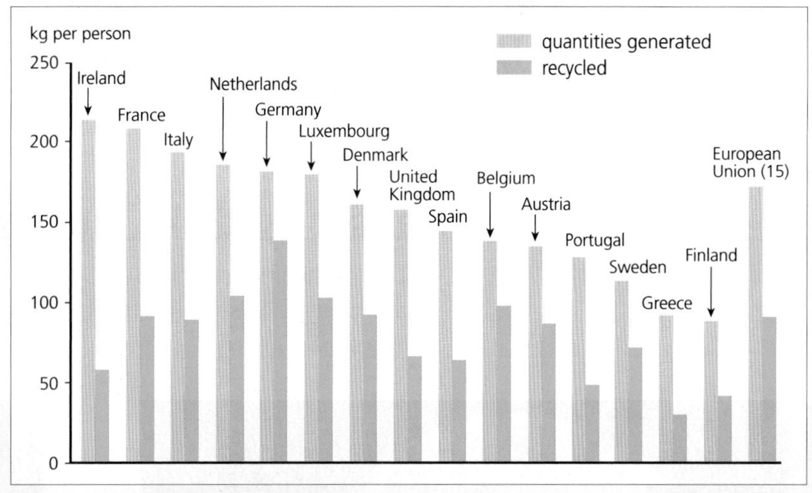

Figure 12.11 Plastic and non-plastic packaging production and recycling data for European countries, 2001.

Look at the data for packaging recycling in European countries and the average for the EU as a whole in Figure 12.11. Note that in Europe about 50% of packaging is plastic.

Questions

1. Which country generates the most packaging per person?
2. Which country recycles the most plastic per person?
3. Which country is 'doing best' on these figures?
4. Explain how you reached your decision in question **1**.
5. Explain why these data might not give an accurate assessment of how well a country was doing with plastic recycling in 2001.
6. Estimate the percentage of packaging that was recycled in the UK in 2001.

Chapter summary

- Crude oil is a mixture of hydrocarbons that was formed over millions of years from the remains of simple marine organisms.
- Crude oil is a finite resource and decisions about its use have a global social, economic and environmental impact.
- Crude oil is separated into less complex mixtures, called fractions, which contain hydrocarbons with boiling points within the same range.
- Some fractions can be further processed by cracking to make small, reactive molecules called monomers, which can be used to make plastics.
- Monomers are joined together to make polymers by a process called polymerisation.
- There are a variety of plastics with different properties, but all have the capability of being shaped or moulded.
- The properties of different plastics make them suitable for different uses.
- There are major advantages to recycling plastic due to its non-biodegradable nature, and also to reduce the use of oil resources to make new plastics.

The ever-changing Earth

Has the Earth always looked like it does today?

The answer to this is no. Over the billions of years of its existence, the Earth has slowly but constantly changed its appearance (see Figure 13.1).

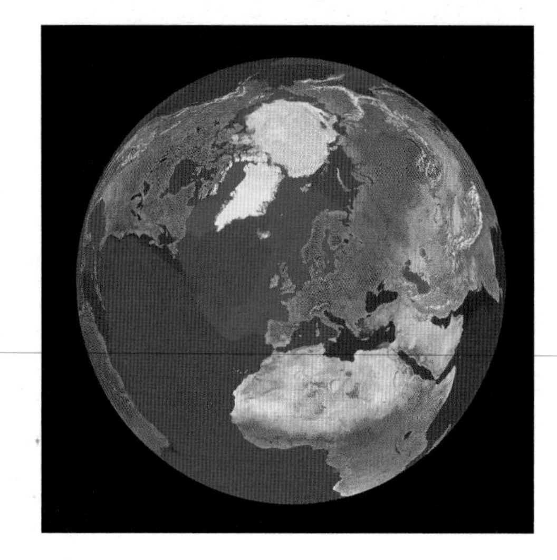

Figure 13.1 The Earth 200 million years ago (left) and as it is today (right).

200 million years ago, the land masses on Earth were all grouped together in one block, which scientists now call **Pangaea**. You might think that the continents stay in one place, but in fact they move across the surface of the planet, and can shift from place to place. In Figure 13.1, the modern picture is a photo taken from outer space, but 200 million years ago humans had not even appeared on the Earth, so no one was taking photos. So, this leads to another question.

How do we know that the continents have moved?

Scientists now know that the surface of the Earth, or lithosphere, is made up of seven large **plates** and some smaller ones, about 70 km thick, which move a few centimetres per year. This movement is called **continental drift**.

The idea of continental drift was put forward by Alfred Wegener (1880–1930), and the problems he had in getting his ideas accepted are a good example of how science works.

Figure 13.2 Alfred Wegener.

The continents of the Earth roughly fit together like a jigsaw. The coastlines of western Africa and eastern South America are a particularly good fit (see Figure 13.3).

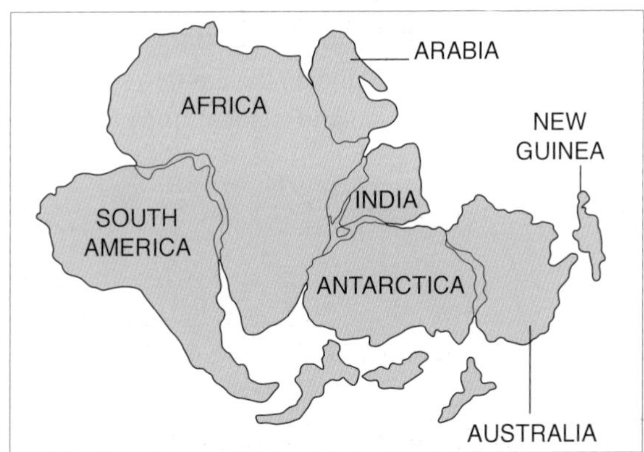

Figure 13.3 The continent 'jigsaw'. This group of continents is known as 'Gondwana'. The original land mass of Pangaea is thought to have split into two, Laurasia and Gondwana.

To explain this, some people had suggested that the continents may have moved, but there was no clear evidence for this apart from the 'jigsaw fit'. Geologists in the nineteenth century actually did believe that continents had moved, but only up and down, not from side to side, as the Earth's surface cooled and contracted (there wasn't any evidence for that, either).

Alfred Wegener looked for evidence of continental drift, and found some:

- Rock formations on both sides of the Atlantic ocean are exactly the same.
- Similar or identical animal and plant fossils are found in areas now widely separated by oceans, e.g. a snail fossil has been found in Sweden, and also in Newfoundland in Canada, and there is no way a snail could swim the Atlantic!
- Certain fossils seem to be in the 'wrong place', e.g. fossil remains of semi-tropical species in northern Norway.

However, there was a weakness in Wegener's theory of continental drift. There was no known mechanism by which continents could plough their way through the Earth's crust without leaving any sort of 'trail'.

Wegener's model was not accepted at the time, and geologists proposed an alternative model to explain the strange distribution of fossils. They suggested that at some stage a 'land bridge' had existed between continents, so that animals and plants could move along them from one continent to another. These land bridges then disappeared (apparently without trace), leaving the continents isolated.

Discussion Point

Is the fact that there is no known mechanism to explain a hypothesis a good reason for rejecting it?

TASK WHERE DID THE TRILOBITES GO?

This activity helps you with:

★ judging the value of a model

★ using evidence to support a conclusion.

Consider the following piece of evidence. There is a fossil trilobite species that is commonly found in Europe, which is also found in Newfoundland on the other side of the Atlantic Ocean, more than 3000 km away. However, it is only found on the Eastern side of the island, never on the western side. Newfoundland is about 300 km wide.

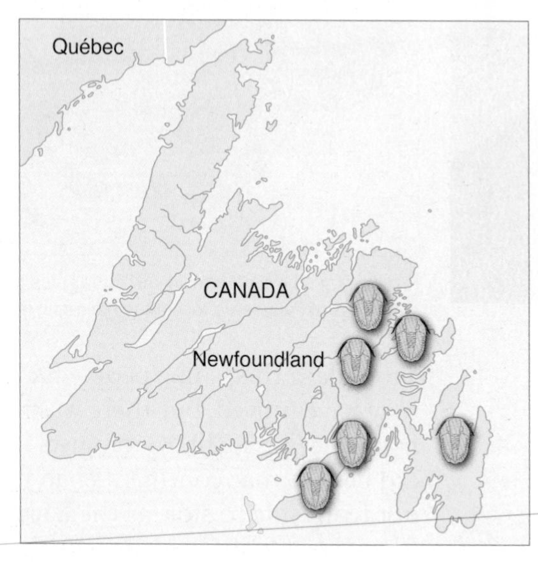

Figure 13.4 Distribution of European Trilobite fossils in Newfoundland, Canada.

Question

How does this evidence fit with the two models mentioned on pages 140–141 (continental drift vs. land bridges)?

How did Wegener's theory become accepted?

In order to convince people that the continents could move across the surface of the Earth, new evidence was needed. Eventually, it was found.

- Studies of the ocean floor found large mountain ranges and canyons. If the ocean floor was ancient it should have been smooth, because of all the sediment coming into it from rivers.
- In 1960 core samples taken from the floor of the Atlantic Ocean were analysed and dated. This showed that rock in the middle of the Atlantic was considerably younger than that from the eastern or western edges.
- No ocean floor was found to be older than about 175 million years, yet rocks on land had been found that were several billion years old.
- Rocks retain a record of the magnetic field of the Earth, which changes from time to time. Analysis of these magnetic records showed that Britain had spun round and moved north in the past, and that the patterns exactly matched those from North America.

It became clear that new ocean floor was forming all the time and spreading outwards, and in places near the borders of the continents it was sinking back into the crust. The Earth's crust was shown to be 'mobile'.

By the 1960s the idea of continental drift became generally accepted, and the theory was renamed **plate tectonics**.

TASK WHAT CAUSES EARTHQUAKES AND VOLCANOES?

This activity helps you with:
★ developing a hypothesis.

Seismology is the study of earthquakes and volcanoes. Seismic studies have shown that earthquakes occur in a pattern, which is now known to indicate the boundaries between the Earth's plates (see Figure 13.6).

This means that there is a chance of predicting where new earthquakes and volcanoes may arise, and that the position of known volcanoes and earthquakes can be used to trace the boundaries of the plates.

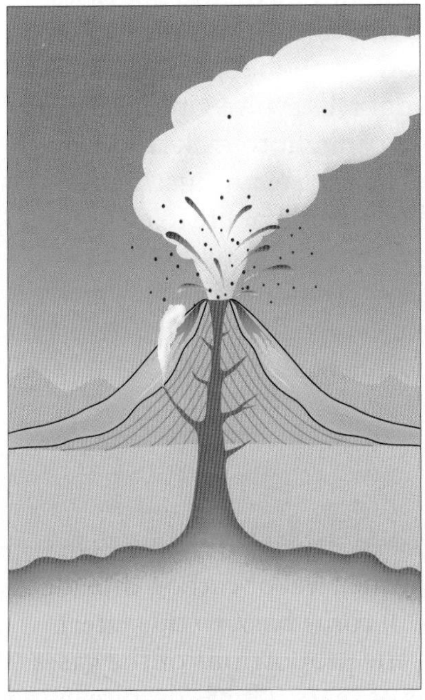

Figure 13.5 Volcanoes occur when molten rock (magma) comes through the surface under pressure. Layers of rock cool and set to form the cone of the volcano.

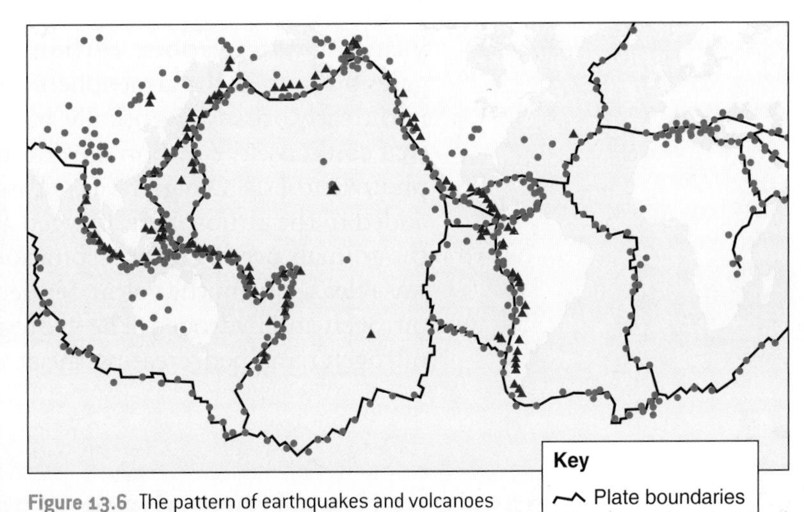

Figure 13.6 The pattern of earthquakes and volcanoes defines plate boundaries. For example, the ring of volcanoes around the Pacific Ocean is the boundary of the Pacific plate.

Where two plates join, there can be four types of movement:

- The plates can move apart – molten rock (magma) below the surface is released. If this happens under pressure, this is a volcanic eruption.
- The plates can collide. This 'crumples' the edges of the plates, forming mountain ranges.
- One plate can slide under the other (this is called **subduction**). Magma is released and volcanoes can occur.
- The plates can slide past one another, neither moving towards nor away from each other.

Any movement of plates can lead to earthquakes. Plate movement causes a build-up of huge quantities of energy in the rock. When the energy is released, vibrations occur that travel through the rock, causing earthquakes of varying severity depending on the amount of energy built up.

Questions

Look at Figure 13.6.

1. Some plate boundaries have earthquakes but no volcanoes. Suggest a reason for this.
2. There are certain places where earthquakes and/or volcanoes occur, which are not on plate boundaries. Suggest a reason for this.

How did volcanoes help life on Earth evolve?

Table 13.1 Composition of the atmosphere today. The atmosphere also contains water vapour, but the amount varies.

Gas	Quantity in dry air, expressed in volumes
Nitrogen (N_2)	78.1%
Oxygen (O_2)	20.9%
Argon (Ar)	0.9%
Carbon dioxide (CO_2)	0.035%
Others: Neon (Ne) Helium (He) Krypton (Kr) Hydrogen (H_2) Xenon (Xe) Ozone (O_3) Radon (Rn)	0.065%

If it wasn't for volcanoes, Earth would not have evolved an atmosphere capable of sustaining life. The original atmosphere of Earth was composed mainly of **hydrogen** and **helium**, but these low-density gases soon escaped Earth's gravity and drifted into space. At this time, the Earth was young and was still cooling down after its formation. There were large numbers of volcanoes on the surface, constantly erupting. The eruptions contained a mixture of gases, including **water vapour**, **carbon dioxide** and **ammonia**. These gases built up in the atmosphere and the carbon dioxide dissolved in the early oceans. Eventually, bacterial cells evolved in the oceans that could use the carbon dioxide and sunlight to make food, by photosynthesis. **Oxygen** was released as a waste product, and so was added to the atmosphere. Oxygen allowed animal life to evolve, as the animals needed it for respiration. The poisonous ammonia that was released from the volcanoes decomposed in sunlight to form nitrogen and hydrogen. The hydrogen escaped the atmosphere, but nitrogen remained, creating the atmosphere as it is today.

Why do the gases in the atmosphere stay constant?

There is concern at the moment that the level of carbon dioxide in the atmosphere is going up (we will come back to this later). This is a concern because it is new – the proportions of the gases have remained constant for millions of years. This is despite living organisms using and producing both oxygen and carbon dioxide. In the past, the processes using oxygen have been balanced by those producing it, and the same has applied to carbon dioxide. Oxygen production and use is still balanced, but carbon dioxide production is now exceeding use.

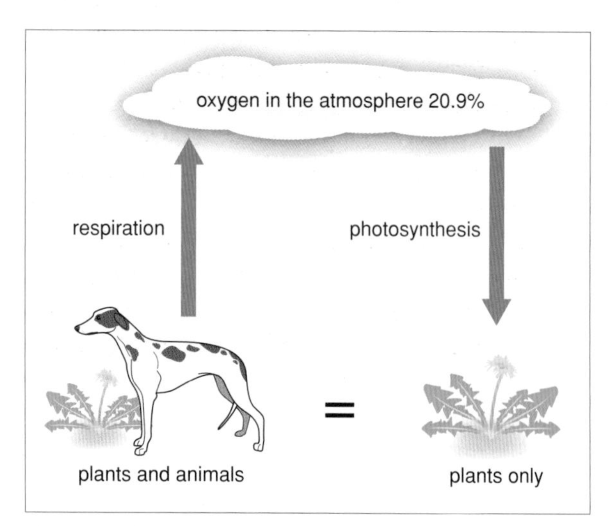

Figure 13.7 The balance of oxygen use and production.

Oxygen

Nearly all living things use oxygen to obtain energy from respiration. Oxygen is produced by plants during photosynthesis and, because they produce more than they need for their own respiration, they add it back to the atmosphere. In terms of oxygen, respiration and photosynthesis balance each other (Figure 13.7).

Carbon dioxide

In the past, the levels of carbon dioxide in the atmosphere have been kept constant by the **carbon cycle**, in which respiration and photosynthesis again play a part. The carbon cycle is summarised in Figure 13.8.

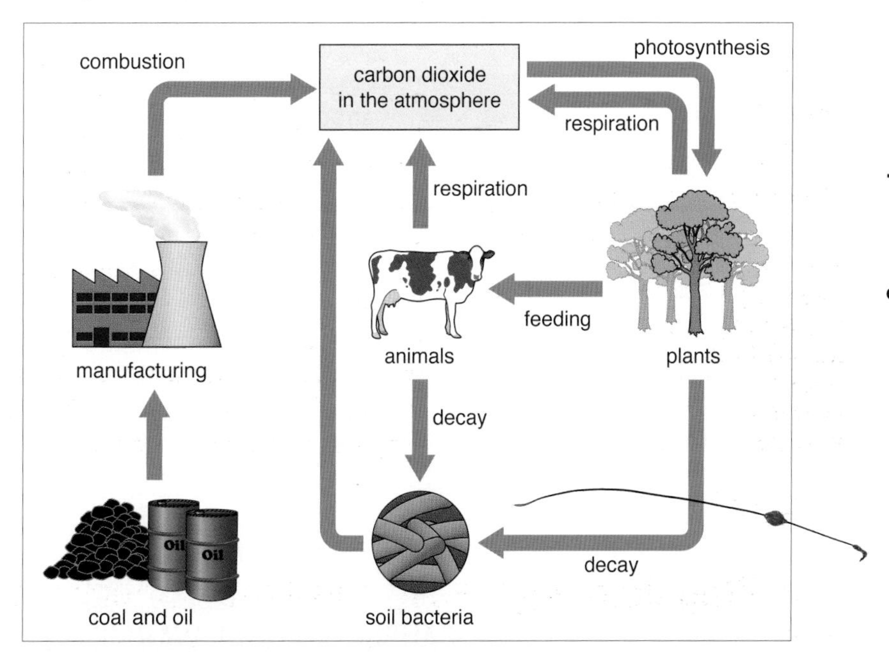

Figure 13.8 The carbon cycle.

In the 'natural' part of the carbon cycle, small amounts of carbon are not recycled, because under certain conditions the dead bodies of plants and animals become fossilised rather than decaying. The carbon in their bodies remains fixed inside them rather than being released back into the atmosphere as carbon dioxide. In this way, fossil fuels (oil, coal and natural gas) have stored carbon for millions of years. Coal has been burned for hundreds of years and this has released some carbon dioxide into the atmosphere. However, in the last 150 years the discovery of oil and natural gas and the huge growth of industry has seen an enormous increase in the burning of fuels, and carbon that has taken millions of years to build up in the Earth has been released rapidly into the atmosphere in the form of carbon dioxide. This **combustion** (burning) of fossil fuels has disturbed the balance that existed before and the level of carbon dioxide in the atmosphere has increased instead of remaining constant. This change is thought to have caused changes in the Earth's climate, as we shall see later.

Discussion Point

The combustion of fossil fuels uses oxygen as well as producing carbon dioxide, yet the levels of oxygen in the atmosphere do not seemed to have significantly decreased. Suggest a possible reason for this.

PRACTICAL HOW CAN WE DEMONSTRATE THE PRODUCTS OF COMBUSTION?

This activity helps you with:
★ designing experiments.

Your teacher will demonstrate this experiment.
The burning of the hydrocarbons in fossil fuels produces carbon dioxide and water. The **cobalt chloride paper** in test tube A detects water, turning from blue to pink. Carbon dioxide can be shown in two ways in test tube B:

- using **lime water**, which turns cloudy in the presence of carbon dioxide
- using **bicarbonate indicator**, which detects a change in pH. Carbon dioxide is an acid gas and this turns the bicarbonate indicator from red to yellow.

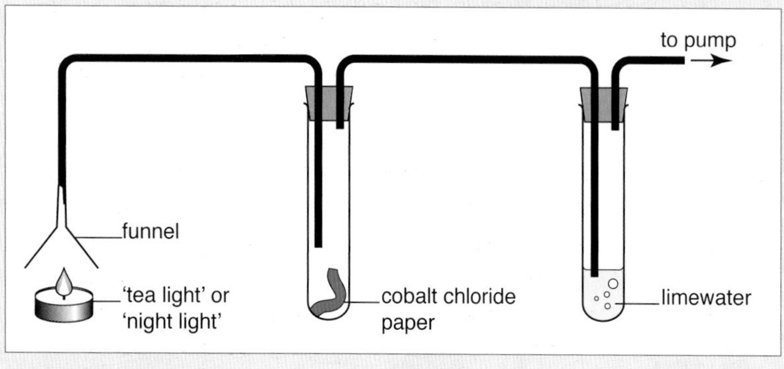

Figure 13.9 Experiment to demonstrate the products of combustion.

Questions

1 Suggest one reason why it might be better to use lime water in this experiment rather than bicarbonate indicator.

2 Design an experiment to test which of the two ways of detecting carbon dioxide is the most sensitive (i.e. detects a smaller quantity of carbon dioxide) using this apparatus.

TASK ARE WE CAUSING GLOBAL WARMING?

This activity helps you with:
★ judging the strength of evidence
★ assessing the effects of bias in secondary data
★ communication skills.

Global warming and its possible causes are constantly in the news. Nearly all the scientists that research global warming believe that its main cause is human activity – mainly the burning of fossil fuels. However, there are some that do not agree, and think that global warming is a natural phenomenon that occurs every so often in Earth's history, or that it isn't really happening, or that it is happening, but is not caused by emissions of carbon dioxide from fossil fuels.

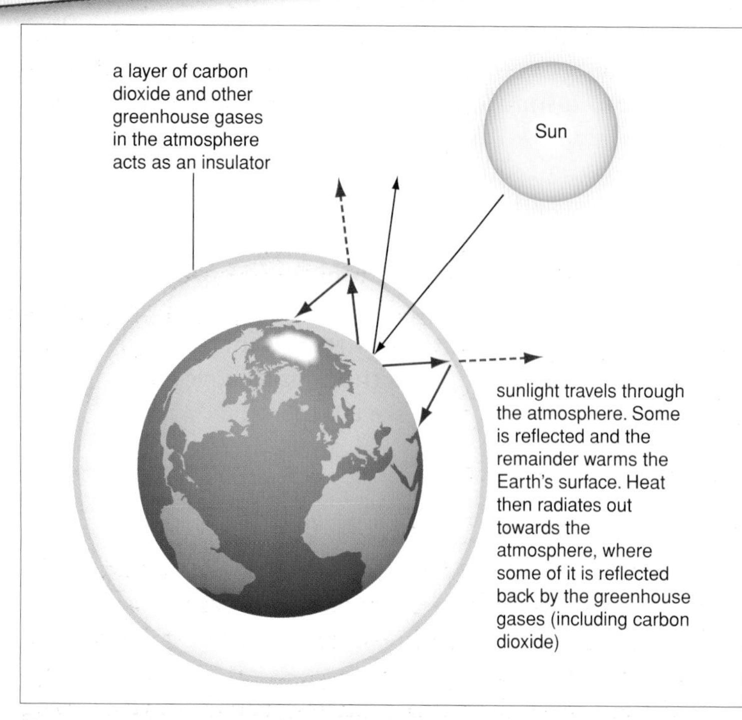

Figure 13.10 Principle of global warming.

So, why can't scientists agree, when looking at similar evidence, and if most of them think that humans are causing global warming, why can't they prove it?

The global warming problem is a good illustration of what most science is like. The world is immensely complex and there are hardly ever simple answers. With global warming, certain facts provide evidence either way.

- The temperature of the Earth's surface has gone up in the last 100 years or so (evidence for global warming).
- The level of carbon dioxide in the atmosphere has risen since around 1750, and the trend seems to match the rise in temperature of the Earth (evidence for carbon dioxide causing global warming).

continued...

- The Earth's temperature does go though natural cycles of hot and cold, and we may well be due for a natural rise in global temperatures (evidence against carbon dioxide causing the rise).
- There is a clear mechanism by which increased carbon dioxide in the atmosphere would cause global warming (this is not exactly *evidence* at all, but a possible explanation).
- The rise in temperature is probably greater than would be expected by the normal temperature cycle (evidence for carbon dioxide causing global warming, but not everyone agrees with this).
- Most scientists believe that the recent activity of the Sun, which causes natural temperature cycles, should be linked with a fall in temperatures not a rise (evidence for a man-made cause, but not everyone agrees because the data are complex).

There are many other pieces of evidence that people who believe in climate change or those that don't will quote to back up their case. The evidence for humans causing global warming (mainly though burning fossil fuels) is strong, but not absolutely conclusive, and it probably never will be. People have to make up their own minds by judging the quality of the evidence. The **Intergovernmental Panel on Climate Change**, a group of well-qualified scientists set up by the United Nations in 1988 to examine the evidence for climate change, has concluded that there is more than a 90% chance that climate change is the result of an increase in greenhouse gases caused by human activity.

Question

Research using the internet and try to find evidence linking global warming with increasing carbon dioxide levels. Decide how strong you think this evidence is, and justify your opinion.

Extension

Find an internet site about global warming that appears biased. These sites will usually put forward only one point of view and will try to make out that the other point of view is completely without foundation. Look at one of their claims, either backing their own opinion, or against those who hold a different opinion. Research other, more balanced sites to see whether there is any real justification for this claim.

Does burning fossil fuels have other harmful effects?

Apart from producing greenhouse gases, the burning of fossil fuels is also a major cause of **acid rain**. Sulfur dioxide and nitrous oxides (other gases produced by burning fossil fuels) are acidic gases that react with water to form acids.

oxides of nitrogen + water → nitric acid

sulfur dioxide + water → sulfuric acid

These gases dissolve in the water vapour in the atmosphere and condense into clouds, which then produce acid rain. The acid rain can kill wildlife, with fish and conifer trees being particularly affected. The acid rain also causes damage to limestone buildings because acids react with limestone, causing it to dissolve in water.

Figure 13.11 Acid rain has caused the damage to this sculpture.

PRACTICAL HOW ACIDIC IS THE RAIN IN YOUR AREA?

This activity helps you with:
- ★ planning and carrying out an investigation
- ★ analysing data
- ★ drawing conclusions.

Apparatus
- * Universal Indicator paper or solution
- * other items depending on individual plan

Unpolluted rain is always slightly acidic (pH 5–6) because of the effect of natural levels of carbon dioxide (which is an acid gas) in the atmosphere. If the rain has a pH of 4 or less, it indicates that this has been caused by pollution.

Risk assessment

- **You will need to prepare a risk assessment when you design your experiment.**

Procedure

1. Devise a method of collecting rain that falls in your neighbourhood.
2. Each day that rain falls, use Universal Indicator solution or paper to test the pH. Construct a suitable table to record your results in.
3. On each day record the direction of the prevailing wind. A forecast of this for the day is usually given with local weather forecasts, which can be obtained from the internet. It is more difficult to find what direction the prevailing wind was in the near past, so look up the forecast early in the day or on the previous day.
4. Continue the survey for at least 2 weeks (longer if there is not much rainfall).

Analysing your results

1. Was the rain in your area:
   - **a** unpolluted
   - **b** polluted
   - **c** polluted on certain days and 'clean' on others?
2. If you detected some pollution but the level was variable, was the acid rain associated with any particular wind direction? If so, can you suggest a reason for this link?

Can we clean up fossil fuel emissions?

It is possible to remove some of the harmful gases produced by the burning of fossil fuels before they enter the atmosphere. At the moment it is only practical to do this on a large scale, such as in power plants.

Carbon capture can reduce the carbon dioxide emissions from power stations by around 90%. It is a three-step process:

- capturing the CO_2 from power plants and other industrial sources
- transporting it, usually via pipelines, to storage points
- storing it safely in geological sites such as depleted oil and gas fields.

The most commonly used form of carbon capture is post-combustion capture, which involves capturing the carbon dioxide from the gases given off by burning. A chemical solvent is used to separate carbon dioxide from the waste gases.

There are also techniques being developed for removing the sulfur dioxide from the waste gases produced by power stations. Such processes are referred to as **sulfur scrubbing**, and can reduce the levels of sulfur dioxide by more than 95%.

Chapter summary

- Alfred Wegener developed a theory that continents move across the surface of the planet, known as the theory of continental drift.
- This developed into the modern theory of plate tectonics.
- Plate boundaries can be identified by the distribution of major earthquakes and volcanoes.
- The atmosphere was formed by gases, including carbon dioxide and water vapour, being expelled from volcanoes.
- The composition of the atmosphere has changed over geological time.
- The atmosphere is composed mostly of nitrogen (78.1%), oxygen (20.9%) and small amounts of other gases.
- The balance between respiration, combustion and photosynthesis maintains the levels of oxygen and carbon dioxide in the atmosphere.
- Increased combustion of fossil fuels in the last 200 years has added extra carbon dioxide to the atmosphere.
- There is much media debate on the issue of global warming, but the vast majority of scientists attribute the main cause of global warming to the increase in carbon dioxide in the atmosphere caused by the combustion of fossil fuels.
- Sulfur dioxide and oxides of nitrogen in polluted air cause acid rain, which can kill animals and plants and damage limestone buildings.
- The removal of carbon dioxide (by carbon capture) and sulfur dioxide (by sulfur scrubbing), can reduce the pollution caused by burning fossil fuels.

14 Generating and transmitting electricity

Electricity – the 'Swiss army knife' of energy?

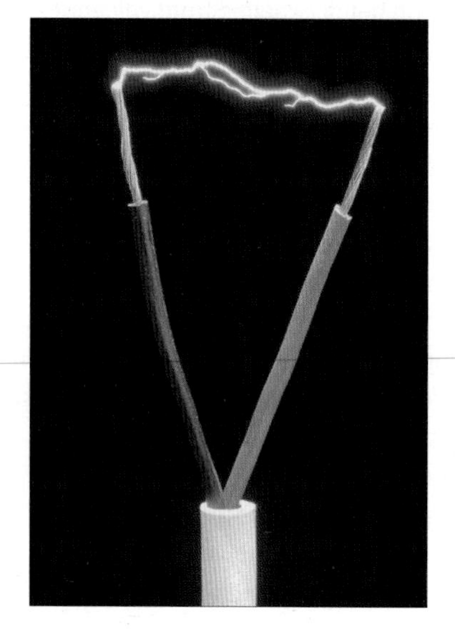

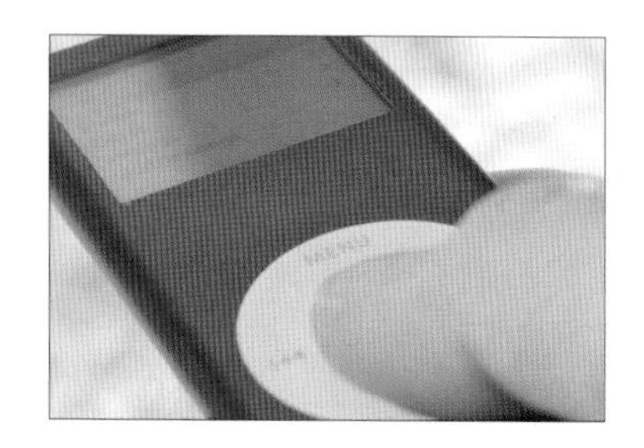

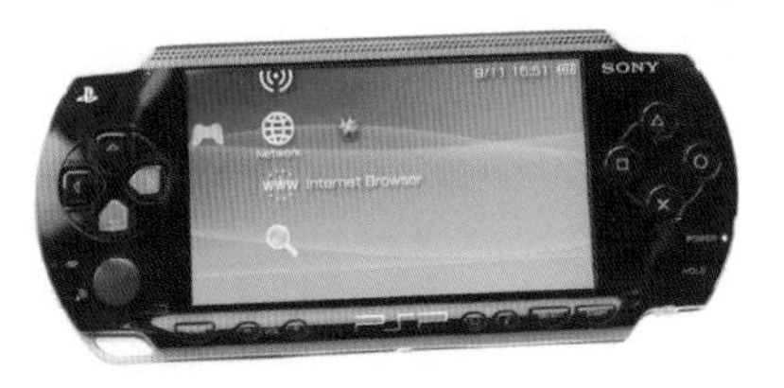

Figure 14.1

Why is electricity in such demand? Why are modern lives dominated by the use of electricity?

Reason 1

Electricity is a form of energy, like heat and light, but unlike these other forms of energy it is quite easy to transform (change) it into other forms. As such, it's easy to use electricity to make other more useful forms like kinetic (movement) energy and sound energy.

Reason 2

Electricity is easy to move over long distances - electric current will travel easily through metal wires - which makes getting electricity from a place where it is generated to a place where it is needed, very simple.

Reason 3

Electricity is easy to generate from other forms of energy. Power stations burn a fuel (which is a concentrated store of chemical energy) such as oil, producing heat, which turns water into steam. The moving steam turns a turbine, which turns a generator, producing electricity. Relatively small amounts of electricity can also be produced when certain chemicals react together in a

QUESTIONS

1. What are the three main reasons why electricity is so useful to us?
2. What fuels, apart from oil, can be used to make electricity in a power station?
3. Describe how energy is transformed from one form to another inside a power station.

battery. Although battery technology is improving due primarily to the development of electric cars, batteries still cannot generate large quantities of electricity capable of supplying houses or businesses. However, they are brilliant at powering small, portable machines like laptops, mobile phones and ipods.

QUESTIONS

4 When electric current travels through a wire it causes the wire to heat up. Explain why this is a problem for electricity supply companies.
5 Electricity is normally transmitted at very low current but very high voltage. Why might this make electricity cheaper for us as consumers?
6 Inside a battery an electrochemical reaction occurs between a chemical, like sulfuric acid, and metal or carbon electrodes. Why do you think that this makes it difficult to design and make batteries that can supply large amounts of electricity for long periods of time?

Discussion Point

Mobile smartphones push battery technology to the limit. What do you think are the main considerations when designing a battery for a new mobile smartphone?

TASK: INVESTIGATING BATTERIES

This activity helps you with:
★ planning an experiment.

Why do batteries come in different sizes? Common batteries include AAA, AA, C and D.

Procedure

Design an experiment to compare the effectiveness of two different batteries at powering an electric device. If you are then going on to do the experiment that you have planned, you will need to:

- construct a list of suitable apparatus
- order the apparatus in conjunction with your teacher and science technician
- produce a suitable risk assessment for the activity. Your teacher will supply you with a suitable blank risk assessment form.

How do we make electricity?

Figure 14.2 A wind farm and a conventional power station.

Every year the International Energy Agency (IEA) and the UK Government Department of Energy and Climate Change (DECC) collect data about the amount of electricity generated from different sources. These sources are broken down into two main groups: **renewables** and **non-renewables**. A non-renewable source of energy is defined as a source that, once it is used, cannot be created again. Fossil fuels and nuclear fuel are non-renewables – the physical conditions on Earth will not allow these fuels to be created again. Renewables are sources of energy that are continuously being produced, mostly as a result of the action of the Sun.

QUESTIONS

7 Construct a table listing the different types of renewable and non-renewable sources of energy.
8 For each of the renewable sources of energy that you have listed, briefly explain how it is related to the action of the Sun.

Are wind and water an answer to our energy needs?

The charts shown in Figure 14.3 were produced by the International Energy Agency and the Department of Energy and Climate Change. They show the proportions of electricity produced by the different types of energy source.

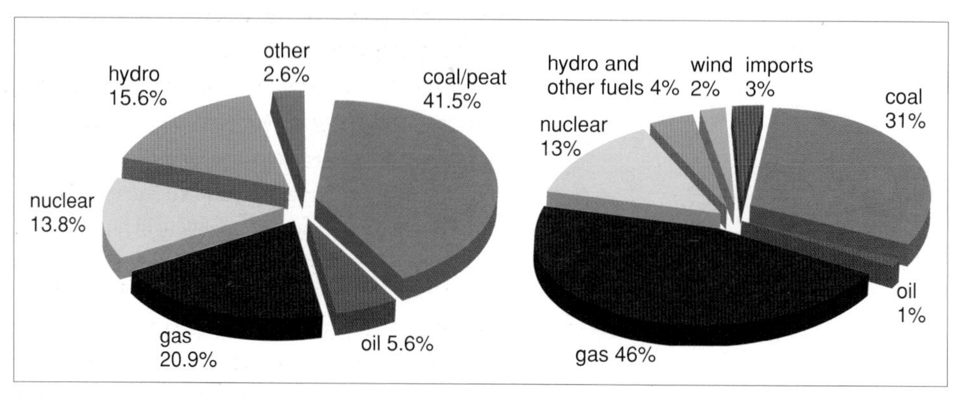

Figure 14.3 World (left) and UK (right) electricity generation by fuel type.

Discussion Point

The need for 'energy security' is much talked about in the press. What do you think this means? Come up with some reasons why the World as a whole and the UK on its own should have different proportions of electricity generated by different energy sources.

QUESTIONS

9 For both the World and the UK, calculate the percentage of electricity generated from:

a renewables

b non-renewables.

10 Calculate the differences between the proportions generated across the World compared with the UK.

Could all our homes become micro power stations?

Figure 14.4 A micro turbine and solar panel on a house roof.

We all rely on instant access to electricity and gas at home. Just turning on a kettle can draw about 3 kilowatts of electric power (3000 joules of energy per second) – the equivalent of about 400 low-energy light bulbs! However, this instant access to energy comes at a price – both economic and environmental. Huge power stations, fossil-fuel powered or nuclear, massive hydroelectric dams and turbines and hundreds of large wind turbines are constantly producing the millions of kilowatts of electricity needed to secure a constant supply. The large-scale power stations are at best only about 40% efficient , which means that for every tonne of coal or oil burned in a power station about 600 kg is wasted heating up the power station and the air around it. In addition, 1500 kg of carbon dioxide is released into the atmosphere, adding to global warming.

Putting this very crudely in economic terms, it costs us £100 to generate £40 of electricity!

Is there an alternative to large-scale power stations?

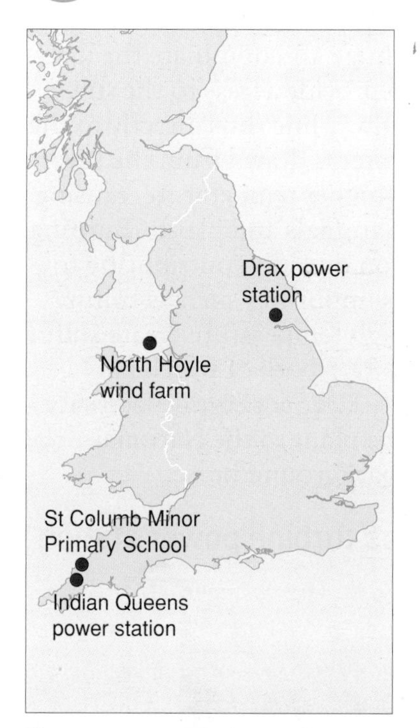

Figure 14.5 Location of power stations in case studies 1–4.

Well, yes and no! On the one hand, individual homes, schools, businesses and government buildings could be fitted with solar panels (that work when it is daylight and don't necessarily need it to be sunny) and micro-turbines could be fitted wherever it is consistently windy enough. Combine this with a programme of building insulation and we would reduce demand and generate a considerable amount of our domestic and commercial consumption. However, there are limitations to this. It's not always daylight, the wind does not always blow, and large-scale industry needs secure, guaranteed, large amounts of power that could not be delivered by local micro-generation.

The answer could be a complex combination of many different sources - large-scale, fossil-fuel burning and nuclear power stations, together with a mixed bag of renewable sources of energy, improved insulation and more energy-efficient devices. A good way to understand the problem is to look at some different case studies.

Case study 1 – Drax coal-fuelled power station

Drax power station is a coal-fuelled conventional power station in North Yorkshire (see Figure 14.6). When online it can produce up to 3960 **megawatts (MW)** of electricity continuously 24 hours a day, 7 days a week - approximately 7% of the whole UK's electricity! 36 000 tonnes of coal are brought to the power station every day via railway from UK coal mines and imports from Russia, Columbia and the USA. The coal is burnt in a furnace at several thousand degrees celsius, together with air/oxygen, producing enough thermal heat energy to turn nearly 60 tonnes of water into steam per second. The steam is then superheated to 568 °C and pressurised to 166 times atmospheric pressure.

Figure 14.6 Drax power station.

The superheated steam turns turbines, causing them to spin at 3000 rpm. Each of the six turbines is connected to an electric generator producing 660 MW of electricity which is output to the National Grid.

Drax factfile

Type: Conventional fossil fuel power station

Built: 1974

Primary energy source: coal

Electricity output: 3960 MW

Energy input: 11 250 MW

Carbon footprint: 22.8 million tonnes per year

Setup cost: estimated £1 billion to build similar power station

2008 cost prices: average achieved electricity price: £58.30/MWh; average fuel costs: £25.10/MWh

Projected lifetime: early 2020s if carbon capture and storage systems not fitted

Projected decommissioning cost: £10 million

Start-up time: always on

Environmental impact: very high

Environmental impact

Drax covers 750 hectares of previously agricultural land which will be very difficult to return to either land for building houses on or farmland. The building structures are huge and dominate the local skyline. Major roads have been built to provide access to the site, with the associated level of traffic. A railway line runs directly to the power station carrying coal. Cooling water is drawn from the River Ouse, and is returned, but at a slightly higher temperature, causing local heating which affects the aquatic animals and plants. Burning coal produces the greenhouse gas carbon dioxide, contributing to global warming. The coal also contains impurities such as sulfur and nitrogen, which when burned at high temperature create sulfur dioxide (SO_2) and oxides of nitrogen (NO_x), both of which are soluble and contribute to acid rain. Very large supergrid electricity pylons take the electricity away from the plant to the National Grid. The plant produces a continual background noise.

Case study 2 – Indian Queens gas turbine power station

Figure 14.7 Indian Queens gas turbine power station.

Indian Queens power station is effectively a huge jet-engine situated on Goss Moor, a government designated SSSI (Site of Special Scientific Interest) in mid-Cornwall (see Figure 14.5). It can generate 140 MW of electricity for a maximum of 24 hours at a time and only operates for on average 450 hours per year.

At full power, 44 000 litres per hour of kerosene (jet fuel) or diesel fuel is injected under pressure along with air and purified water into a huge jet engine. The fuel mixture is ignited and the resulting continuous controlled explosion produces an exhaust that spins a turbine, that turns a 140 MW generator.

Indian Queens power station factfile

Type: open cycle gas turbine

Built: 1996

Primary energy source: kerosene/diesel

Electricity output: 140 MW

Energy input: 425 MW

Carbon footprint: 57 000 tonnes per year

Setup cost: £60 million

2010 cost prices: electricity sold to Grid at approximately £270/MWh; diesel costs approximately 45p/litre.

Projected lifetime: 30 years

Projected decommissioning cost: £2 million

Start-up time: 14 minutes

Environmental impact: high

Environmental impact

Indian Queens power station covers several hectares of moorland immediately next to an SSSI nature reserve. The tall exhaust chimney is clearly visible locally, and the fuel is brought to the site by large tankers via a specially built access road, although tankers are only needed after the plant has been operational, which is only about 20 days per year. Burning kerosine or diesel produces the greenhouse gas carbon dioxide, contributing to global warming, and the fuel contains impurities such as sulfur and nitrogen, which when burned at high temperatures create sulfur dioxide (SO_2) and oxides of nitrogen (NO_x), both of which are soluble and contribute to acid rain. Large electricity pylons take the electricity away from the plant to the National Grid. When operational, the gas turbine engine produces a considerable amount of noise.

Case Study 3 – North Hoyle offshore wind farm

North Hoyle wind farm is a 30 turbine wind farm situated 5 miles offshore from Prestatyn in North Wales.

Figure 14.8 North Hoyle offshore wind farm.

North Hoyle is one of the windiest places in the northern half of the UK. The mean average annual wind speed is 9 m/s! When the wind is blowing the turbine blades on *each* of the wind turbines turn a generator producing 2 MW of electricity. The 30 wind turbines collectively produce 60 MW of electricity – enough to power 50 000 homes per year.

North Hoyle offshore wind farm factfile

Type: wind turbine

Built: 2003

Primary energy source: wind

Electricity output: 60 MW

Energy input: dependent on the strength of the wind

Carbon Footprint: 160 000 tonnes per year **saved**

Setup cost: £80 million

2010 cost prices: £60/MWh (subsidised – actual generating costs = £144/MWh)

Projected lifetime: 25 years

Projected decommissioning cost: included in set-up cost

Start-up time: Immediate

Environmental impact: medium/low.

Environmental impact

Each of the 30 wind turbines is 67 m high, and the whole site covers an area of 10 km^2. On a clear day it can be seen from over 16 miles away. Fishing is restricted in and around the wind farm, but this has led to an increase in the number of species locally. The spinning turbine blades create a hazard for seabirds. An undersea cable connects the wind farm to pylons onshore which then connect to the National Grid. The 30 turbines produce background noise when operational.

Case study 4 – St Columb Minor Primary School, Newquay

St Columb Minor Primary School is an unusual primary school in Newquay, North Cornwall.

Figure 14.9 St Columb Minor Primary School's PV panels and wind turbine.

Not only does the school boast a 6 kW micro-wind turbine on the school playing fields, but it also has a large photovoltaic solar panel array on the roof generating 13.8 kW during daylight and a 4 kW solar thermal water system. Since 2008 electricity consumption has fallen by 37%, due to the installation of the solar PV and wind turbine, coupled with savings from reduced use of lighting and other electrical equipment. Improved insulation and the installation of energy-efficient boilers have also reduced gas consumption by 6%. As a result, the school's energy costs fell by 10% in 2008/09 and a further 20% in 2009/10, despite energy supply cost increases of 50%.

Environmental impact

The micro-wind turbine is 15 m high and is clearly visible in the local residential area. It produces a background noise when operational. The PV and solar thermal panels are fitted to the flat and pitched roofs of the school and can be seen locally. The school is connected to the National Grid through its normal power lines.

St Columb Minor Primary School factfile

Type: single micro-wind turbine, photovoltaic (PV) solar panels and solar thermal water panels

Built: 2008

Primary energy source: wind and solar

Electricity output: 19.8 kW (plus 4 kW saving from solar thermal panels)

Energy input: 23.8 kW **saved**

Carbon footprint: 11 tonnes per year **saved**

Setup cost: £118 000

2010 cost prices: approx. £2300 saved per year; approx. 40p/kWh paid by Grid for electricity generated from photovoltaic panels and 27p/kWh paid by Grid for the electricity generated from the wind turbine.

Projected lifetime: 20 years

Projected decommissioning cost: unknown

Start-up time: immediate

Environmental impact: low

TASK

GOVERNMENT REPORT

This activity helps you with:

- ★ presenting information graphically
- ★ producing a written report
- ★ researching scientific information
- ★ selecting scientific information
- ★ referencing scientific information.

1. Using the information contained in the four case studies, and other information researched (and referenced) from the internet, produce a report for the DECC (Department of Energy and Climate Change), explaining how the UK could use a mixture of different types of electricity supply to meet present and future electricity consumption levels.
2. Detail the environmental impact of your chosen options. You need to discuss the efficiencies of the different methods of generating electricity *and* you need to discuss the carbon footprints of each method together with its effect on global warming. Remember to fully reference any information that you use in your report.

TASK

GREEN SCHOOL

This activity helps you with:

- ★ presenting information diagrammatically
- ★ researching scientific information
- ★ selecting scientific information
- ★ referencing scientific information
- ★ using and analysing data.

Imagine your school is about to be knocked down and rebuilt as a 'state of the art', 'minimal carbon footprint' new school.

1. Produce a series of labelled design sketches for your head teacher showing your new design and the energy saving/generating measures that you would include.
2. For each energy saving or generating idea, include researched and referenced details of a suitable commercially available system and an analysis of costs and projected savings, both in terms of money and in carbon dioxide.

National electricity consumption patterns

The patterns in the UK national consumption of electricity are shown in Figure 14.10.

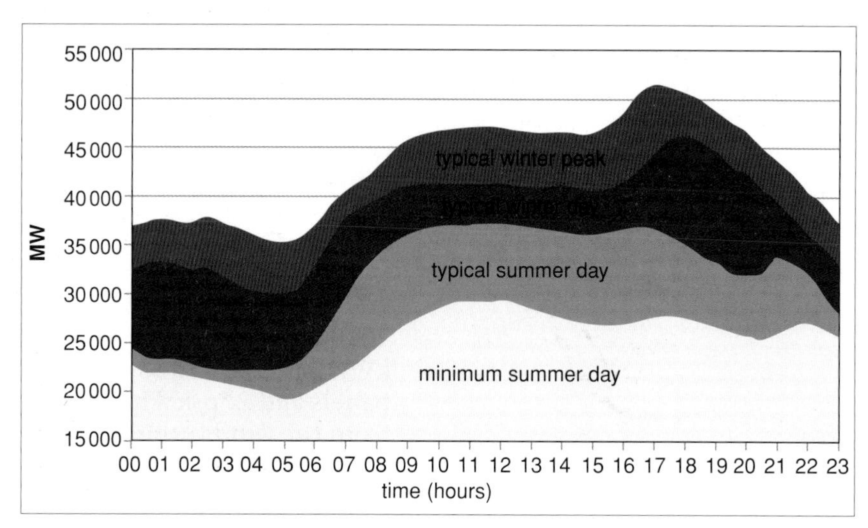

Figure 14.10 Seasonal changes in electricity consumption.

The National Grid is responsible for matching the annual and daily consumption of electricity with production in the UK. It is based in the National Control Centre in Wokingham. The two competing processes have to be balanced exactly. Sometimes there is not enough production capacity in the UK to fully match the consumption and the National Grid has to buy extra capacity from the French National Grid. The National Grid is constantly predicting and forecasting consumption so that the generators can be prepared to produce exactly the right amount of electricity.

How do we move electricity around?

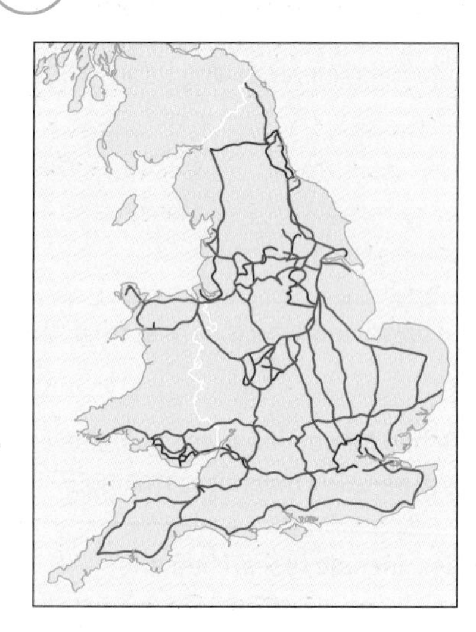

Figure 14.11 The National Grid.

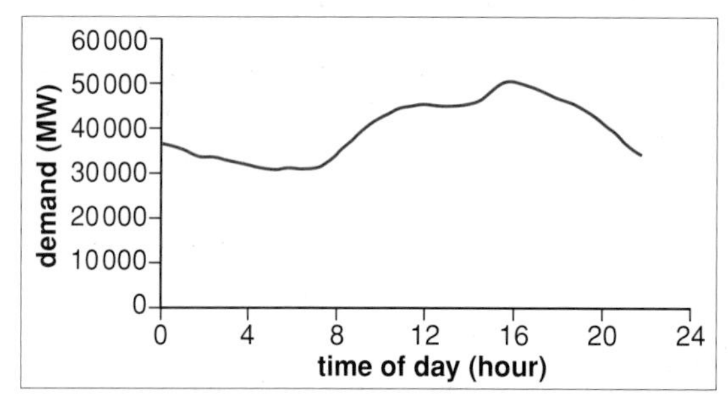

Figure 14.12 UK electricity demand on 4 January 2009.

The fourth of January 2009 was not a particularly significant date (apart from the fact that it was the 366th anniversary of the birth of Isaac Newton, arguably the greatest scientist that ever lived!). The graph in Figure 14.12 shows the UK national demand for electricity, in megawatts (MW), ranging from midnight to midnight. During that 24-hour period the demand for electricity varied from a minimum of 30 894 MW at 5.30 am to a maximum of 50 599 MW at 5.00 pm – a difference of 19 705 MW!

Discussion Point

During half-time of the 2010 World Cup Final the country's demand for electricity rose substantially – as millions of kettles were turned on to make a cup of tea! This event produced a huge spike in electricity demand. How do you think that the National Grid copes with very sudden spikes in demand?

The UK's electricity is generated in a vast network of fossil fuel power stations, wind farms, hydroelectric and nuclear power stations, all connected together and to us, and to the continent, by a network of cables, wires and pylons that stretches like a spider's web across the country. National Grid operators are constantly predicting the demand for electricity in 30 minute blocks, and then directing power generators to supply the relevant amount. If the amount generated is too little then parts of the UK start to experience power cuts.

Some power stations, like Drax in North Yorkshire, are constantly producing electricity (about 7% of the total in Drax's case) and they are 'on' all the time. Other generators, like Dinorwig pumped storage power station in North Wales, are only needed at peak times – they can be given very little notice to be switched on and off as the demand fluctuates. We also import and export electricity across the English Channel. The combined effect of this complex system is to match the electricity demand pattern illustrated by the graph for 4 January 2009. Without this system, the security of our electricity supply would be compromised and we would spend significant amounts of time in the dark, cold, and unable to use our computers, TVs and phones!

QUESTIONS

11 What factors do you think affect the amount of electricity demand?

12 Sketch a copy of the graph for 4 January 2009. On 4 July 2009, the minimum demand was 21 756 MW at 5.00 am and peaked at 34 755 MW at 11.30 am. On your sketch draw the demand line for 4 July 2009.

13 Why does the electricity demand vary during the day?

14 Why does the electricity demand vary during the year?

How big is a megawatt?

The megawatt is a unit of electrical power. 1 MW = 1 000 000 W = 1 million watts. Power is the word we use for the energy transferred per second. High power means a lot of energy per second. Engineers measure the power output of power stations in MW. At home we also use electrical power. A standard low-energy light bulb has a power of about 8 W. An electric kettle has a power of about 3 kW (3000 W). Drax power station can produce enough electricity to power nearly 1.3 million electric kettles at once!

Electrical power depends on the supply voltage and the current flowing. Current, voltage and power are linked by the following equation:

power (W) = voltage (V) × current (A) or $P = VI$

Example

Q An electric oven has an electric current of 13 A supplied with a voltage of 230 V. What is the power of the oven?

A power (W) = voltage (V) × current (A)

power = 230 × 13

= 2990 W

QUESTIONS

15 A hairdryer has an electric current of 2.5 A and a supply voltage of 230 V. What is its power?

16 A travel kettle draws 1.8 A when it is used in the USA where the voltage is 110 V. What is the power of the travel kettle?

17 An electric oven operates off the mains with a voltage of 230 V and has a power output of 3 kW (3000 W). What current does the oven draw?

How is electricity moved around the country?

Figure 14.13 Electricity pylons.

When electric current flows through a wire, the wire heats up. If electricity was transmitted around the country at high current, the amount of energy wasted as heat would be colossal and the price would be so high that nobody would be able to afford it. So how is it moved around?

Remember the electrical power equation:

power = voltage × current

So, a 250 kW wind turbine could generate electricity at 10 A and 25 000 V or at 1 A and 250 000 V, both of which would produce a power output of 250 kW.

The major power stations in the UK are connected together by part of the National Grid, operating at 400 kV, called the Supergrid, where the electrical energy wasted as heat works out at about 1%. The part of the National Grid that connects homes and small businesses to the Supergrid operates at 275 kV or 132 kV. If the whole grid operated at a lower voltage of 25 kV, the electrical energy wasted would be about 40%.

Therefore, electricity is transmitted around the country at very high voltage but very low current to minimise the energy wasted as heat.

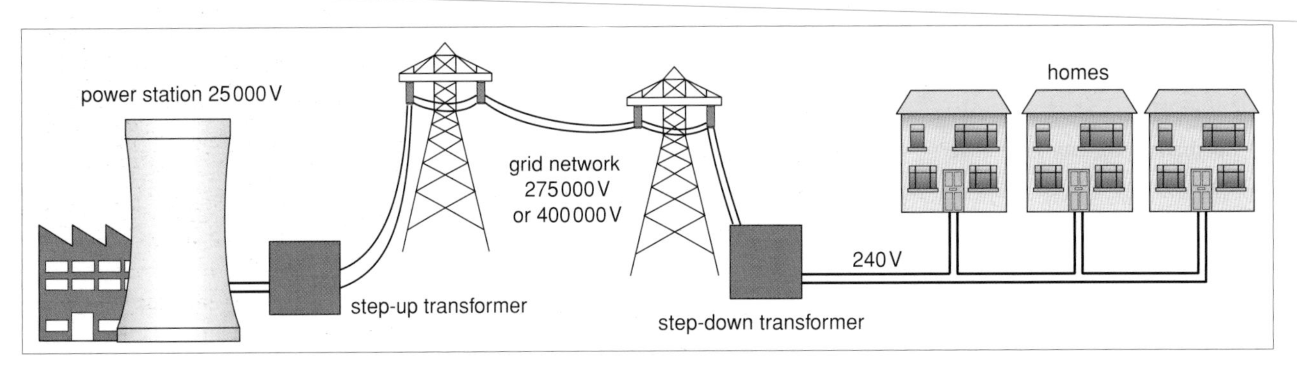

Figure 14.14 The National Grid transmission system.

The trouble with high voltage/low current is that the high voltages are very dangerous and cannot be used by household devices such as hairdryers, mowers, TVs and computers. The electricity needs to be changed by a **transformer** before it enters our homes. **Step-up transformers** are found at power stations. They transform the electrical energy into high voltage/low current to allow it to be transmitted around the country on the National Grid, minimising heat loss. **Step-down transformers** are found at the user end of the National Grid. They convert the electrical energy into low voltage/high current so it can be used safely in electrical devices. In total the National Grid is about 92% efficient, which is pretty good.

QUESTIONS

18 Write down the units of power, voltage and current.

19 What is the equation that links power, voltage and current?

20 Barry power station is a gas-fired power station in South Wales. It can generate electricity with a voltage of 25 000 V at a current of 10 000 A. What is the power of Barry power station in megawatts, MW?

21 Why is electricity transmitted around the National Grid at very high voltage?

22 Why do we not use high-voltage electrical devices at home?

23 What is the name of the device that changes the voltage and current of electricity?

PRACTICAL INVESTIGATING TRANSFORMERS

This activity helps you with:

★ working as a team
★ handling apparatus
★ organising your work
★ measuring and recording data
★ calculating values from data.

You can make a simple transformer using the following apparatus:

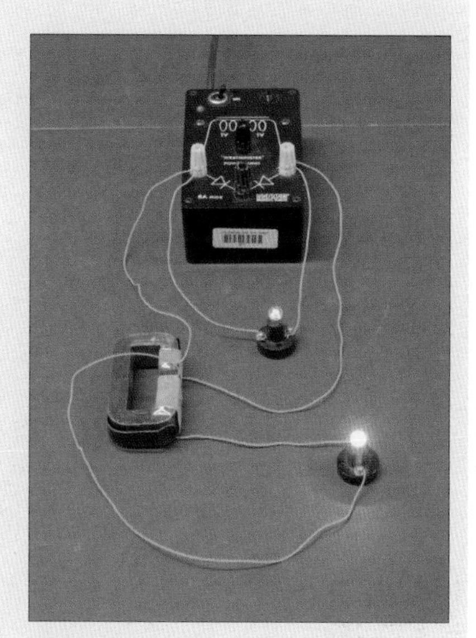

Figure 14.15 A C-core transmission experiment.

Apparatus

* lengths of plastic clad wire
* 2 C-cores
* low-voltage ac power supply
* 2 low-voltage light bulbs
* 2 ac multimeters

Risk assessment

- **Your teacher will supply you with a suitable risk assessment for this experiment.**

Procedure

1 Connect up the circuit shown in Figure 14.16.

2 To make a step-up transformer, wind 20 turns onto the primary coil and 40 turns onto the secondary coil.

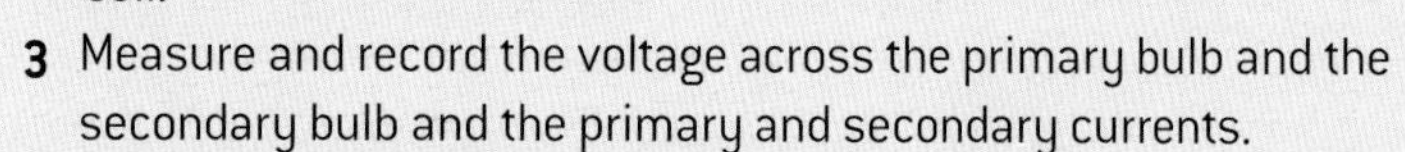

3 Measure and record the voltage across the primary bulb and the secondary bulb and the primary and secondary currents.

4 Compare the primary bulb brightness to the secondary bulb brightness.

5 Use the primary voltage and current to calculate the primary power, and the secondary voltage and current to calculate the secondary power.

6 Use these values to calculate the efficiency of the transformer.

7 Disconnect the transformer and turn it around. Use it to make a step-down transformer.

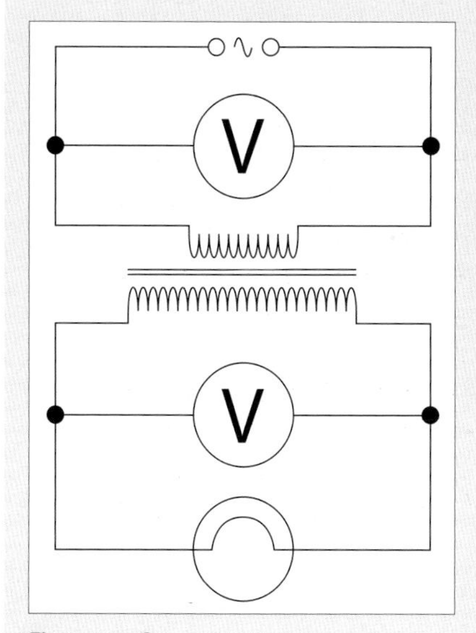

Figure 14.16

Analysing your results

1 Compare the efficiencies of the two types of transformer. Is the step-down transformer more or less efficient than the step-up transformer?

2 Investigate what happens when you change the number of turns on the primary and secondary coils.

Chapter summary

- Electricity is a really useful form of energy because it is easy to produce and easy to transfer into other useful forms of energy.
- Power stations (renewable, non-renewable and nuclear) have significant but very different commissioning costs, running costs (including fuel) and decommissioning costs that need to be considered when planning a national energy strategy.
- Large-scale power generation by power stations and micro-generation of electricity, e.g. using domestic wind turbines and roof-top photovoltaic cells, have different advantages and disadvantages. They all have very different environmental impacts.
- Data can be used to determine the efficiency and power output of power stations and micro-generators.
- There is a need for a national electricity distribution system (the National Grid), in order to maintain a reliable energy supply that is capable of responding to a fluctuating demand.
- The National Grid consists of power stations, substations and power lines.
- Electricity is transmitted across the country at high voltages because it is more efficient, but low voltages are used at home because it is safer.
- Transformers are needed to change the voltage and current within the National Grid.
- It is possible to experimentally investigate the operation of step-up and step-down transformers, in terms of the input and output voltage, current and power.
- power = voltage × current; $P = VI$

Energy

Peak oil – how will you ride the slide?

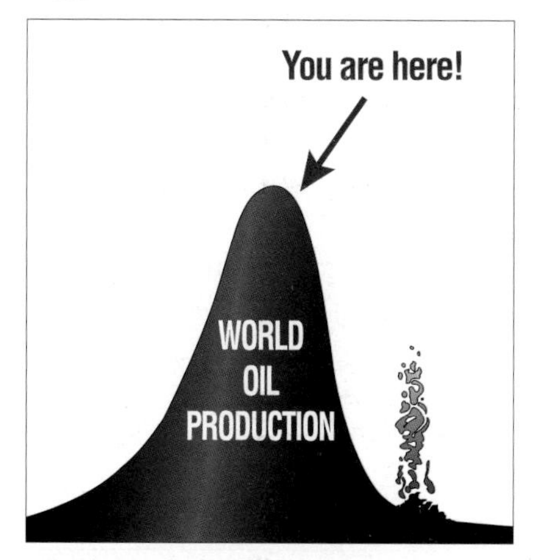

Figure 15.1 Peak oil has happened!

Approximately 34% of the world's total energy supply is from oil. Sometime during 2010, world oil production peaked. From now on, the amount of oil processed by the world's refineries will only ever get less. How will you ride the slide?

Oil is a hugely valuable resource. We burn it in large quantities in power stations to produce electricity and it is refined to make petrol, diesel and kerosene to power our cars, lorries, ships and aeroplanes. It is also used as the raw material for most plastics. How would you manage without cheap transport and all the myriad of things that are made from plastic?

The world's demand for electricity is increasing. Countries like China and India are increasing their consumption of electricity substantially as more and more of their population are connected to mains electricity. In 2011 approximately 6% of the world's electricity is generated as a result of burning oil (nearly 42% is generated from coal and 21% from gas).

TASK MAKING THINGS CLEARER

This activity helps you with:
- ★ working as a team
- ★ presenting information graphically
- ★ evaluating your work.

Discussion Point

Do graphics matter when you are trying to get a scientific argument over to the general public? What are the advantages of graphics? Can you think of examples of scientific graphics that have really helped you to understand some science?

The problem of 'peak oil' is so important that it is absolutely vital that everyone is aware of it. Many people find it much easier to access information if it is presented in a graph or a picture. In this task you will examine a table of data and decide on the best way to present the data in a graphical way.

1. Study Table 15.1. It lists world oil production since 1900 and projected to the year 2080. You could plot these data as a bar chart or a line graph. (If you are going to use a spreadsheet such as Excel to plot the data, Excel calls this an XY scattergraph.)

   Work with a partner – one of you will plot the data as a bar chart and the other will plot the data as a line graph.
2. Which of the two ways of plotting the data do you think is best in this case – the bar chart or the line graph?
3. Why is a line of best fit not a good idea on the bar chart?

Table 15.1 Data showing world production of oil since 1900 and projected to 2080.

Year	World oil production (million barrels per day; Mbl/day)
1900	1
1910	2
1920	3
1930	4
1940	5
1950	11
1960	20
1970	48
1980	67
1990	66
2000	74
2010	82
2020	58
2030	30
2040	16
2050	9
2060	6
2070	5
2080	3

Why we all need to know about efficiency and energy transfer

When an iPod plays a music track through some earphones, chemical energy stored in the iPod's rechargeable battery is transferred into useful electrical energy (later converted into sound by the earphones). Some energy is wasted as thermal (heat) energy, causing the battery and the iPod to heat up. IPod rechargeable batteries are very good at doing their job, and for every 100 J of chemical energy stored in the battery, 98 J are transferred into electrical energy and only 2 J are wasted as heat. As the batteries convert such a large amount of their available stored chemical energy into useful electrical energy (and little is wasted) we say that the batteries are very **efficient**. The efficiency of a device or a process is normally expressed as a percentage (%). A device that converts all its available input energy into useful output energy is 100% efficient.

Efficiency is calculated using the following mathematical formula:

$$\text{efficiency} = \frac{\text{useful energy (or power) output}}{\text{total energy (or power) input}} \times 100\%$$

Calculating efficiencies – examples

Q The Li-ion battery in a mobile phone can hold 18 000 J of chemical energy. If the battery converts 16 000 J into useful electrical energy and 2000 J is wasted as heat energy, what is the efficiency of the battery?

A Total energy input = 18 000 J
Useful energy output = 16 000 J

$$\text{efficiency} = \frac{\text{useful energy (or power) output}}{\text{total energy (or power) input}} \times 100\%$$

$$\text{efficiency} = \frac{16\,000}{18\,000} \times 100\%$$

$$= 89\%$$

Q A power station supplies electricity to the National Grid at a power output of 60 MW. In order to do this it burns coal at a rate with an equivalent power input of 200 MW. What is the efficiency of the power station?

A Total power input = 200 MW
Useful power output = 60 MW

$$\text{efficiency} = \frac{\text{useful power output}}{\text{total power input}} \times 100\%$$

$$\text{efficiency} = \frac{60}{200} \times 100\%$$

$$= 30\%$$

Q A solar panel is rated as being 30 % efficient. The panel outputs electrical power at a rate of 180 W. What is the input power of sunlight on the panel?

A Efficiency = 30 %
Useful power output = 180 W

$$\text{efficiency} = \frac{\text{useful power output}}{\text{total power input}} \times 100\,\%$$

Rearranged:

$$\text{total power input} = \frac{\text{useful power output}}{\text{efficiency}} \times 100\,\%$$

$$\text{total power input} = \frac{180}{30} \times 100\,\%$$

$$= 600\text{ W}$$

QUESTIONS

1 Low-energy light bulbs are very efficient – much more efficient than the older, tungsten filament bulbs of the same light output, which waste a lot of the electrical energy input as heat. Calculate the efficiency of the following bulbs:

- **a** low-energy 1.6 W output, 8 W input
- **b** tungsten filament 1.6 W output, 60 W input.

Figure 15.2 An energy-efficient light bulb.

2 The most efficient type of light bulb available today is manufactured from high-intensity LEDs (light emitting diodes). These bulbs can be 90% efficient. If one such LED bulb gives out 18 W of light, how much electrical power is input?

3 A large wind turbine can output 0.50 MW of electrical power from a maximum wind power of 0.75 MW. Calculate:

- **a** the efficiency of the wind turbine
- **b** the amount of wind power wasted as heat and sound energy if the turbine is working at maximum power.

Figure 15.3 An LED light bulb.

continued...

Discussion Point

The electricity generated by Dinorwig is produced by falling water – a form of hydroelectric energy. Would you consider the electricity produced by Dinorwig to be 'renewable' or 'non-renewable'? Do we really need power stations like Dinorwig?

QUESTIONS *contd.*

4 Dinorwig pumped storage power station is located in North Wales. During the night, when demand for electricity is low and there is 'spare' capacity (as not all the big conventional power stations can be 'switched off' at night), water is pumped from a lake at the bottom of Elidir Mountain through a huge water tunnel, up to another lake, Marchlyn Mawr, 70 m up the mountain. During the day or early evening when there is a sudden rise in demand for electricity (e.g. during half-time of the FA Cup Final – when millions of people decide they all want a cup of tea at the same time, and millions of kettles are switched on), the water in the top lake is sent back down the water tunnel through the pumps, this time acting as generators, producing 1800 MW of electricity.

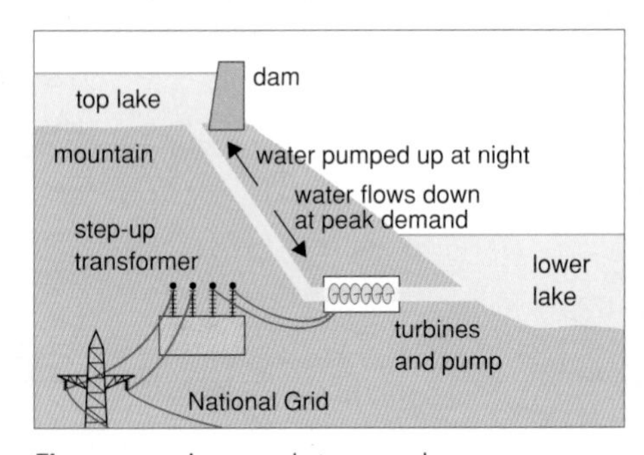

Figure 15.4 A pumped storage scheme.

Figure 15.5 Dinorwig pumped storage power station.

Dinorwig can be up and running at full power in 12 seconds, but there is only enough energy stored in the top lake to last for 5 hours. This is enough to cover a surge in demand during peak times. At night the pumps work at 2400 MW.

a Calculate the efficiency of the power station.

b When all the water has run from the top lake to the bottom lake, 32.4 TJ (TJ = Terajoules, 32.4×10^{12} J) of electrical energy have been generated by the pump-generators. Use the efficiency of the power station that you calculated in part **a** to calculate the total energy (in TJ) needed at night to pump the water back up to the top lake.

c Electricity produced by Dinorwig costs the National Grid approximately £200 per MW. Electricity produced by a conventional coal or oil power station costs the National Grid only approximately £30 per MW. Why does the National Grid need Dinorwig when its electricity costs so much more?

TASK THE AMPAIR UW100

This activity helps you with:
- ★ working with tabulated data
- ★ presenting information graphically
- ★ calculating values from data
- ★ looking for patterns in data
- ★ drawing best-fit lines
- ★ extrapolating data.

The Ampair UW100 is a small micro-water-turbine generator that individual households can buy and use if they live next to a river. The unit will locally generate a small amount of electricity – usually to charge batteries.

Figure 15.6 The Ampair UW100 water turbine.

The datasheet supplied by the company gives the following data about the performance of the UW100:

- minimum flow rate needed to turn turbine = 1 m/s
- maximum safe operating flow rate = 6 m/s

Table 15.2 Data for calculating the efficiency of the UW100.

Water flow speed m/s	Potential input power (W)	Electrical output power (W)	% efficiency
1	19	12	
2	152	24	
3	513	50	
4	1216	72	
5	2375	84	
6	4104	94	

Questions

1. Copy and complete the table by calculating the efficiency of the UW100 for each water flow speed.
2. Plot a graph showing how the efficiency of the UW100 varies with water flow speed and draw a best-fit line through the data.
3. Describe in words the pattern shown by the best-fit line.
4. Why do you think the efficiency varies in this way?
5. How do you think the efficiency varies for very low and very high water flow speeds (down to minimum flow rate and up to maximum flow rate)?
   - a Draw dashed lines on your graph, extrapolating (continuing) your best-fit line to show how you think the efficiency changes for low and high water flow speeds.
   - b Give explanations for your extrapolations.
6. What would be the consequences of increasing the turbine's efficiency too far.
7. Why might the designer wish to limit the electrical output power?

PRACTICAL MEASURING THE EFFICIENCY OF AN ELECTRIC HEATER

This activity helps you with:
★ working as a team
★ handling apparatus
★ organising your work
★ measuring and recording data
★ calculating values from data
★ analysing the results of an experiment.

Apparatus
* 12 V power supply unit
* 12 V electric heater
* joulemeter
* thermometer
* 100 cm³ glass beaker
* stirring rod
* 4 mm connecting wires
* 100 cm³ measuring cylinder

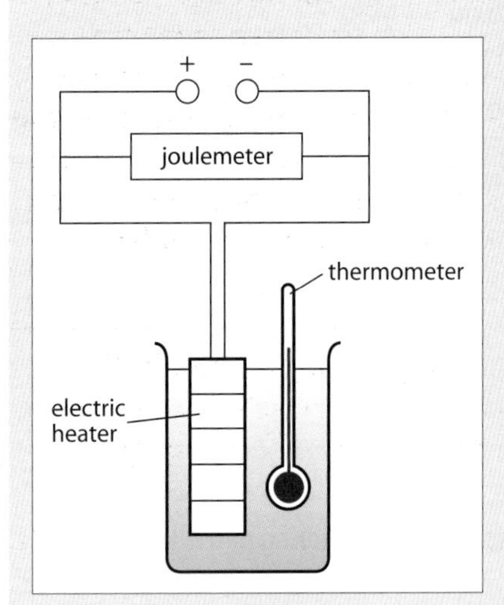

Figure 15.7

In this experiment you will investigate how efficient a small 12 V electric heater is at heating water. Some of the electrical energy supplied by the heater will be wasted, heating the air and glass around the heater. The rest will be usefully used to heat the water. You need to know that it takes 420 J to raise the temperature of 100 g of water by 1 °C.

Risk assessment

- **Take care – the electric heater will get hot.**
- **You will be supplied with a risk assessment by your teacher.**

Procedure

1 Use the measuring cylinder to measure 100 cm³ of water and pour it into the beaker.
2 Connect the power supply to the joulemeter and then the heater to the joulemeter.
3 Put the heater into the water together with the thermometer. Measure and record the temperature of the water, T_{start}.
4 Make sure the joulemeter reads zero (reset if needed) then turn on the 12 V power supply.
5 Every 30 seconds, stir the water with the stirring rod and monitor the temperature of the water until the temperature has increased by 10 °C. Then turn off the power supply and record the reading on the joulemeter, E_{input}.
6 Keep measuring the temperature of the water with the heater switched off and record the maximum temperature reached by the water, T_{max}.
7 Calculate the temperature change of the water, $T_{change} = T_{max} - T_{start}$
8 Calculate the useful energy output by the heater, $E_{useful} = 420 \times T_{change}$
9 Calculate the efficiency of the heater $= \dfrac{E_{useful}}{E_{input}} \times 100\%$
10 Pack away all your apparatus neatly when it has cooled down.

Analysing your results

1 Apart from the air, where else is heat energy wasted? (*Hint*: what else heats up along with the water?)
2 How much energy is wasted in total?
3 Why is it important to keep measuring the temperature of the water even after the power supply unit has been switched off?
4 Do you think the heater will be more or less efficient at heating 250 cm³ of water rather than 100 cm³? Explain your answer.

Sankey diagrams

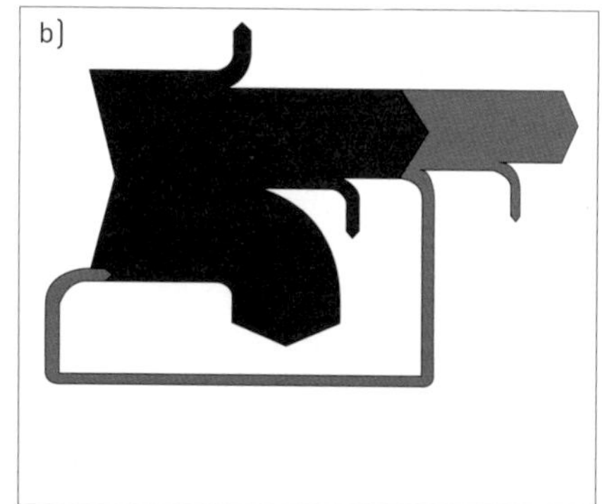

Figure 15.8 a) A steamship; b) Sankey diagram for a steamship.

Figure 15.9 Captain Matthew Sankey

Sankey diagrams are a clever way of illustrating both energy (or power) transfers and efficiency. Sankey diagrams were invented by an English sea captain, Captain Matthew Sankey. Captain Sankey was really interested in the steam engines that powered his ship, and he invented Sankey diagrams as a way of showing the energy transfers that happened as the engines on his ship worked – he was trying to make the engines more efficient.

The good thing about Sankey diagrams is that not only do they show the different energy transfers that happen, but because they are always drawn to scale, they also show the relative or percentage amount of energy being transformed at each stage of the transfer. The width of the bars on the Sankey diagram at any point show the amount of energy involved, and because of this the efficiency can be shown by the width of the useful energy (or energies) compared with the total input energy width or the wasted energy width.

Example

As an example, we will look at an energy-efficient light bulb.

Every second, 10 J of electricity is input to the bulb. Only 2 J is output as useful light energy so 8 J is wasted as heat energy.

Figure 15.10 An energy efficient light bulb.

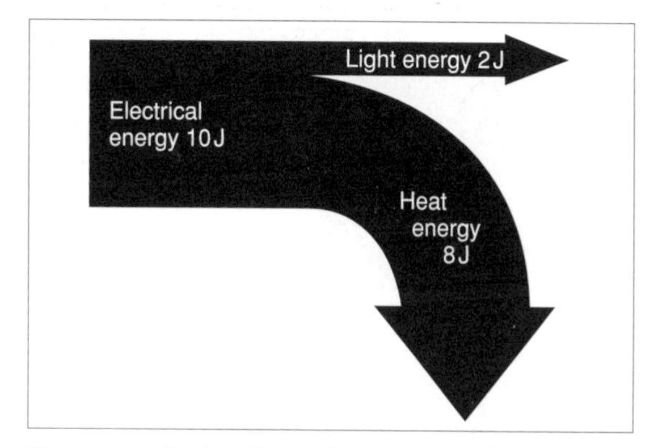

Figure 15.11 Sankey diagram for an energy-efficient lightbulb.

When this Sankey diagram was drawn the electrical energy bar was drawn so that it is 10 units wide, the light energy bar was drawn 2 units wide and the heat energy bar was drawn 8 units wide. Usually (but not always) the useful energy transfer runs along the top of the diagram (as a straight bar) and the wasted ones curve away below. In this case it is easy to see that of the 10 J of electrical energy input only 2 J is output as useful light energy. This makes 'low-energy' light bulbs only 20% efficient – not so good until you find out that standard incandescent filament bulbs are only about 2% efficient.

QUESTIONS

5 LED light bulbs can be up to 80% efficient. Draw a Sankey diagram for an LED light bulb, assuming 100 J of electrical energy is input.

6 Some internal combustion engines are 25% efficient. Here is a Sankey diagram showing one in action.

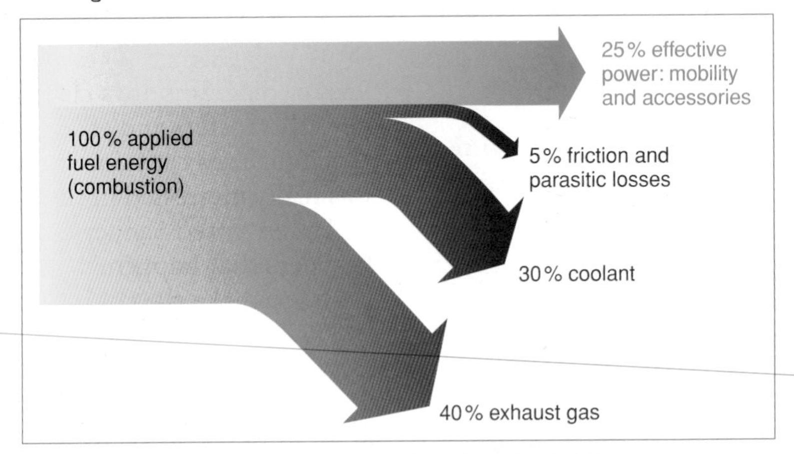

Fig 15.12 Sankey diagram for an internal combustion engine.

Each litre of petrol produces approximately 35 MJ of heat energy when it is burnt in the engine.

a How much of this energy is transformed into useful kinetic energy (effective power)?

b How much is wasted as exhaust gas?

TASK AN INCONVENIENT TRUTH?

Figure 15.13 Al Gore.

In his 2006 documentary film *An Inconvenient Truth*, former US Vice-President Al Gore championed the need for humanity to address the problem of global warming. He outlined in graphic detail the consequences of continued burning of fossil fuels and their resulting carbon footprint. Global warming over the next 100 years could raise the sea level by up to 2 m. The consequences of this are devastating for mankind.

continued...

TASK contd.

Discussion Points

Discuss all or some of the following questions:

1. Can we afford to continue the uncontrolled increase in the use of fossil fuels?
2. Is it right that we should continue to run some power stations that are 30% efficient when others of a similar but more modern design are 50–60% efficient?
3. Can we morally defend the use of incandescent light bulbs that are only 2% efficient?
4. Is it morally right that high-efficiency LED light bulbs should be so expensive?
5. Do you leave electrical appliances on stand-by at night? This causes the National Grid to run the equivalent of one whole fossil-fuelled power station, purely to have a little light on while you are asleep, even though you cannot see it.
6. Is it right that manufacturers of electrical appliances do not have (by law) to build in an isolating on/off switch for appliances? This could easily be done.
7. Is it right that the average European manufactured motor car has an engine that is capable of a fuel consumption of 5 litres per 100 km, yet an average US manufactured motor car is only capable of 11 litres per 100 km?
8. What do you think about building the world's largest snow dome ski slope in the Middle East desert, where the average daytime temperature is well over 30 °C, and the air-conditioning alone would require immense amounts of electrical energy?
9. Is it fair that power stations built in the UK have to comply with stringent carbon-emission targets, yet the same power station built in China would require no carbon-emission targets?
10. Why do some people object to having a wind farm built near them, yet then complain about the price of electricity?
11. The world is facing 'an inconvenient truth'. Too many people make too much money as a result of fossil-fuel exploitation. Three of the world's top four most profitable companies are oil companies. The average American consumes twice as much energy as the average UK citizen and nearly 20 times more than the average citizen of India. Can we really allow this exploitation to continue?

Making things better – generating 'free' electricity

Figure 15.14 Elan Valley dam, West Wales.

Electricity is produced by a generator. A large magnet (or electromagnet) spins inside a coil of wire, generating a current. In a conventional fossil-fuel powered power station, the generator is spun by a turbine, which is driven by high-pressure steam. The steam is produced when the fossil fuel is burned and the heat is used to boil water.

Any moving fluid (a liquid or a gas) will spin a turbine, so running water and the wind can be just as effective as steam at spinning the turbine. In fact, we have been using running water and the wind for centuries to provide us with energy – waterwheels and windmills.

Modern water and wind turbines have low-friction turbine blades that are easily turned by the moving fluid and are connected directly to the generator that produces the electricity. The important thing about this, however, is that there must be enough kinetic energy in the water or the wind to turn the turbine blades. This depends on the

Figure 15.15 a) Waterwheels and b) windmills have been generating energy for centuries.

speed of the moving wind/water *and* the mass of the air/water moving through the turbine blades. We can calculate the mass of the wind or water by knowing its **density**.

Density is a measure of how much mass (matter) is present in a given volume of a material – usually given as 1 cm^3 or 1 m^3. Water has a density of 1 g/cm^3 or 1000 kg/m^3. Air (at sea level and 15 °C) has a density of 0.0012 g/cm^3 or 1.2 kg/m^3.

Density can be calculated using the following equation:

$$\text{density} = \frac{\text{mass}}{\text{volume}} \quad \text{or} \quad \rho = \frac{m}{V}$$

Why is density so important for electricity generation?

Water is approximately 1000 times denser than air. This means that when water moves through a turbine blade, there is more mass of material moving so more kinetic energy. Moving water is a very efficient way of generating electricity. Large amounts of water, moving at speed through a hydroelectric power station, can generate huge amounts of electricity. The generators at the foot of the Hoover Dam in Nevada provide more than enough electricity for the city of Las Vegas with all its bright lights.

Figure 15.16 The Hoover Dam in Nevada, USA.

Figure 15.17 The bright lights of Las Vegas.

QUESTIONS

7 Seawater is more dense than fresh water (due to the salt dissolved in it). Calculate the density of seawater if 2000 cm^3 has a mass of 2060 g.

8 The Farr wind farm near Inverness is an unusually high site for a wind farm, with an average elevation of 500 m. Although air moves faster at altitude it is significantly less dense than at sea level. At 500 m, 10 m^3 of air has a mass of 9.5 kg. Calculate the density of air at the Farr wind farm.

9 The Elan Valley dams and hydroelectric scheme in West Wales supply a total of 11 000 homes with electricity and 300 000 m^3 of water to Birmingham each day. If the density of water is 1000 kg/m^3, calculate the mass of water delivered to Birmingham each day from the Elan Valley.

Examples – calculating density

Q A student measures the mass of 100 cm^3 of water in a measuring cylinder to be 101.05 g. What is the density of the water?

A

$$\text{density} = \frac{\text{mass}}{\text{volume}}$$

$$\text{density} = \frac{101.05}{100}$$

$$= 1.0105\ g/cm^3$$

Q A clown blows 0.5 g of air into a balloon, which expands to occupy a volume of 400 cm^3. What is the density of the air?

A

$$\text{density} = \frac{\text{mass}}{\text{volume}}$$

$$\text{density} = \frac{0.5}{400}$$

$$= 0.00125\ g/cm^3$$

Q A wind turbine requires 24 kg of air per second, with a density of 1.2 kg/m^3, in order turn at maximum efficiency. What volume of air is needed?

A

$$\text{density} = \frac{\text{mass}}{\text{volume}}$$

Rearranged:

$$\text{volume} = \frac{\text{mass}}{\text{density}}$$

$$\text{volume} = \frac{24}{1.2}$$

$$= 20\ m^3$$

PRACTICAL MEASURING THE DENSITY OF WATER

This activity helps you with:

- ★ working as a team
- ★ handling apparatus
- ★ making a prediction
- ★ organising your work
- ★ measuring and recording data
- ★ calculating values from data
- ★ analysing the uncertainty in experiments
- ★ analysing the results of an experiment.

In this experiment you will measure the density of water using a variety of different methods. Your task is to decide which of the methods will give you the most accurate value. The density of water is determined by measuring the mass of a given volume. This requires you to measure the mass and the volume separately. Look at Table 15.3.

Table 15.3 Methods of measuring mass and volume of liquids.

Ways of measuring the mass of a liquid	Ways of measuring the volume of a liquid
Electronic laboratory balance	Beaker
Kitchen scales	Measuring cylinder
Laboratory mechanical balance	Volumetric flask

1 Predict which combination of apparatus you think will produce the most accurate value of the density of water. Explain why.

PRACTICAL *contd.*

Apparatus

* lab balance
* kitchen scales
* mechanical lab balance
* 100 cm³ beaker
* 100 cm³ measuring cylinder
* 100 cm³ volumetric flask

Risk assessment

- **Your teacher will supply you with a suitable risk assessment.**

Procedure

1 Use each piece of volume measuring apparatus to measure out 100 cm³ of water.

2 Then use each piece of mass measuring apparatus to measure the mass of each 100 cm³ of water. Record your measurements in a suitable table. Calculate the density of the water for each combination of volume–mass apparatus, using the equation:

$$\text{density} = \frac{\text{mass}}{\text{volume}}$$

Evaluating your results

1 The given value of the density of water is 1.00 g/cm³. Which combination of apparatus produces:

   a the most accurate value (the closest to the given value)

   b the least accurate value?

2 Give reasons why you think that there is such a difference in the accuracy of the calculated densities of water.

3 If you repeated each measurement do you think it would make a difference? Explain your answer.

4 Does it matter which way a physical quantity such as density is measured?

Extension – measuring the density of air

An excellent method of measuring the density of air, produced by the Institute of Physics (IOP), using a balloon and a bucket of water, can be found at:
www.practicalphysics.org/go/Experiment_168.html

Chapter summary

- Energy transfer (Sankey) diagrams can be used to make it easier to understand the idea of energy efficiency in terms of input energy, useful output energy and wasted energy.
- It is possible to investigate the efficiency of energy transfer in an electrical context by measuring the energy transfers in and out of an appliance.
- $$\text{efficiency} = \frac{\text{useful energy or power output}}{\text{total energy or power input}} \times 100\%$$
- The efficiency of fossil-fuel power stations and the National Grid are absolutely vital to the environmental debate on carbon footprint and global warming.
- $$\text{density} = \frac{\text{mass}}{\text{volume}} \text{ or } \rho = \frac{m}{V}$$
- The density of materials, including air and water, can be experimentally determined by measuring volumes and masses.
- Knowledge of density can be used to inform discussion of the energy available from moving water and air.

How much could you save?

Figure 16.1 Thermal imaging shows varying heat loss in a row of houses.

On 14 April 2011, the *Guardian* newspaper had the following headline:

At least 10% of new homes fail energy efficiency test

Official figures show that a high number of new homes don't comply with legal standards to cut carbon emissions and utility bills.

At least one in 10 new homes in Britain does not meet legal requirements for energy efficiency, condemning tens of thousands of householders to higher energy bills, and exacerbating climate change. The government has identified improving households' energy efficiency as the best way to reduce carbon emissions at the same time as keeping a lid on rising utility bills. Since April 2008, all new homes have had to meet tough standards on draught proofing, lighting and heating. All homes require an Energy Performance Certificate (EPC) indicating how they rate. But at least 30,000 of the 300,000 homes built since then do not meet these legal standards, according to official figures just released. Andrew Warren, director of the Association for the Conservation of Energy, said: "Buying a home is the biggest single purchase people will make in their lives. With energy costs mounting – never mind the environmental issues – it's perfectly respectable to expect that buildings meet the minimum legal standards for energy efficiency".

Figure 16.2 The world needs YOU!

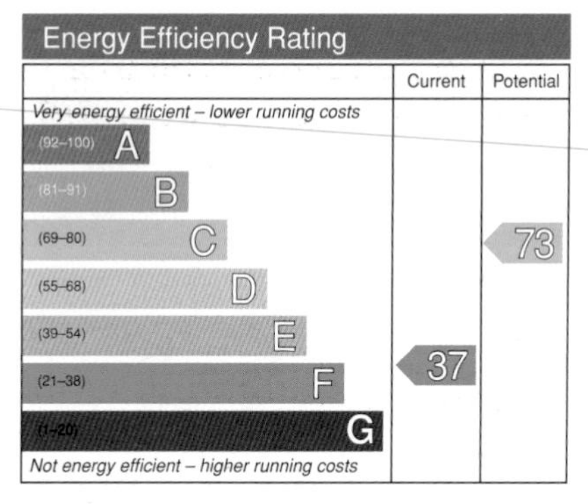

Figure 16.3 An Energy Performance Certificate.

The world needs YOU! It is your generation that must take and make hard decisions about the future. The world cannot carry on using energy with such recklessness. It must become socially unacceptable to waste energy, in the same way that it is now socially unacceptable to drink and drive. Energy efficiency, insulation and greater use of renewable energy *must* become the accepted way that we use energy, and individuals through to local, national and international governments must all do their bit. What is *your* contribution going to be?

In the *Guardian* article, the author refers to the Energy Performance Certificate or EPC. The EPC produced for a house when it is sold contains an energy efficiency rating – Figure 16.3 shows an example of one of these.

The energy efficiency rating of this house shows that the house is 37% efficient, an F rating. The efficiency is calculated by using a computer model based on factors such as insulation, heating and hot water systems, ventilation and fuels used in the house. The average energy efficiency rating for a dwelling in England and Wales is band E (rating 46%). The EPC then goes on to suggest ways that the householders could improve the overall efficiency of the house. In this case, the EPC recommended that the householders could:

- install cavity-wall insulation — saving £411 per year
- install low-energy lighting — saving £11 per year
- install a hot-water cylinder thermostat — saving £102 per year
- replace the boiler with a Band A condensing boiler — saving £323 per year
- replace single-glazed windows with double glazing — saving £30 per year
- install solar photovoltaic panels (25% of roof) — saving £49 per year
- **Overall effect** — **saving £926 per year**

The EPC then gives details about each type of energy saving measure, suggesting ways to go about installing them and where to find further information.

Some of the energy-saving measures are quite expensive to install. Replacing boilers and installing double glazing and solar photovoltaic panels is not cheap, but some simple measures are surprisingly cheap. Many older houses suffer from excessive draughts, mostly through the sides of windows and ill-fitting doors. Fitting weather strips and draught excluders (that cost only a few pounds to buy and can be installed as DIY) to doors and windows can substantially improve the insulation of a house.

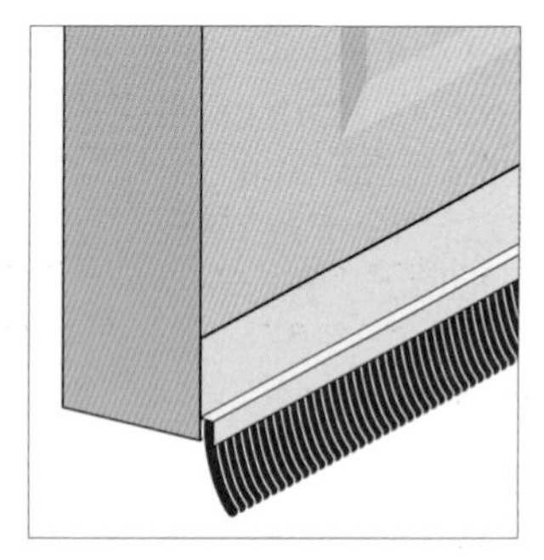

Figure 16.4 A draught excluder fitted to the bottom of a door.

How do draught excluders work?

All forms of insulation effectively work in the same way – they prevent heat energy moving from somewhere hot to somewhere cold.

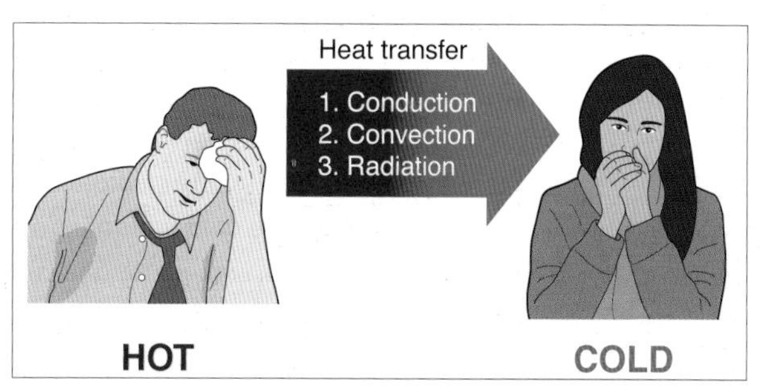

Figure 16.5 Heat transfer.

Insulation reduces the flow of heat energy from hot to cold by reducing the effect of the three mechanisms of heat transfer, namely:

- conduction
- convection
- radiation.

Draught excluders work by reducing convection currents under a door or through the gaps in the frame – they can save between 10% and 20% of a household's heating costs.

PRACTICAL INVESTIGATING CONVECTION

This activity helps you with:
★ making experimental observations
★ analysing experimental observations using a scientific model
★ drawing an experimental diagram.

Your teacher will demonstrate (or let you investigate) convection currents in action in a model chimney.

Apparatus
* chimney demonstration apparatus
* candle
* splint

Risk assessment

- **Wear eye protection and tie back any loose hair/clothing.**
- **Your teacher will supply you with a suitable risk assessment.**
- **The chimney above the candle will get very hot, as will the air leaving it. Do not touch the glass or put your hand above the hot chimney until the candle is put out and the chimney has cooled.**

Procedure

1. Remove the front glass panel from the apparatus.
2. Light the candle under one of the glass chimneys.
3. Replace the glass panel.
4. Light a wooden splint, let it flame for a moment and then blow it out. The splint will produce smoke.
5. Hold the smoking splint in the top of the other glass chimney, and observe the motion of the smoke.
6. Draw a diagram of the apparatus and add arrows showing the motion of the smoke.

Figure 16.6 Smoking chimney apparatus.

continued...

PRACTICAL contd.

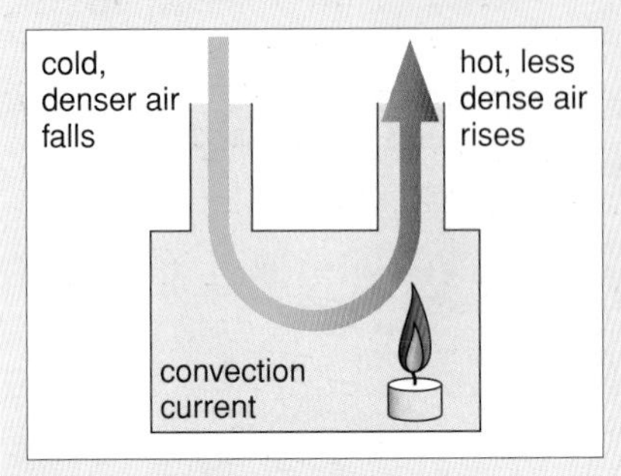

Figure 16.7 How it works.

Explaining the motion of the smoke

The heat energy produced by the candle heats the air immediately above the flame. The heated air particles move faster. As they move faster they get (on average) further apart. This means that 1 cubic centimetre of hot air above the candle flame contains fewer particles than the cold air around it. Fewer air particles means less mass of air per cubic centimetre – this means that the **density** (mass per unit volume) of the hot air is less than the density of the colder air around it. The less dense, hot air then rises above the colder, denser air. As the hot air rises it is replaced directly above the flame by colder air that is sucked in from around it. Any smoke particles within the chimney system are carried on this **convection current** of air as it moves down the cold chimney and up the hot chimney (see Figure 16.7).

PRACTICAL — PRACTICAL CONVECTION

This activity helps you with:
★ working as part of a team
★ working safely in science
★ making scientific observations.

You can also investigate convection currents in water.

Apparatus

* flat-bottomed, round flask
* Bunsen burner
* gauze
* tripod
* heatproof mat
* potassium permanganate crystal
* tweezers
* eye protection

Risk assessment

- **Wear eye protection.**
- **You will be supplied with a risk assessment by your teacher.**

Procedure

1. Your teacher will provide you with a small crystal of potassium manganate(VII) (permanganate). Drop it carefully into a flask of cold water.
2. Use a very low flame to gently heat the water near to the crystal.
3. Observe what happens.

As the water near the crystal is heated it becomes less dense and rises. The coloured water shows the path of the convection current in the water. It carries the heat energy along with it.

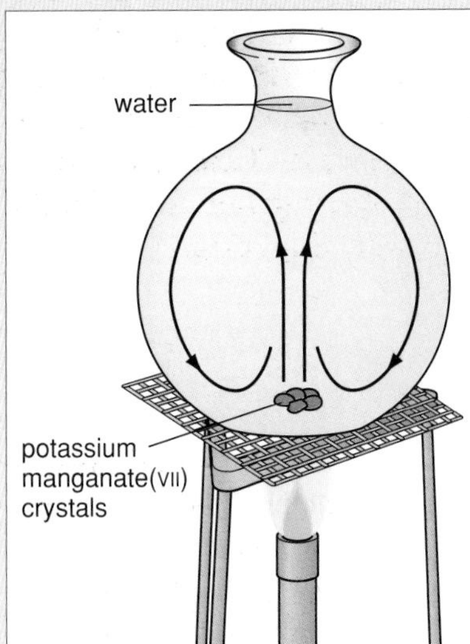

Figure 16.8 Apparatus to show convection currents.

Draught excluders act as barriers between the places that are hot and the places that are cold. The cold air outside a room is prevented from being sucked through the air gap under a door as the hot air inside the room rises. The draught excluder physically prevents the colder air particles from entering the room.

Convection currents are everywhere. All **fluids** (gases or liquids) will undergo convection if they are heated. Even very thick liquids like the molten rock inside the Earth's mantle will undergo convection (in fact it's these convection currents that drive plate tectonics – the movement of the Earth's crustal plates). The Earth's prevailing winds, called the Trade Winds, are also governed by convection. The Sun is the heat source and the atmospheric convection currents that it causes are called Hadley cells.

Figure 16.9 Convection currents in the mantle of the Earth.

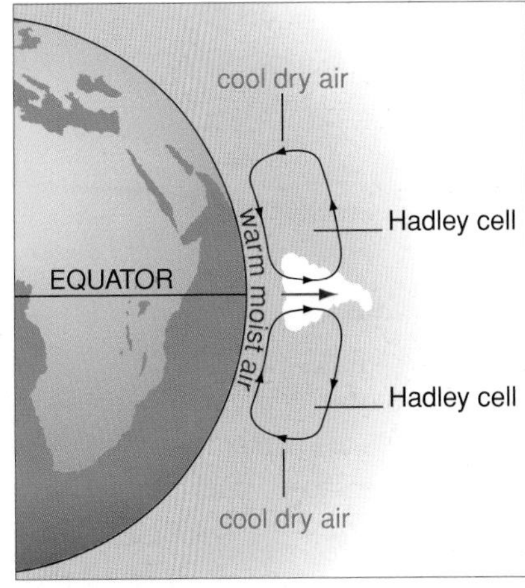

Figure 16.10 Hadley cells at the equator.

Most central heating systems work as a result of convection currents. Hot water is pumped around a system of pipes and radiators. The air inside a room is heated by the radiator, setting up convection currents. The hot air above the radiator gets less dense and rises, causing colder, denser air to replace it. Whilst the radiator is hot there will be a continuous current of rising hot air and falling colder air circulating around the room, warming it up.

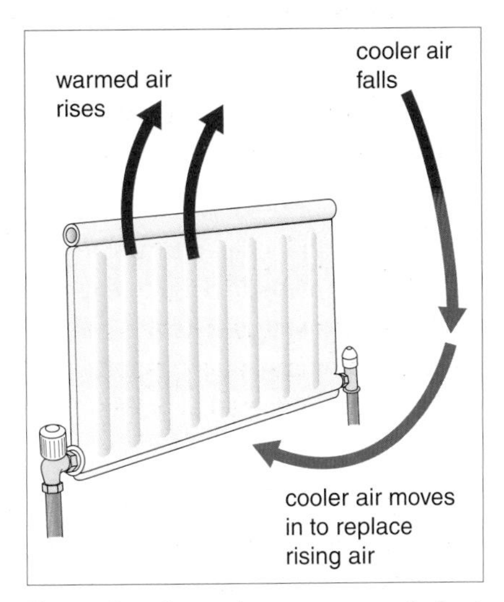

Figure 16.11 Convection currents transfer heat from the radiator to your room.

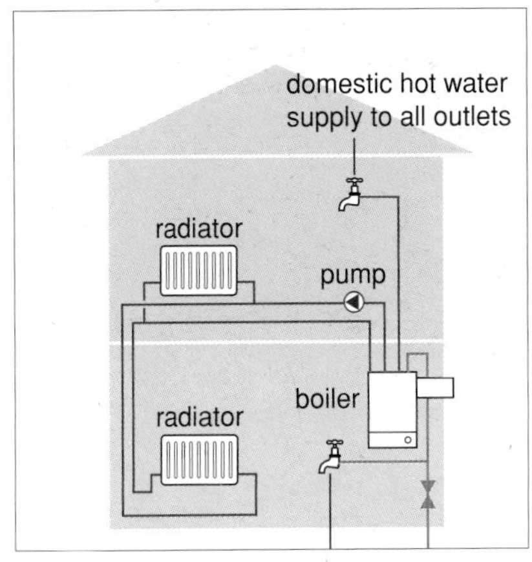

Figure 16.12 A domestic central heating system.

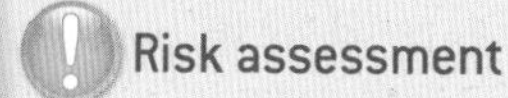

PRACTICAL INVESTIGATING CONVECTION CURRENTS

This activity helps you with:
- ★ using a data logger to monitor temperatures
- ★ plotting graphs using data-logging software or Excel
- ★ determining patterns in data plotted graphically
- ★ using a scientific model to explain experimental patterns.

Apparatus
- * data logger and four thermometer probes
- * access to PC in order to download data from data logger

You can measure the temperature variations within a convection current using a data logger and temperature probes. You will need to have access to a room with a radiator, and you will need to have three or four temperature probes attached to the data logger.

Risk assessment

- **You will be supplied with a risk assessment by your teacher.**

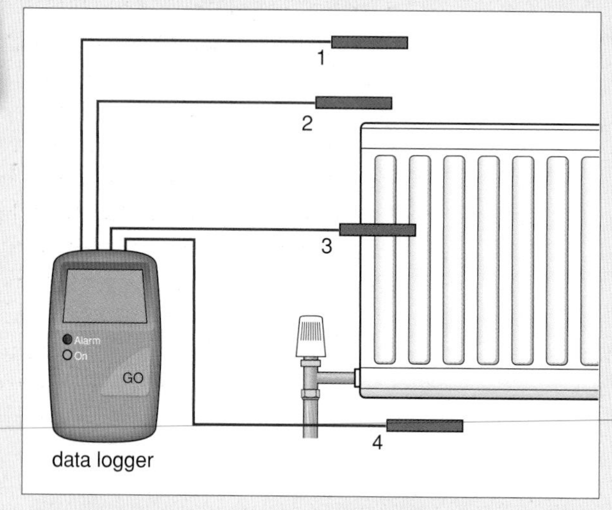

Figure 16.13 Data logger measuring temperatures near a radiator.

Procedure

1. Set the data logger up in a similar way to the diagram in Figure 16.13.
2. Program the data logger to record the temperature of each probe every 10–20 minutes over the course of 24 hours. The 24-hour period must include a proportion of the time where the radiator is on.
3. When the data logger has finished you can download the data to a graph-plotting program, such as the programs supplied with the data logger. Alternatively you can transfer it to Excel.
4. Plot a graph of the temperature of each probe with time over the course of the 24 hour period.

Analysing your results

1. How does the temperature of the probe attached to the radiator vary? This probe has measured the true temperature of the radiator and has mapped the pattern of on/off for the radiator over the 24-hour period.
2. Pick one time when the radiator was on. How did the temperatures measured by the other probes compare with the temperature of the probe attached to the radiator?
3. Explain your answer to Question **2** in terms of a convection current of air.
4. Compare the temperature profiles (over time) of each of the other probes with the probe attached to the radiator.
5. Explain the differences in the patterns that you identified in Question **4**.

How else can I save energy?

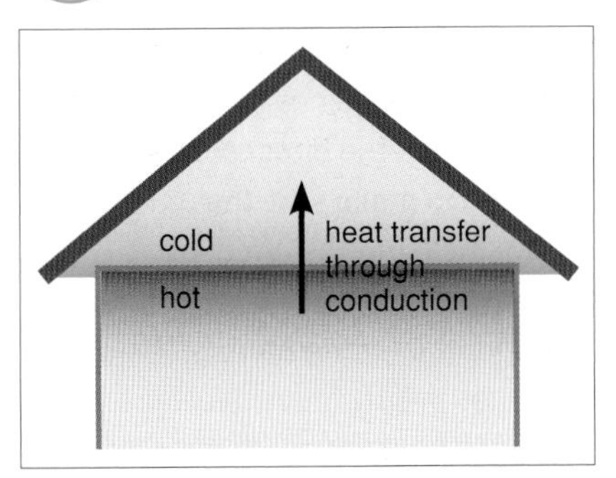

Figure 16.14 Heat transfer in a house with a loft.

Another cheap way to save energy on a home heating bill is to install a thick layer of insulation into the loft. The materials used in loft insulation are poor conductors of heat. Generally the air in rooms in houses tends to be quite warm, and the air at the top of the room tends to be hottest because of convection currents. The air inside of loft spaces tends to be quite cold, and the coldest parts of the loft tend to be near the floor of the loft. This means that there is quite a temperature difference between the room-side and the loft-side of the ceiling. The heat energy will move through the material of the ceiling, from hot (the room) to cold (the loft).

The heat energy moves through the material of the ceiling by vibrating the particles that make up the ceiling board. This process is called **conduction**. Metals are very good conductors of heat, so it would not make good sense to have metal ceilings in houses. Non-metals are good insulators. Heat does not conduct through them very well. The materials used for loft insulation are generally made out of non-metals such as wool, glass or mineral fibres, and the fibres also trap air between them because air is a good insulator. (This is why birds fluff their feathers up in cold weather.) The thicker the layer of insulation, the better the insulation works. Thicker insulation, however, costs more to install.

PRACTICAL INVESTIGATING CONDUCTION

This activity helps you with:
★ investigating a scientific model
★ comparing scientific observations
★ making scientific measurements.

Apparatus
* clamp stand and boss
* metal bar
* match heads
* Vaseline
* Bunsen burner
* heatproof mat

Risk assessment
- **You will be supplied with a risk assessment by your teacher.**

Procedure
1 Set up the apparatus as shown in Figure 16.15.
2 Heat one end of the bar and observe what happens.

The end nearest the Bunsen burner becomes hotter than the rest of the bar. This temperature difference causes heat energy to be passed along the metal bar by conduction. Metals are good conductors of heat. The match heads burst into flame as the heat travels along the bar.

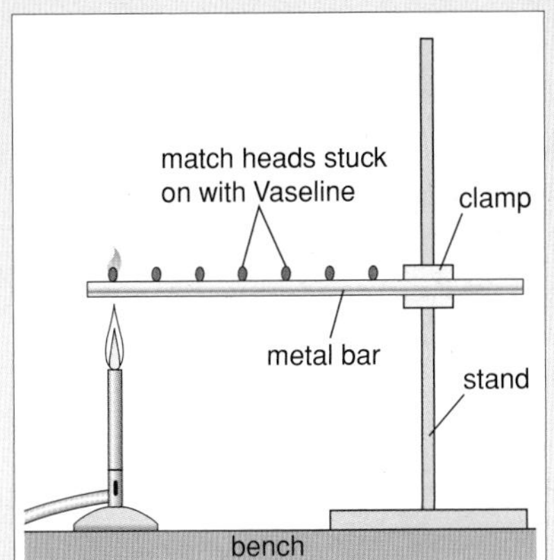

Figure 16.15 The matches burst into flame as the heat travels along the bar.

PRACTICAL INVESTIGATING LOFT INSULATION

This activity helps you with:

★ designing a scientific experiment
★ writing a suitable risk assessment
★ carrying out a scientific investigation
★ collecting and recording scientific measurements
★ analysing the results of a scientific investigation
★ evaluating a scientific investigation.

Does thicker insulation reduce the heat loss? In this investigation you will model a warm room with a container of hot water. Copper calorimeters work best for this, but you could use beakers. To reduce the effect of heat loss through convection and evaporation you need to make sure that each container that you use has a suitable lid. You can measure the effect of insulation by measuring the change in temperature of the hot water in the containers after 10 minutes, or you could measure the temperature of the water every minute for 10 minutes. A simple investigation could involve comparing the temperature change of an un-lagged container with the temperature change of containers lagged with layers of insulation of different thicknesses.

Design and plan an investigation to determine the effect of increasing the thickness of insulation around a container of hot water.

Apparatus

You will need to select suitable apparatus to carry out this experiment and order it in conjunction with your teacher and science technician.

Risk assessment

You will need to produce a suitable risk assessment for this experiment. Your teacher will supply you with a blank risk assessment sheet.

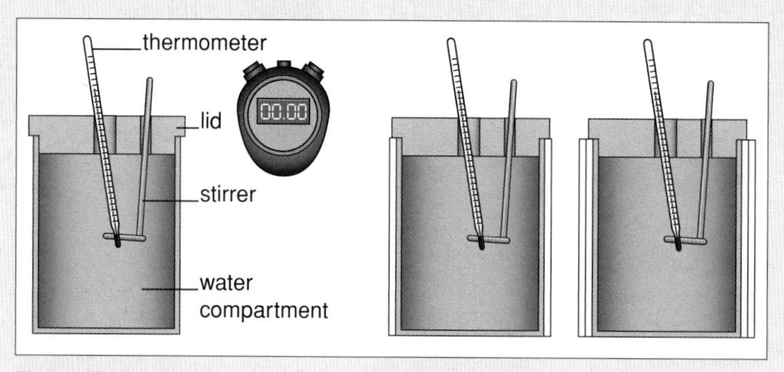

Figure 16.16 Modelling loft insulation using a calorimeter.

Analysing and evaluating your results

How will you present the data from your investigation?

1 Analyse the measurements that you have taken.
2 What is the effect of increasing the thickness of the insulation?

Evaluate your experiment

3 How good are the data that you have taken?
4 How could the data be improved?
5 How well did you do the investigation?
6 What could you do to improve the way that you did the experiment?

Can I save money by using 'free' energy from the Sun?

Figure 16.17 A solar water panel.

You can use solar panels. There are two types:

- Solar photovoltaic panels convert sunlight directly into electricity using solar cells.
- Solar water panels heat cold water as it passes through pipes on the roof that are designed to capture as much of the heat from sunlight as possible.

Figure 16.18 shows how a solar water heating system works.

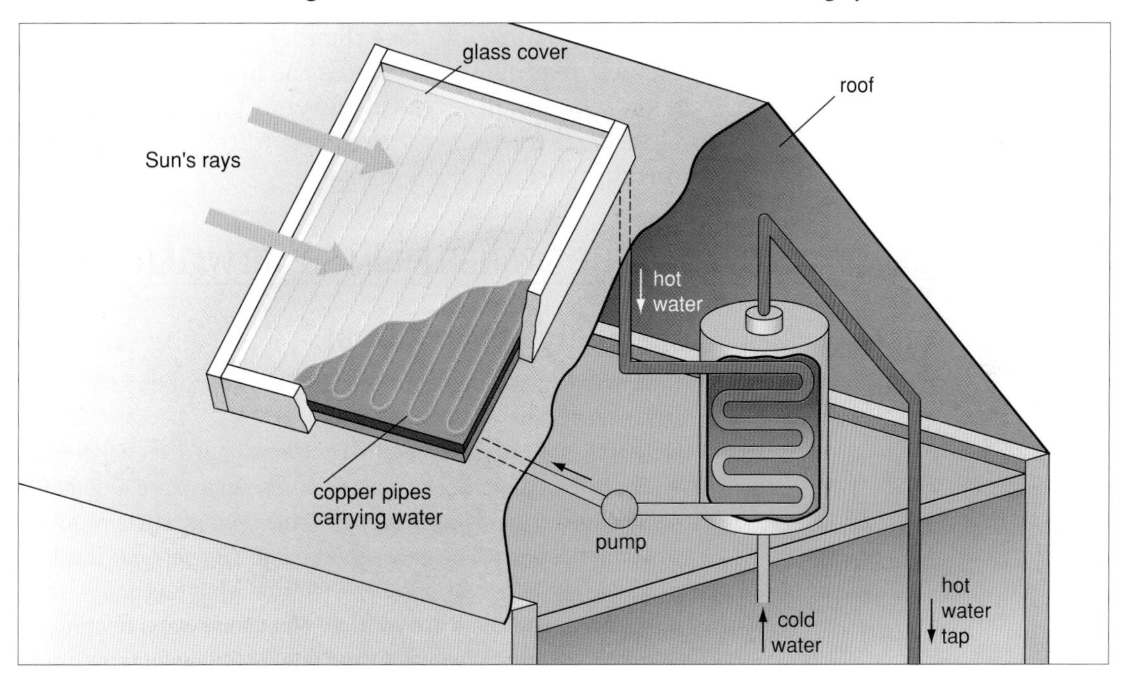

Figure 16.18 Flat plate thermal collector on house roof.

In the UK, solar water panels are normally installed into hot-water systems together with an efficient boiler system. The solar panels are used to pre-heat the water entering the boiler, reducing the amount of electricity, oil or gas needed to heat the water to operating temperature. On dull, overcast, cloudy days and at night, the boiler does all the work and the solar panels are shut off. In countries with very sunny climates, solar water panels are one of the main ways of heating water. Solar photovoltaic panels do not need direct sunlight in order to generate electricity, just normal daylight.

Solar water panels work by capturing **radiation** in the form of **infrared** electromagnetic energy from the Sun. All objects emit infrared radiation but hot objects, like the Sun, emit lots of higher-energy infrared. You will learn about infrared radiation in Chapter 18. As an introduction, your teacher may show you the demonstration shown in Figure 16.19.

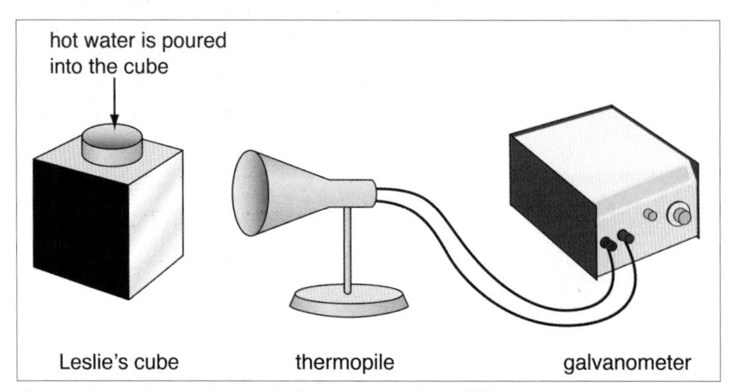

Figure 16.19 Comparing radiation emission from different surfaces.

The thermopile shown in Figure 16.19 produces a small electric current when infrared (heat) radiation falls on it. Each surface of the cube is turned to face the thermopile. The meter shows the greatest reading when it faces the matt (dull) black surface. It shows the smallest reading when it faces the shiny, silvery surface. Dull, black surfaces are the best emitters of radiation and shiny metallic surfaces the worst.

TASK CALCULATING THE SAVINGS FROM SOLAR WATER HEATING

This activity helps you with:
- ★ accessing information from text
- ★ doing simple calculations
- ★ comparing two sets of data.

The Energy Saving Trust

The Energy Saving Trust produces lots of information for homeowners about how to save energy at home.

Here is the information that they publish about solar water heating:

- **Costs** for a typical solar water heating system are around £4800.
- **Savings** are moderate – a solar water heating system can reduce your water heating bill by between £50 and £85 per year. It will also save up to 570 kg of CO_2 emissions, depending on what fuel you will be replacing.
- **Maintenance** costs are very low. Most solar water heating systems come with a 5–10 year warranty and require little maintenance. You should take a look at your panels every year and have them checked more thoroughly by an accredited installer every 3–5 years, or as specified by your installer.

Table 16.1 Savings from solar water heating depending on type of fuel used.

Fuel displaced	Saving per year	CO_2 saving per year (kg)
Gas	£50	250
Electricity	£80	570
Oil	£55	310
Solid	£60	520

All savings are approximate and are based on the hot-water heating requirements of a 3-bed semidetached home with a 3.4 m² panel.

continued...

The Energy Saving Trust also produces information about draught-proofing:

- **Cost:** DIY draught-proofing costs around £100 for materials. Professional draught-proofing costs around £200 for the full service
- **Benefits:** Draught-proofing saves money and makes your home snug and pleasant.
- **Savings:** Full draught-proofing will save on average £25 per year. Blocking gaps around skirting boards and floor boards could save another £20 per year. Draught-free homes are comfortable at lower temperatures so you'll be able to turn down your thermostat. This could save you another £55 per year.

How the savings add up

If every household in the UK used the best possible draught-proofing, every year we would save:

- almost £200 million
- enough CO_2 to fill nearly 225 000 hot air balloons
- enough energy to heat over 260 000 homes.

Questions

1 If a typical solar water system costs £4800 to install, calculate the payback time for each of the fuel systems in Table 16.1.

2 Use this information to calculate how much carbon dioxide you would save during the payback period.

3 Use the information given to compare the financial savings from draught-proofing with the savings from solar water heating.

Discussion Point

Which energy-saving system is best? Is solar water heating worth it?

TASK ADDING IT ALL UP AT HOME

This activity helps you with:

★ analysing scientific information presented graphically and in text format

★ designing a graphic to display your scientific ideas

★ making simple scientific calculations

★ making decisions based on scientific data.

Figure 16.20 on the next page shows a typical house. The numbered labels show the different energy-saving measures that can be carried out to save money and reduce the house's carbon footprint.

Table 16.2 summarises the energy savings from the most common insulation systems.

Table 16.2 Costs and savings for insulation in a modern semidetached house.

Type of insulation	Cost of fitting (£)	Annual saving on fuel bills (£)
Cavity wall insulation	260–380	100–120
250 mm loft insulation where none at present	220–250	140–180
Fit a jacket to hot water tank (DIY)	10+	10–20
Draught-proofing (DIY)	40+	10–20
Floor insulation (DIY)	100+	15–25
Filling gaps between skirting board and floor	25	5–10
Double glazing	3000+	40

1 On a sheet of A3 paper, draw a schematic diagram of your home.

2 Use the information above to audit the energy-saving measures that could be carried out on your house. Indicate these measures on your schematic diagram.

continued...

3 Calculate the total cost to install all the measures. Calculate the total annual savings from all the measures, the payback time and the total carbon dioxide saved per year from all the measures.

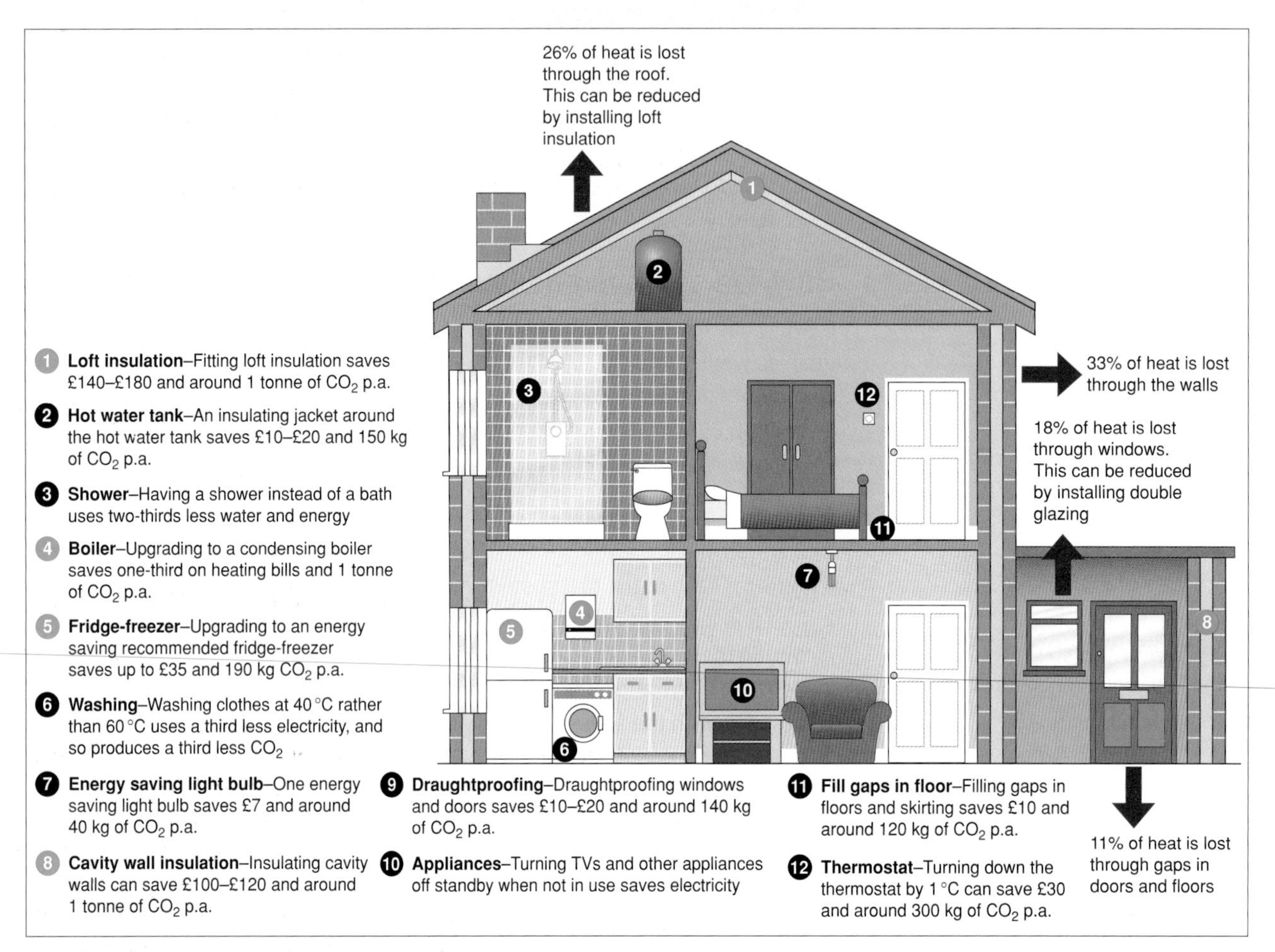

Figure 16.20 How to save energy in your house (the blue circles show that grants and advice are available through the Energy SavingTrust, and the black cicles show no-cost or low-cost measures).

Questions

1 Which is the most cost-efficient energy saving measure to install? Explain how you decided this. Show all your calculations on your schematic diagram.
2 Which energy-saving measure saves the most money per year?
3 Which energy-saving measure saves the most CO2 per year?
4 Which part(s) of a typical house lose the most heat?
5 How much money could be saved per year if all the energy-saving measures in Figure 16.20 were acted on?
6 Why do you think that some energy-saving measures have a grant available from the Energy Saving Trust?
7 Calculate the payback time for each energy-saving measure in Table 16.2.

Discussion Point

If you were building your own house from scratch, what systems would you install to save energy, reduce carbon dioxide emissions and minimise your energy bills?

How much does it all cost to run?

Most people have a whole host of electrical appliances at home. From televisions to cookers, from computers to washing machines, each device or appliance costs money to run, but some devices and appliances are more efficient than others and so cost less.

The run cost of an electrical device depends on how much electrical energy the device uses and the electricity tariff (the cost per unit of electrical energy).

The electrical energy used by an appliance is calculated using the equation:

energy transfer = power × time

or

$$E = P \times t$$

When calculating the cost of domestic electricity, electricity supply companies use units of electrical energy called kilowatt-hours (kWh) or simply units.

Units of electrical energy are calculated using the equation:

units used (kWh) = power (kW) × time (h)

The cost of the electrical energy is then calculated by multiplying the number of units used by the cost per unit:

cost = units used × cost per unit

The cost per unit of electrical energy depends on the electricity supply company and the type of electricity plan that the user is on, but an average UK cost per unit is between 10p and 15p.

Each mains electrical device or appliance has an electrical information plate somewhere on it. Sometimes it is a separate plate screwed or stuck onto the device, but most devices with plastic housings have the electrical information moulded into the plastic somewhere (usually on the back or underneath the device). The electrical information tells the user:

- the **power** of the device in watts or kilowatts
- the **frequency** of the electrical supply in hertz (50 Hz in the EU and 60 Hz in the USA)
- the supply **voltage** in volts (220 V in the EU and 110 V in the USA)
- (sometimes) the **current** drawn by the appliance in amps (usually on the transformer power supplies of computers, games consoles, etc.).

Examples

Q A 100 W lamp is left on for 10 minutes. How much electrical energy is transferred?

A energy transferred (J) = power (W) × time (s)

Put the numbers in and convert the minutes to seconds.

energy transferred (J) = 100 × (10 × 60)

= 60 000 J

= 60 kJ

Q Suppose that you leave a 3 kW heater on in your room. You put it on at 8 a.m. and forget about it until you get home at 4 p.m. If a unit (1 kWh) costs 12.5p, how much will it have cost to leave the heater on?

A number of units used (kWh) = power (kW) × time (h)

Put the numbers in:

number of units = 3 kW × 8 h
= 24

cost = number of units × cost per unit

Put the numbers in:

cost = 24 × 12.5p
= £3.00

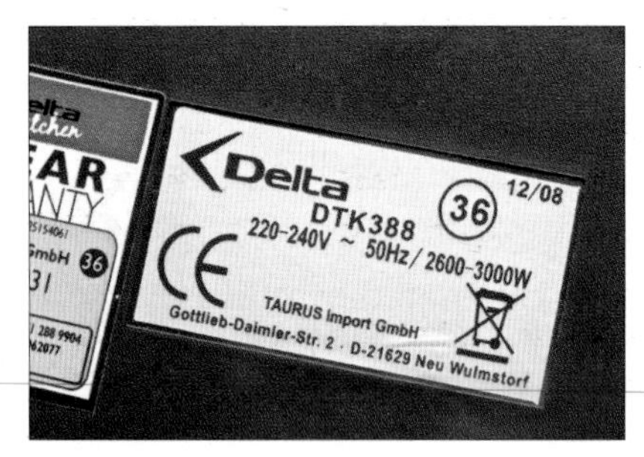

Figure 16.21 This label tells you the rate at which the appliance transfers electrical energy.

QUESTIONS

4 Beth is worried. She left the heater in her room on from 7.00 a.m. until 5.00 p.m. It is a 3 kW heater.

a How many hours was it on for?

b How many units of electricity did it use?

c If the electricity cost 12.5p per unit, how much did her mistake add to the family electricity bill?

5 Which of the following appliances costs most to run per day?

A 4 kW cooker on for 1 hour

B ten 60 W light bulbs on for 4 hours

C a 1 kW washing machine on for 45 minutes

D a 45 W Playstation on for 3 hours

6 It is no longer possible to buy 100 W tungsten filament light bulbs in the UK. The equivalent low-energy light bulb that gives out the same amount of light as a 100 W filament bulb is a 25 W compact fluorescent bulb. If you have four 100 W tungsten filament bulbs in your house that are on for an average of 2 hours each per day, how much do you save per year by replacing them with 25 W compact fluorescent bulbs, if electricity is charged at 12.5p per unit by your electricity supplier?

7 Aled was last out of the house when they went on holiday for a whole week. He left the hall light on, which had a 60 W light bulb in it. How much did that mistake cost? (Assume electricity costs 12.5p per unit.)

8 Aled's dad gets an electricity bill. The current reading on the electricity meter is 34 231 units, and the previous meter reading was 33 571 units.

a How many units have been used?

The electricity company charges 12.5p per unit for the first 250 units used, and 10p per unit after that.

b On the electricity bill, how much have the first 250 units cost in total?

c How much have the remaining units used cost?

d What is the total bill?

TASK HOW MUCH DOES MY ENERGY COST ME?

This activity helps you with:

★ analysing scientific data presented in a table
★ making choices based on scientific data
★ presenting scientific data as pie charts.

Table 16.3 Relative cost of different types of domestic fuel source.

Fuel source	Cost per kWh (p)	CO_2 per kWh (kg)
Domestic electricity	12	0.527
Domestic gas	4	0.185
Domestic coal	4	0.966
Domestic fuel oil	6	0.245

Questions

1 Which is the most expensive way to heat your house?
2 If you had to choose between a gas fire or a coal fire to heat your living room, which one would you choose and why?
3 People who live in houses a long way from the mains gas system often choose domestic fuel oil as their energy source. Why do you think that this is often a good choice?
4 Many older tower blocks use electricity as the main source of heat energy. The flats are frequently fitted with night-storage heaters that only come on at night. They contain large concrete blocks that heat up overnight and then release the heat energy slowly during the day. The electricity tariff at night can be as low as 7p per unit. Why is this a good alternative if gas is unavailable?
5 Draw pie charts showing the relative costs and carbon footprints of the four different types of domestic fuel source.

Could I use renewable technology in my home?

The Energy Saving Trust has an excellent online home energy generation selector that you can use to determine which types of renewable energy would be suitable for your home.

You can access the home energy generation selector at: www.energysavingtrust.org.uk/renewableselector/start/

Check out the short animations that show how each type of renewable energy technology works.

QUESTION

9 Copy and complete the following table. Use the online selector to work out whether the renewable energy technologies would or would not be suitable for your home.

Renewable energy source	Typical installation cost	Reason why it is/is not suitable for my home	Cost saving per year	CO_2 saving per year
Wood-fuelled boiler				
Wood-fuelled stove				
Air-source heat pump				
Solar electricity panel				
Solar water heating				
Ground-source heat pump				
Wind turbine				
Hydroelectricity				

Can we save it? Yes, we can!

What contribution to energy saving can you make? What simple, everyday steps could you take to reduce the world's addiction to energy? If we all made some simple little steps, then the cumulative effect would be massive. We could reduce the number of fossil-fuelled power stations and the subsequent emissions of CO_2. But... it does require effort, and it does require co-operation and it does require good science. Without it...well...just take a look at the planet Venus – a thick atmosphere of carbon dioxide, nitrogen and clouds of sulfuric acid. The daytime temperature is over 450 °C and the pressure is 95 times greater than on Earth.

This is all the result of a huge, runaway greenhouse effect caused by volcanic activity sometime in Venus's past, whereby enormous amounts of carbon dioxide were released from the planet's interior. How do you fancy living on a planet like that?

Figure 16.22 The surface of Venus.

Chapter summary

- Energy transfer can be calculated using the equation:

  energy transfer = power × time; $E = Pt$

- The cost of domestic electricity can be calculated by the equations:

  units used (kWh) = power (kW) × time (h)

  cost = units used × cost per unit

- Different sources of domestic energy can be compared, including:
  - cost comparisons of traditional sources, e.g. electricity, gas, oil and coal
  - cost-effectiveness and environmental implications of introducing alternative energy sources, e.g. domestic solar and wind energy equipment.
- Temperature differences lead to the thermal transfer of energy by conduction, convection and radiation. Density changes in fluids result in natural convection.
- Energy loss from houses can be restricted by various means, such as draught-exclusion, loft insulation and double glazing.

Waves

Surfing the wave – life in the 'Green Room'

Figure 17.1 A surfer in a wave tube.

Figure 17.2 Surfers at Fistral beach, Newquay.

In big surf, as the waves start to break they can curve over the top of a surfer, forming what surfers call a 'tube'. As a surfer moves along the tube it's often referred to as being in the 'Green Room' due to the colour of the sea. This is the ultimate rush for a surfer. It only happens in big surf, and only the best surfers can get into this special place.

Waves at sea are formed by a number of different factors, like the topography (shape) of the shoreline and the underlying seabed, but the most dominant factor is the direction and strength of the wind. The wind far out to sea causes the water to 'peak' and 'trough', forming a **swell**. As the swell moves onshore and breaks it forms surf. The best surfing beaches, like Fistral Beach in Newquay, face into the prevailing wind and swell.

When surfers are assessing the surf at a beach, they are actually doing some basic physics. The height of the surf is a measure of the **amplitude** of the waves. More amplitude means more energy, bigger surf and more fun. The time between each wave is related to the **frequency** of the waves. If the frequency is too high the surf becomes messy, with the wave-fronts interfering with each other. The best surf happens when the frequency is very low and the time between the waves is very long – typically 12 to 18 seconds in the UK. The distance between the waves is called the **wavelength**, which is related to the frequency. High frequency usually means short wavelength and vice versa. In the best surf the distance between the waves can be up to 50 m. The frequency and the wavelength of the waves are always related to the speed of the waves.

How do we describe waves?

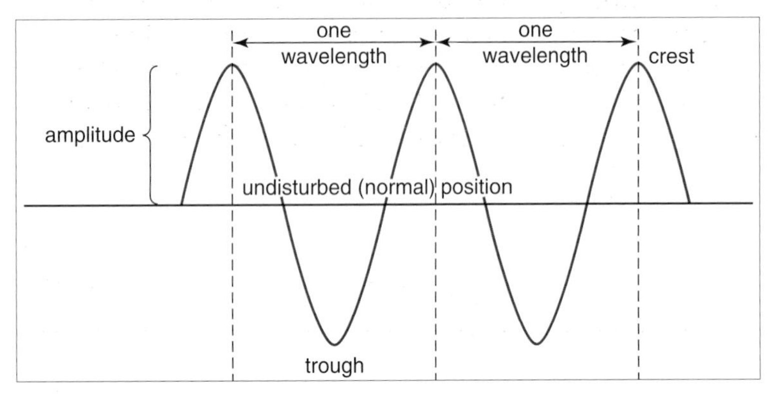

Figure 17.3 Wave measurements.

Figure 17.4 Artists impression of space around a black hole.

The frequency, *f*, of any wave is the number of waves that pass a point in 1 second. Frequency is measured with a unit called hertz (Hz). Waves at sea have a very low frequency, typically 0.1 Hz – which means you only get one every 10 s or so. X-rays and gamma rays have incredibly high frequencies. The X-rays given out by the Black Hole Cygnus X1, for example, have a frequency of about 10^{18} Hz, i.e. 1 000 000 000 000 000 000 of them arrive every second!

Wavelength, λ, is the distance a wave takes to repeat itself over one cycle. It can be measured from one wave crest to the next wave crest or from one trough to the next trough (see Figure 17.3). Because wavelength is a distance it is measured in metres, m. The wavelength of radio waves used to transmit Radio Five Live on AM is 909 m or 693 m, depending on where you live, whereas the wavelength of the gamma rays used to treat a cancerous tumour can be 10^{-12} m, about a hundred times smaller than the radius of one atom!

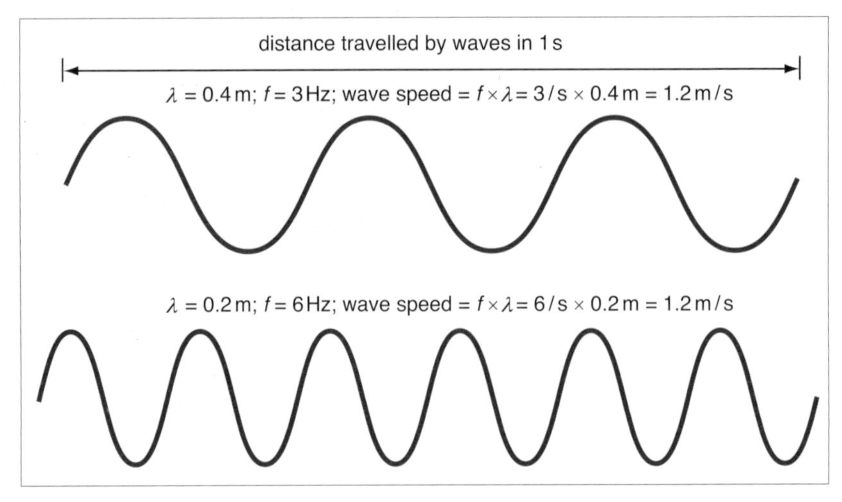

Figure 17.5 These two waves are travelling at the same speed, so the one with a higher frequency has a shorter wavelength.

The amplitude of a wave is a measure of the energy carried by the wave – more energy, more amplitude. The amplitude is measured from the undisturbed (normal) position of the wave to the top of a crest or the bottom of a trough for transverse waves like water waves or the waves in the electromagnetic spectrum. The wave produced during the Boxing Day Tsunami in 2004 had an amplitude of 24 m when it hit the shoreline at Aceh in Indonesia!

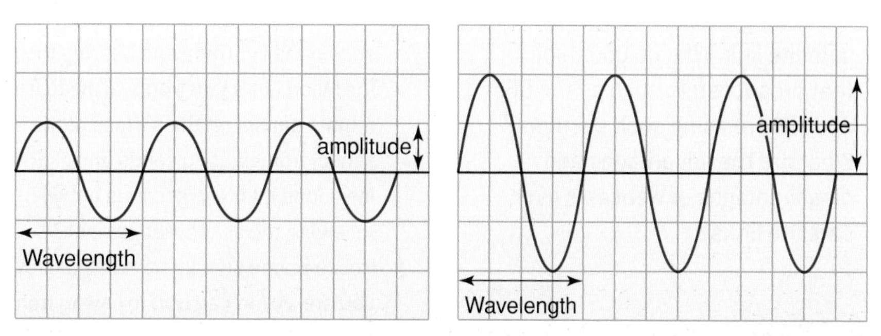

Figure 17.6 These two waves have the same frequency and wavelength but different amplitudes.

How fast can you go on a water wave?

Do all water waves travel at the same speed or do they go faster near the beach than they do out at sea? Do water waves in the laboratory behave any differently?

The speed of a water wave is easy to measure. Like the speed of objects such as cars and runners, wave speed can be determined by measuring the distance travelled in a given time, and then calculated using the speed equation:

$$\text{speed} = \frac{\text{distance}}{\text{time}}$$

Wave speed is a general property of all waves. All electromagnetic spectrum waves travel at exactly the same speed – the speed of light, which is 300 000 000 m/s
(3×10^8 m/s). Sound waves travel at about 330 m/s at sea-level, and ultrasound travels at about 1500 m/s through flesh. The seismic waves produced during an earthquake can travel as fast as 5000 m/s (5 km/s) through hard rock like granite.

Example

Q A surf canoeist takes 12 s to travels 48 m on a wave crest coming onto a beach. What is her speed?

A $$\text{speed} = \frac{\text{distance}}{\text{time}} = \frac{48}{12} = 4\ \text{m/s}$$

Discussion Point

In recent years, news programmes have started to feature 'live' interviews with correspondents in faraway places. More frequently, these live interviews are being carried out via a webcam link rather than using a full, high-quality satellite link. Why do you think that broadcasters such as the BBC and ITN are using such systems? What are the advantages and disadvantages of webcasts over satellite links?

QUESTIONS

1 A water wave takes 20 s to travel 90 m between two buoys. What is the speed of the water wave?
2 An earthquake occurs 16 km (16 000 m) away and it takes 4 s for the first seismic wave to arrive. How fast is the seismic wave travelling?
3 In a thunderstorm, the lightning is seen almost immediately. The thunder, however, travels much more slowly, at 330 m/s. If the time delay between seeing the lightning and hearing the thunder is 6 s, how far away is the storm?
4 The Moon is 384 403 000 m away from the Earth. A radio signal is sent to a remote sensor on the surface of the Moon from a transmitter on Earth. The sensor immediately sends an acknowledgement back to the transmitter. How long does it take in seconds between the transmitter emitting the signal and receiving the acknowledgement?
5 Mobile phone signals travel as microwaves at the speed of light, 3×10^8 m/s. If you are 20 km (20 000 m) away from the nearest phone mast:
   a how long does it take for your signal to get from your handset to the nearest mast
   b what implications does this have for mobile phone communications
   c how do mobile phone companies get around this problem?

Measuring the speed of waves

PRACTICAL SPEED OF WAVES ALONG A SPRING

This activity helps you with:
★ working as part of a team
★ handling apparatus
★ organising your tasks
★ taking and recording measurements
★ using equations to calculate answers
★ plotting graphs
★ looking for patterns in results
★ making and testing predictions

Risk assessment
- **Your teacher will supply you with a suitable risk assessment for this experiment.**

Apparatus
* slinky spring
* stopwatch
* tape measure
* newtonmeter
* calculator

Procedure
1 Work in a group of three or more.
2 Stretch the spring between you and your partner.
3 Give the spring a single flick to one side to check that the wave can reach the fixed end and be reflected back in a reasonable time.
4 Measure the extended length of the spring.
5 On a given signal, start the wave and time it over as long a path as possible.
6 Calculate the wave speed from the following formula:

$$\text{speed (m/s)} = \frac{\text{distance (m)}}{\text{time (s)}}$$

7 It is better to time the wave over several reflected paths. Multiply the distance by the number of times the wave has travelled this distance.
8 Now find out how, if at all, the wave speed changes when you stretch the spring.
9 Decide how you are going to present your results.
10 Plot a graph to find out if there is a relationship between speed and tension (amount of stretch).
11 What do you think would happen to the wave if you put the slinky on a carpet? Make a prediction and, if you have time, test your prediction.

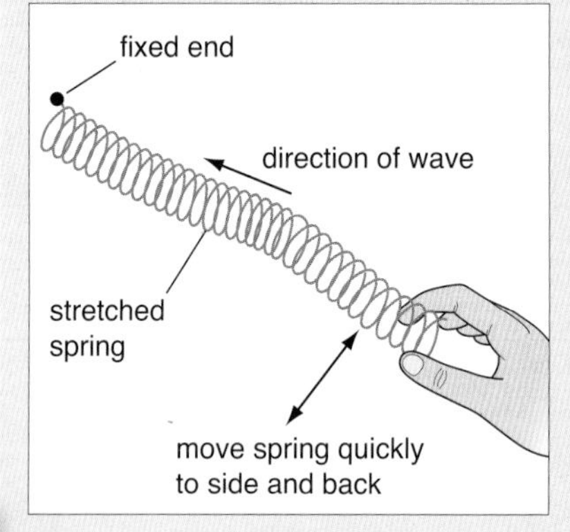

Figure 17.7 Making a wave in a slinky spring.

PRACTICAL

FINDING WHAT AFFECTS THE SPEED OF WAVES IN WATER

This activity helps you with:

* ★ working as part of a team
* ★ planning an investigation
* ★ producing a risk assessment
* ★ handling apparatus
* ★ organising your tasks
* ★ taking and recording measurements
* ★ reaching conclusions
* ★ producing a scientific report.

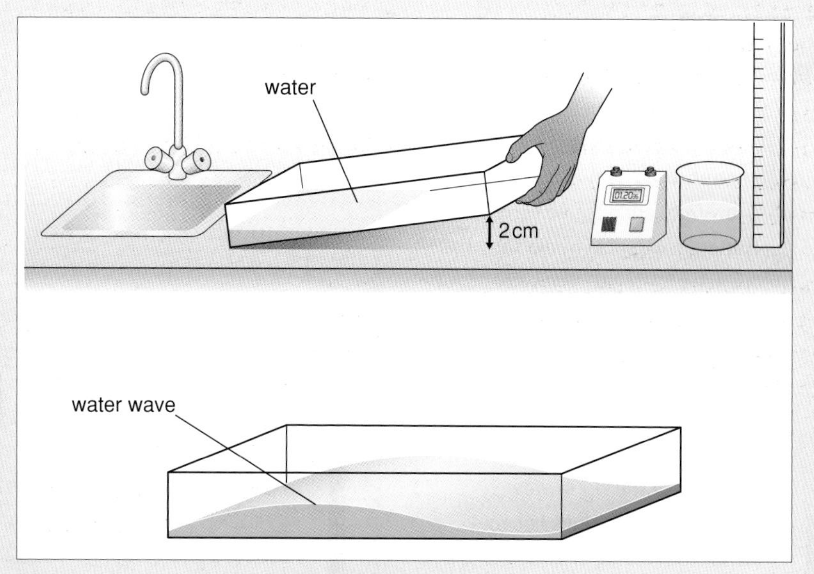

Figure 17.8 Apparatus for measuring waves.

Apparatus

* tray
* stopwatch
* ruler
* beaker

Risk assessment

You will need to produce a risk assessment for this experiment. Your teacher will supply you with a suitable blank risk assessment sheet.

Procedure

1. In your group, think about the results of your slinky wave speed investigation and make a list of things that might affect the speed of waves on water.
2. Decide what apparatus to use.
3. Think about:
   - **a** the sort of container you need to hold the water (the water in your container must not be deeper than 1 cm)
   - **b** how you will make the waves
   - **c** what sort of measurements you will take on the waves
   - **d** how you will take the measurements
   - **e** what apparatus you will need to make the measurements
   - **f** what things you are going to change or keep constant
   - **g** what things you are going to measure as a result of your changes
   - **h** how you are going to record your measurements
   - **i** how you will display and present your findings.
4. Write a report on how you carried out the investigation. What were your conclusions?

The wave equation

A slinky spring can be used to demonstrate waves (see Figure 17.7). If you move the spring quickly from side to side you can set up a wave where the peaks and troughs do not appear to move along the slinky. If you increase the frequency of moving the spring, setting up another wave, then you can clearly see that the peaks and troughs get closer together – there is a direct link between the frequency and the wavelength of the waves.

The wave equation directly links wave speed, frequency and wavelength:

wave speed (m/s) = frequency (Hz) × wavelength (m)

Electromagnetic waves all travel at the same speed – the speed of light, which is 3×10^8 m/s. This special number is given its own symbol, *c*. As wavelength has the symbol λ, and frequency, *f*, then the wave equation (for electromagnetic waves) becomes:

$$c = f \times \lambda$$

Examples

Q A slinky produces waves with a frequency of 2 Hz and a wavelength of 0.75 m. What is the speed of the waves on the slinky?

A wave speed = frequency × wavelength
= 2 × 0.75
= 1.5 m/s

Q A submarine uses sonar with a frequency of 7500 Hz to echolocate objects on the seabed. If the speed of sound in seawater is 1500 m/s, what is the wavelength of the sonar waves?

A wave speed = frequency × wavelength
Rearranged:

$$\text{wavelength} = \frac{\text{wave speed}}{\text{frequency}} = \frac{1500}{7500}$$

wavelength = 0.2 m

QUESTIONS

6 Whales can communicate across vast oceans, often over thousands of kilometres. They do this by generating very low-frequency sound waves at high energy. A typical whale song has a frequency of 3 Hz but a wavelength of 500 m. What is the speed of the whale song in seawater?

7 An oscilloscope measures the frequency of an electrical signal to be 50 Hz with a wavelength of 0.2 m. What is the speed of the signal wave?

8 An oboe makes a musical note with a frequency of 200 Hz and a wavelength of 1.65 m. What is the speed of the sound?

9 The bright red light produced by a laser pointer has a wavelength of 6×10^{-7} m and a frequency of 5×10^{14} Hz. What is the speed of the light?

10 Test-match cricket is transmitted by BBC Radio 4 Longwave with a wavelength of 1500 m. The radio waves travel at the speed of light ($c = 3 \times 10^8$ m/s; 300 000 000 m/s). What is the frequency of Radio 4 Longwave?

11 A surfer is watching the surf on a beach. She counts 20 waves hitting the shore in 5 minutes. She estimates that the waves are travelling at a wave speed of 4.5 m/s. Estimate the wavelength of the waves.

12 Seismic waves can have low frequencies, typically 25 to 40 Hz. The speed of seismic waves in granite is 5000 m/s, but only 3000 m/s in sandstone. During an earthquake in a region of sandstone and granite, what would be the shortest and longest seismic wavelengths recorded?

The electromagnetic spectrum

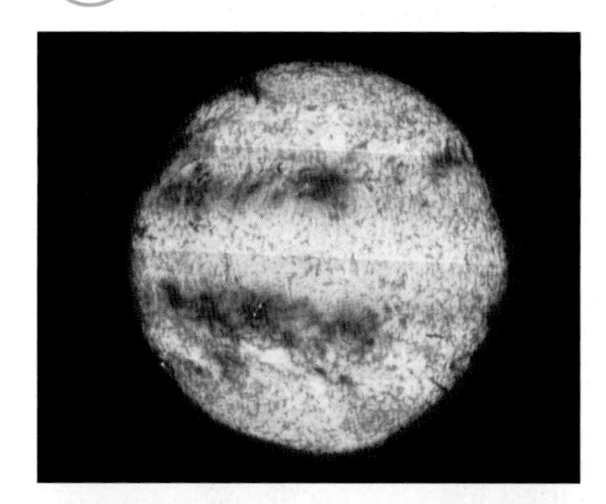

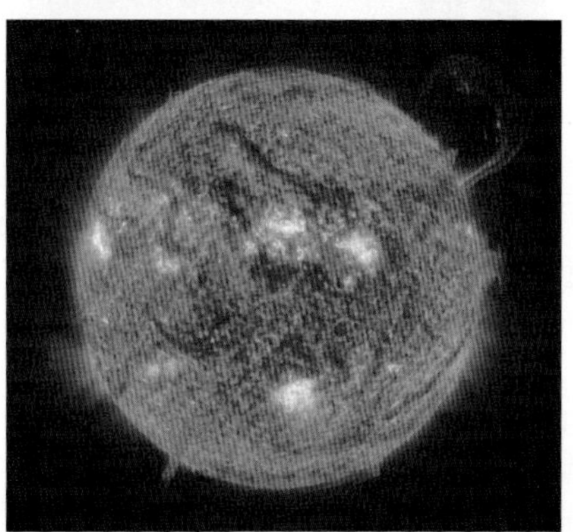

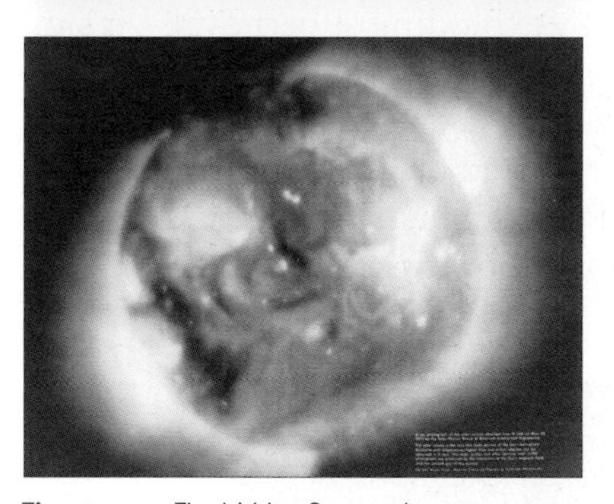

Figure 17.9 The hidden Sun – what we can see and what we can't see!

The photos in Figure 17.9 are all of exactly the same thing – the Sun. They have been taken with a range of different terrestrial and space-based telescopes and cameras, using different parts of the electromagnetic spectrum. We cannot see most of the electromagnetic spectrum – our eyes can only pick up and detect visible light, and our skin can detect some infrared. The photos show the intensities of the different parts of the spectrum in **false colour**. Higher-energy parts tend to be brighter colours than lower-energy parts.

The electromagnetic spectrum is a family of waves with several things in common. All electromagnetic waves:

- travel at the same speed, c, the speed of light (3×10^8 m/s in the vacuum of space)
- transfer energy from place to place
- raise the temperature of the material that absorbs them
- can be reflected and refracted.

The vast reaches of the Universe are continually bathed in all parts of the spectrum. Very hot, super-massive objects like stars, black holes, neutron stars and galactic centres, all produce huge amounts of all the different parts of the spectrum. In fact, the hotter and more energetic the object is, the higher the energy of the electromagnetic waves that can be emitted. Colder lower-energy objects like planets, nebulae (gas clouds) and the background space of the Universe only emit lower-energy electromagnetic waves like radio waves, microwaves and infrared.

The complete electromagnetic spectrum is shown in Figure 17.10.

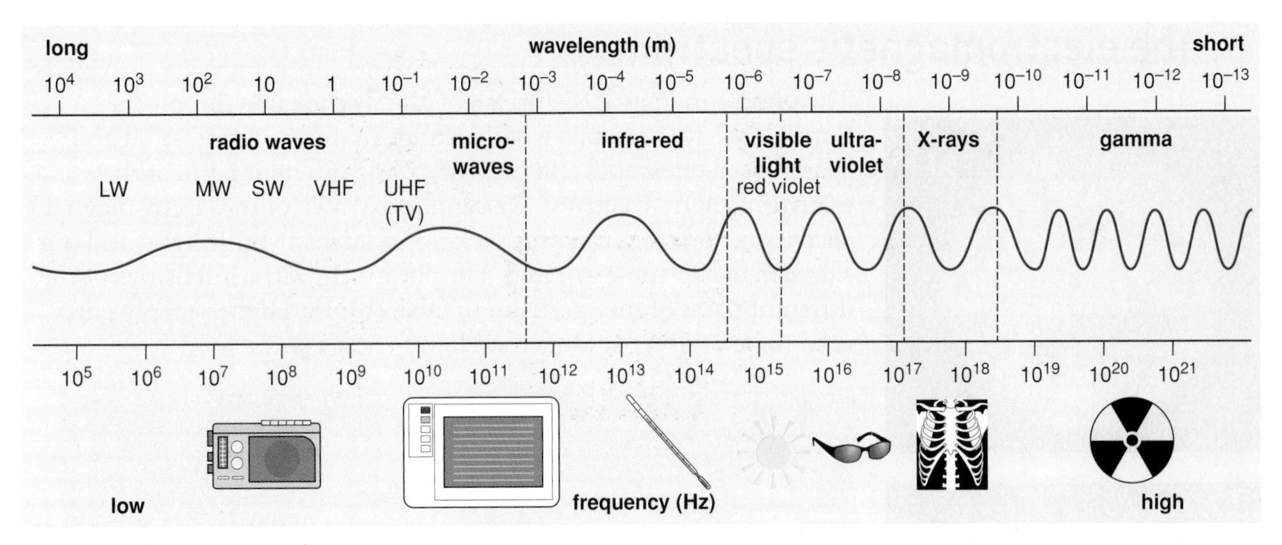

Figure 17.10 The electromagnetic spectrum.

Radio and TV waves

Radio waves have the longest wavelengths, lowest frequencies and lowest energies. They are emitted from a wide range of objects in space. Stars, nebulae (gas clouds), comets, planets and galaxies all emit radio waves. Radio signals from space are particularly useful when astronomers are looking at relatively low-energy and low-temperature objects. They are particularly good for studying the structure of nebulae produced by exploding supernovae – it's in these huge gas clouds that new stars are forming.

Radio waves can be used on Earth to transmit communications signals. TV and radio signals are produced by aerial transmitters and picked up by aerial receivers. They can be relayed across the planet by geostationary satellites, allowing a global communications network.

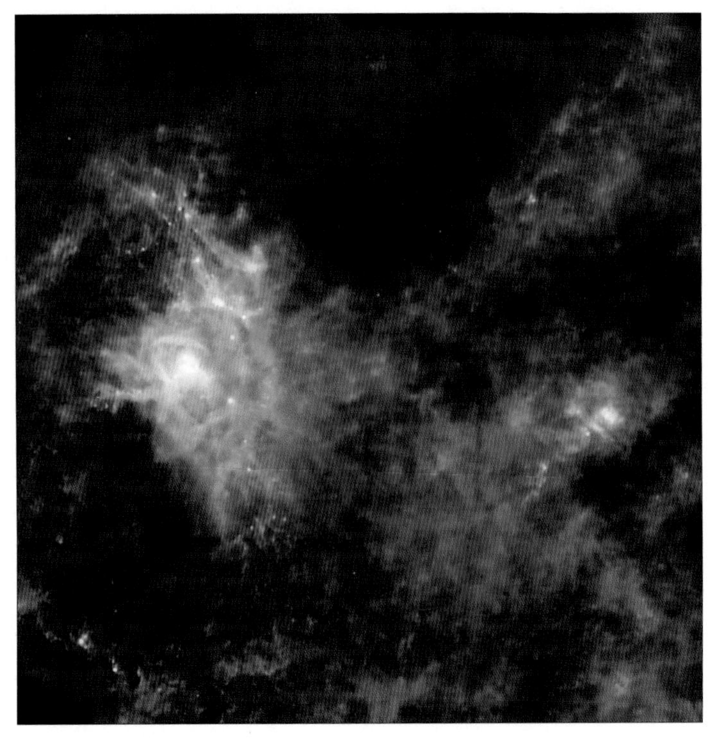

Figure 17.11 The Eagle nebula.

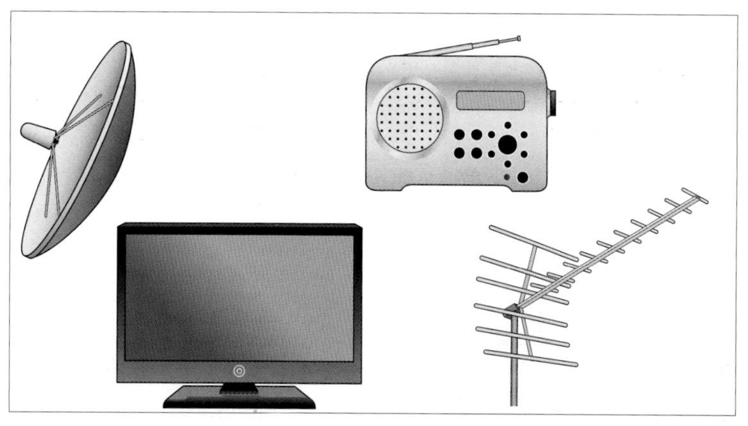

Figure 17.12 These appliances rely on radio and TV signals.

Figure 17.13 NASA's Cosmic Background Explorer searches for microwave radiation from outer space.

Microwaves

The Universe was formed about 13.5 billion years ago, as the result of a huge explosion known as The **Big Bang**. The explosion produced high-energy gamma rays that filled the Universe. Over billions of years, the Universe has expanded and cooled and the gamma rays produced at the time of The Big Bang have also 'cooled' (lost energy). As they have lost energy, the gamma rays have turned progressively into X-rays, then ultraviolet, then visible light, infrared and finally microwaves. When microwave telescopes study the background radiation of the Universe they find huge amounts of microwaves, left over from The Big Bang – the Universe's equivalent of the smoking gun!

Microwaves are also used in microwave ovens (Figure 17.14).

In a microwave cooker the waves come in from the top. They are reflected off the metal sides onto the food to be cooked. The glass door has a metal mesh in it. This stops the waves from escaping, which could be harmful. The frequency is chosen so that the microwaves penetrate the food and energy is transferred to the water molecules in it. As a result the food is cooked quickly and evenly from the inside. Normal cooking heats from the outside and it takes some time for the heat to travel to the centre. Microwaves used in cooking are just short-wavelength (high-frequency) radio waves. So the radio and TV waves used to communicate with satellites are microwaves. Mobile phones use microwaves too, and just like TV transmitters, mobile phone signals need a good line of sight.

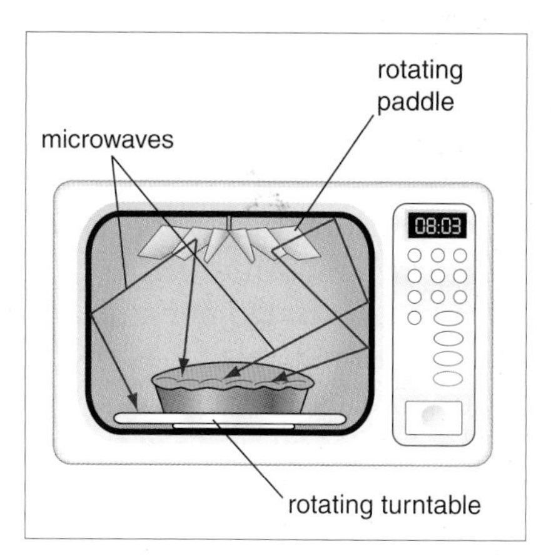

Figure 17.14 A microwave oven.

Figure 17.15 The Herschel Space Observatory.

Infrared waves

Our atmosphere only lets a small amount of infrared through it. Infrared telescopes on Earth are therefore quite limited. In order to get better infrared signals, infrared detectors need to be mounted onto telescopes in low-Earth orbit, to get above the influence of our atmosphere. One example of this is the Herschel Space Observatory.

Infrared detectors need to be cooled to extremely low temperatures and shielded from the infrared radiation produced by the Sun. Infrared can pass through thick dust clouds (nebulae) in space, so infrared telescopes are particularly good at observing star-forming regions and looking into the centre of our galaxy. Cool stars and cold interstellar nebulae, which are invisible in optical light, are also imaged in infrared.

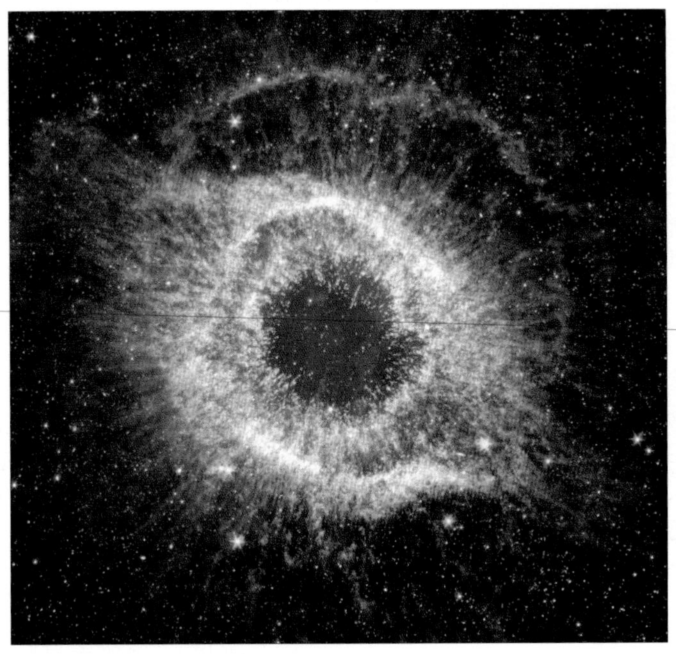

Figure 17.16 Infrared image of the Helix nebula.

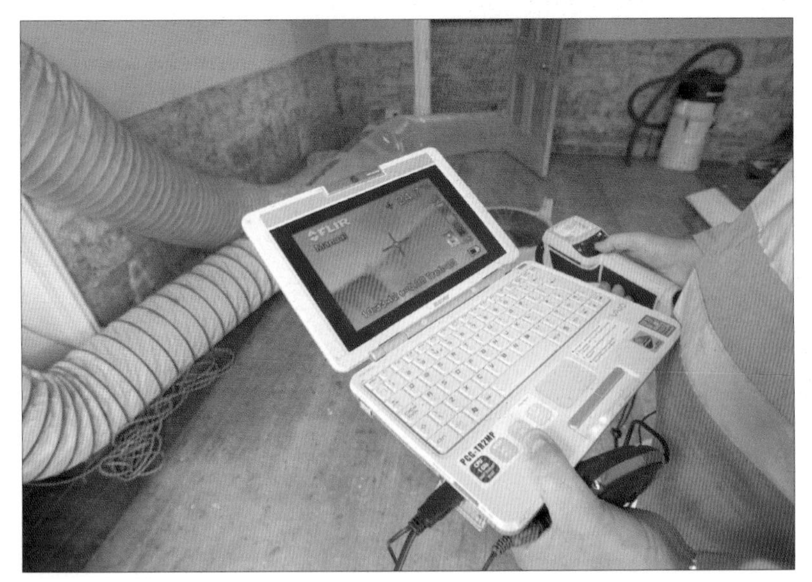

Figure 17.17 Firefighters use infrared cameras to search for people in smoke-filled buildings.

We know and feel infrared radiation as heat radiation, particularly from very hot objects like fires or from the Sun. Everything that is above a temperature of absolute zero (−273 °C) emits infrared radiation. Infrared radiation in itself is not dangerous, so long as you do not get too much of it. If you stand in front of a bonfire for too long, the radiated energy will not only warm you up, it could burn you. Infrared cameras detect heat. They are used by the fire service to find people in smoke-filled buildings and by police helicopters to locate suspects at night. Infrared cameras show which houses are well insulated and which are not.

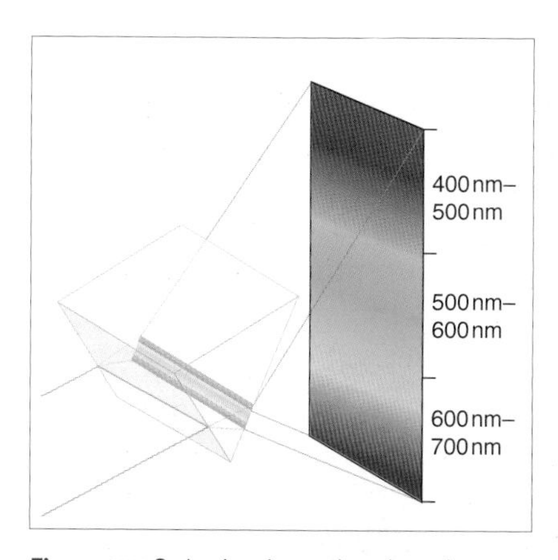

Figure 17.18 A prism is used to show the colours that make up visible light.

Visible light

Our Sun produces vast quantities of visible light from its visible surface called the **photosphere**. The warm, yellow/white light that we see is actually a complete spectrum of colours, first studied scientifically by Isaac Newton in 1704.

Other stars appear to be different colours due to their temperature. The largest, hottest stars are massive blue supergiants like Rigel in the constellation of Orion. In the same constellation is Betelgeuse – a huge red supergiant star.

The Sun is our main source of both light and heat. Its energy keeps us warm and is essential to life. Plants use visible light for photosynthesis to make their own food and oxygen.

It is the only part of the electromagnetic spectrum we can detect with our eyes.

Figure 17.19 The Sun's energy enters our food chain via plants.

Ultraviolet radiation

Ultraviolet (UV) radiation is produced by hot, highly energetic objects such as:

- very massive, bright young stars, like the Pleiades star cluster in the constellation of Taurus
- super-hot white dwarf stars such as Sirius, the Dog Star, in the constellation of Canis Major
- active galaxies like Centaurus A.

Most UV radiation coming from space is absorbed by our atmosphere, so UV astronomers need to put UV telescopes aboard satellites in Earth orbit, such as the Extreme Ultraviolet Explorer (EUVE), which operated from 1992 to 2001.

Figure 17.20 Ultraviolet images of a) the Pleiades cluster and b) the Centaurus A galaxy.

Figure 17.21 The Extreme Ultraviolet Explorer.

At this end of the electromagnetic spectrum the waves become increasingly dangerous. Their wavelengths get shorter and shorter. As their frequencies increase, so does their energy and danger to life. The ultraviolet radiation that does get through our atmosphere damages the skin because the radiation has enough energy to ionise atoms in the skin cells. A sun tan shows that your skin has already been damaged. Sometimes ionising radiation can cause cells to mutate. This can lead to cancer.

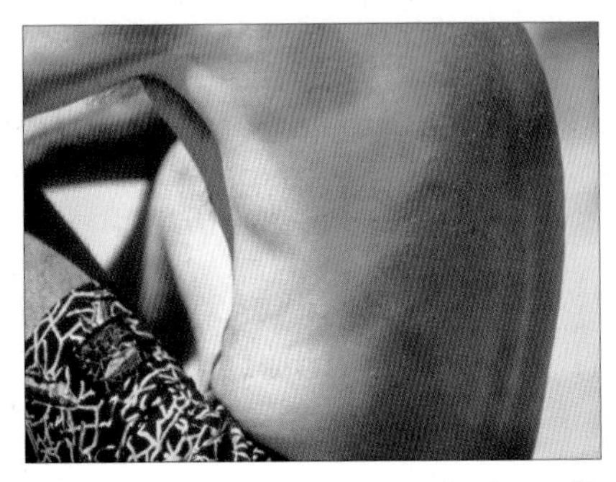

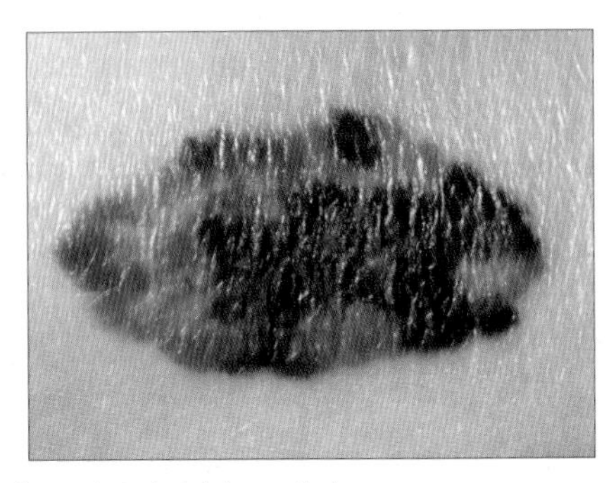

Figure 17.22 Skin cancer is usually triggered by damaging ultraviolet radiation.

X-rays

X-rays are produced by the most energetic and hottest objects in the Universe. Black Holes, neutron stars and the huge explosions of dying super-massive stars (supernovae) are all emitters of X-rays. The X-rays are often produced by material moving at extremely high speed. Black Holes produce lots of X-rays as the matter surrounding the Black Hole is sucked inwards by the huge force of gravity. As the matter accelerates into the Black Hole it emits high-energy X-rays in a beam that is often used to detect the presence of the Black Hole. X-rays are absorbed by our atmosphere, so any X-ray astronomy needs to be carried out on-board orbiting space telescopes such as the Chandra X-ray Observatory, launched by the space shuttle Columbia in 1999.

Figure 17.23 The Chandra X-ray Observatory in orbit.

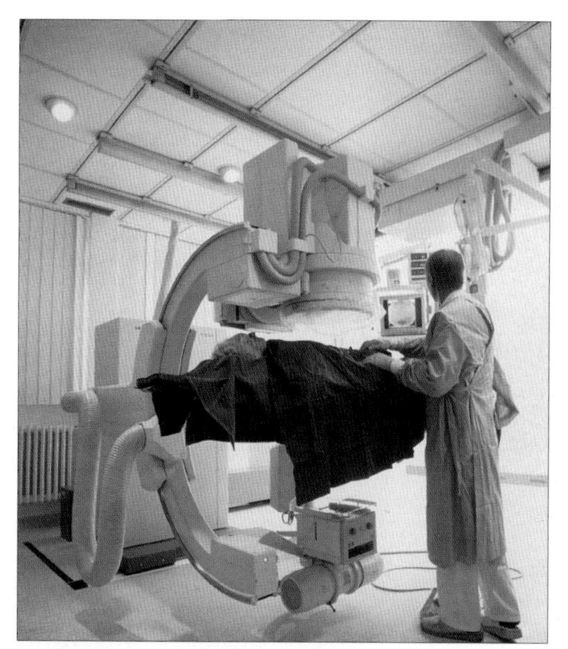

Figure 17.24 An angiography machine uses X-rays to diagnose heart conditions in patients.

X-rays are also ionising (like UV) and overexposure to them can cause cancer. However, they are widely used in medicine where the benefits of the X-ray greatly outweigh the dangers from it (Figure 17.24). They can be used in carefully controlled conditions to cure cancers. Very powerful X-rays are also used to detect flaws and fractures in metals.

Gamma rays

On Thursday 23 April 2009, the Swift Gamma-Ray Burst Telescope, in orbit around the Earth, detected the most distant object every observed. Gamma-Ray Burst (GRB) 090423, a 10-second burst of high-energy gamma rays, was imaged and confirmed by other telescopes to be over 13 billion light years away. In fact, the explosion that produced this burst of gamma rays occurred only 600 million years after the Big Bang – about 5% of the age of the Universe! Astronomers think that GRB090423 was a huge star exploding as a supernova, producing a super-massive Black Hole.

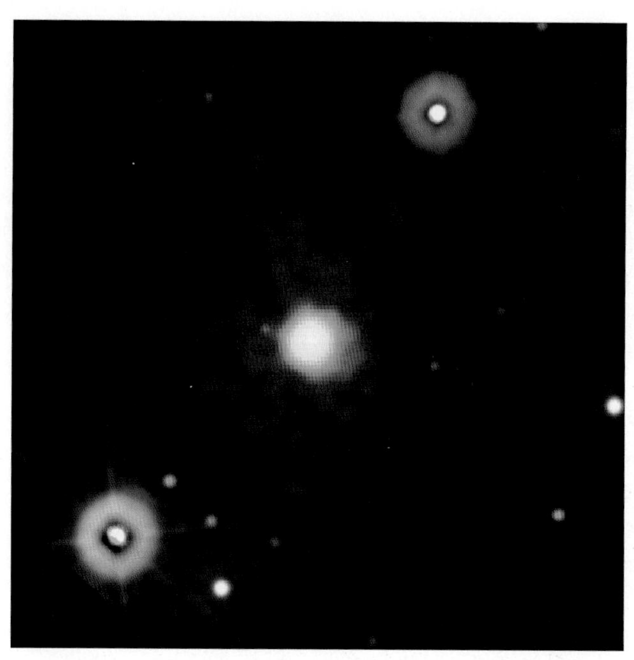

Figure 17.25 Artist's impression of GRB090423.

Figure 17.26 Artist's impression of the Swift GR telescope.

Find out more about the Swift Gamma Ray Mission at: http://swift.gsfc.nasa.gov/docs/swift/swiftsc.html

Gamma radiation also comes from the nuclei of radioactive materials such as uranium. Gamma rays, like X-rays and UV, are ionising and so are very dangerous to all living things. They can cause cancers or kill cells directly. Like X-rays they are used to detect flaws in metals. They can also be used to image and treat cancer, to sterilise medical instruments and to check people and vehicles at ports for illegal imports of radioactive materials.

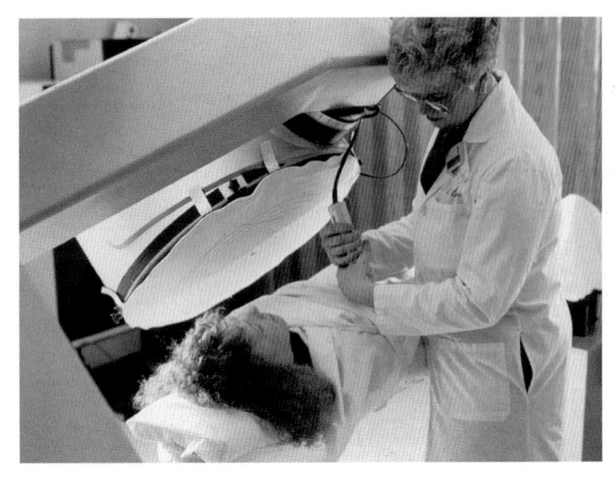

Figure 17.27 A medical gamma ray camera.

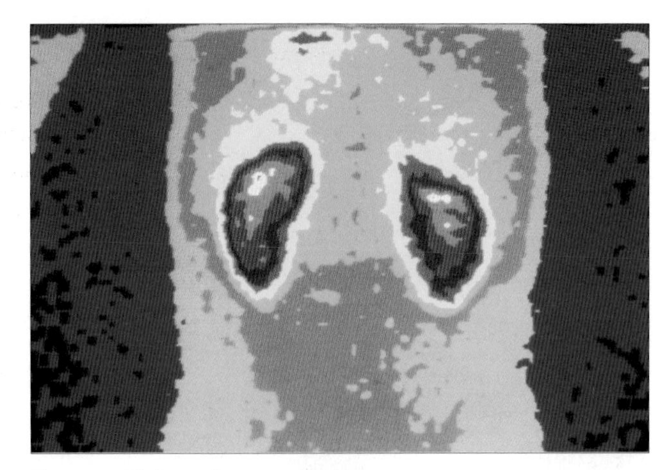

Figure 17.28 Image from a medical gamma ray camera.

QUESTIONS

13 Which part of the electromagnetic spectrum has:

  a the longest wavelength
  b the highest frequency
  c the least energy?

14 Which parts of the electromagnetic spectrum are missing from this list?
gamma UV visible radio

15 State and explain which parts of the electromagnetic spectrum are used in hospitals.

16 The Sun emits all parts of the electromagnetic spectrum. What does this tell you about the temperature of the Sun?

17 Why are some telescopes put into orbit?

18 What information does the electromagnetic spectrum give us about astronomical objects?

Discussion Point

Why are astronomers one of the few groups of scientists that use the complete electromagnetic spectrum?

TASK PRESENTING INFORMATION

This activity helps you with:

★ researching information
★ selecting information
★ presenting written and diagrammatic information
★ referencing information
★ talking about your work.

Scientists are often asked to make presentations at conferences by creating a poster. The posters are then displayed in a hall and the scientists stand in front of them and talk about them to anyone who is interested. Often, poster presentations are made by several scientists who have worked together. Your task is to make a poster presentation about *one* part of the electromagnetic spectrum. You can use the information in this book and information that you have researched on the internet. If you use information from the internet, make sure that you fully reference it. Other students in your class will make similar posters about the other parts of the spectrum. Your poster must include:

- a graphic showing the position of your part of the spectrum in relation to the other parts
- data about wavelength, speed, frequency and energy
- examples of astronomical images taken with your part of the spectrum, together with some brief text about the astronomical object that produced your image
- a picture of the telescope(s) that took your image(s)
- pictures showing a non-astronomical use of your part of the spectrum and some brief text about the use.

Your poster needs to be colourful, clear and easy to read. You may be asked to display your poster and to answer questions about it.

Chapter summary

- Waves can be distinguished in terms of their wavelength, frequency, speed, amplitude (and energy).
- The equations associated with waves are:

$$\text{wave speed} = \text{wavelength} \times \text{frequency};\ c = f \times \lambda$$

$$\text{speed} = \frac{\text{distance}}{\text{time}}$$

- Waves travel at different speeds through different materials.
- The electromagnetic spectrum is a family of waves consisting of radio waves, microwaves, infrared, visible light, ultraviolet radiation, X-rays and gamma rays.
- The electromagnetic spectrum is a continuous spectrum of waves of different wavelengths and frequencies but all the waves travel at the same speed in a vacuum – the speed of light.

Infrared and microwaves

Search and rescue – how we find people when we can't see them!

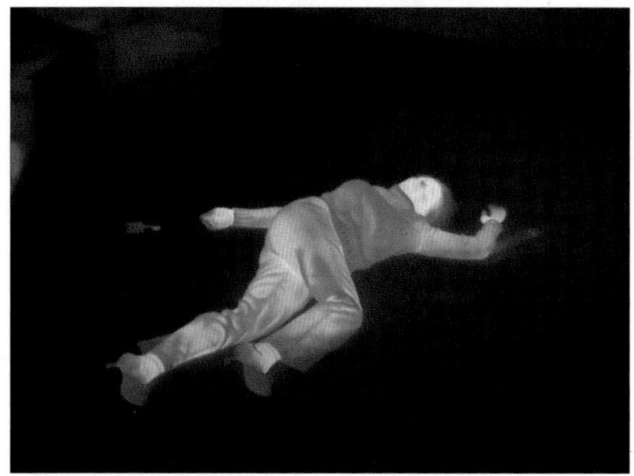

Figure 18.1 Two views of the same thing – a body in a smoke-filled room. The photo on the left is taken with a normal camera and the photo on the right is taken with a special thermal-imaging camera.

Without thermal imaging technology, many people would be dead – the fire and rescue services would simply have missed them in smoke-filled rooms! Thermal imaging also allows 'night-vision' goggles to work. These are often used by the Armed Forces on night-operations. Soldiers and pilots frequently use night-vision goggles to co-ordinate movements of troops and helicopters.

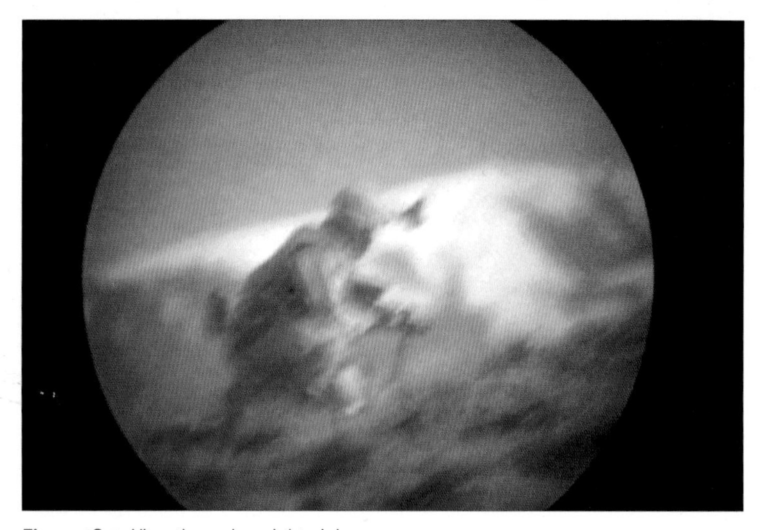

Figure 18.2 View through a night-vision camera.

So what is actually being detected by thermal imaging and night-vision cameras? It's actually infrared radiation. All objects give out some form of electromagnetic radiation. As we have seen in Chapter 17, even the coldest parts of the Universe emit microwaves. Figure 18.3 shows the intensity of electromagnetic radiation emitted by objects at different temperatures, called 'black body radiation'.

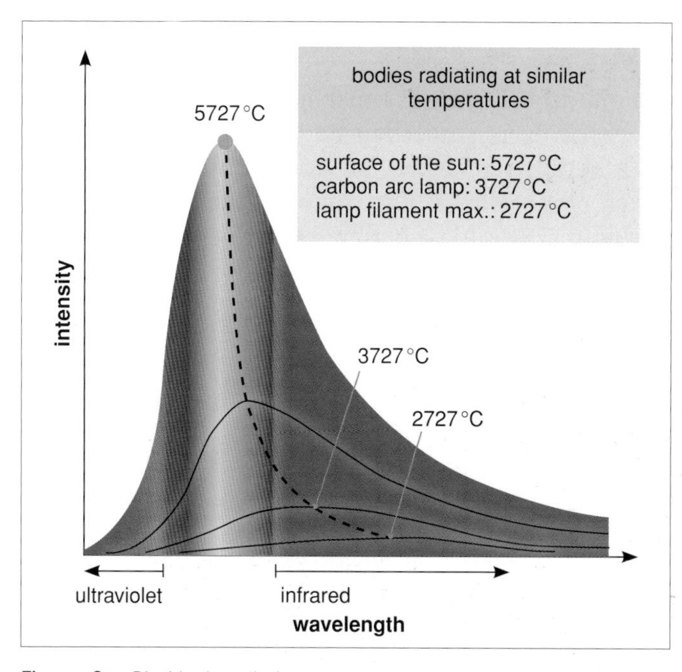

Figure 18.3 Blackbody radiation curves.

In general, the hotter the object, the higher the intensity and the lower the wavelength of the radiation it emits. A human body, with a surface temperature of about 20 °C, will produce very low-intensity infrared radiation. This is invisible to the naked eye, but a thermal imaging camera can easily detect it. It can also distinguish between the infrared produced by different parts of the body at different temperatures – showing the different wavelengths as different 'false colours' on a screen – either different shades of grey or different colours. Figure 18.4 shows a human body imaged in this way.

In addition to the many search and rescue, military and police applications, thermal imaging is also useful in medicine. Infections and the body's response to injury frequently cause localised increases in body temperature that can easily be detected and pinpointed by thermal imaging. This type of technology is now being introduced at ports and airports to detect people with infections and diseases, particularly following epidemics of diseases such as bird flu and swine flu.

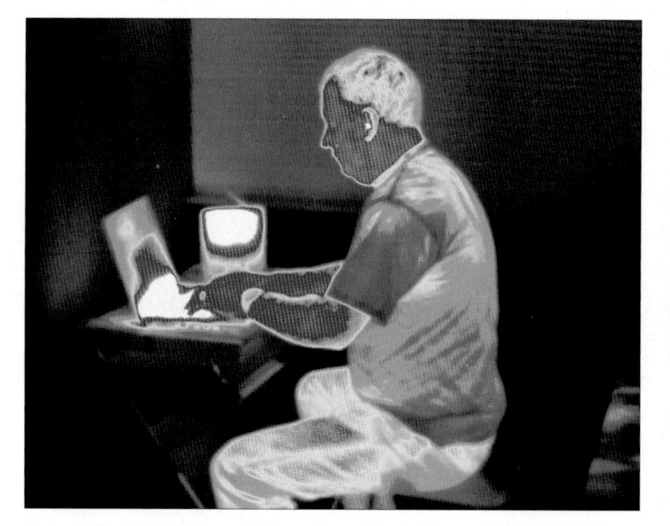

Figure 18.4 False-colour picture of a full-body thermal imaging scan.

Figure 18.5 Thermal imaging body scanner at an international airport.

Discussion Point

People who are unwell generally have higher body temperatures than healthy people. Potential terrorists are usually 'hotter' than the general public, due to their high state of anxiety. Why do you think that thermal imaging technology is being used at major airports? What would be the potential problems in using such systems?

QUESTIONS

1. Why is thermal imaging useful for the fire and rescue services?
2. What are 'night-vision' goggles?
3. How might the pilot of a search-and-rescue helicopter use 'night-vision' goggles to find someone lost at sea?
4. How can thermal imaging cameras pick out the clothes that someone is wearing?
5. What is the difference between the electromagnetic radiation emitted by a tungsten filament lamp and the Sun?

Marathon runners and infrared

Figure 18.6 Why do marathon runners wear aluminium foil when they finish a race?

When marathon runners finish a race they are hot! As they stop exercising, their bodies cool rapidly. To prevent hypothermia, they are covered with a thin 'blanket' of metallised plastic film. Why? Surely a thick fleecy blanket would be better? How does the shiny plastic blanket work?

PRACTICAL MODELLING THE THERMAL RADIATION FROM MARATHON RUNNERS

This activity helps you with:

- ★ working safely
- ★ working as part of a team
- ★ handling apparatus
- ★ using ICT with your scientific work
- ★ organising your tasks
- ★ taking and recording measurements
- ★ plotting graphs
- ★ looking for patterns in results
- ★ reaching conclusions
- ★ understanding uncertainties in your work.

In this experiment you will measure the heat lost by hot water when it is placed into different types of container. The water models a human body and the containers model different clothes worn.

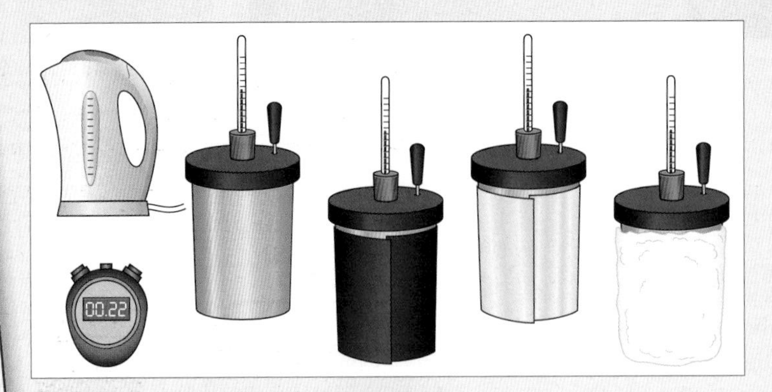

Figure 18.7 Apparatus for the insulation experiment.

Risk assessment

- **Take care with hot water.**
- **You will be supplied with a risk assessment by your teacher.**

continued...

PRACTICAL contd.

Apparatus

* a kettle filled with hot water
* 4 copper calorimeters
* black paper
* aluminium baking foil
* blanket material
* plastic drinking cup lids
* 4 thermometers *or* data logger and 4 temperature probes
* heatproof mat
* stopwatch
* sticky tape
* optional: measuring cylinder

Procedure

Work in a group of two or three students.

1 Fill the kettle with water and allow it to boil.
2 Lag the four calorimeters as follows: **A** black paper; **B** aluminium baking foil; **C** blanket material; **D** no lagging.
3 Fill each calorimeter with same amount of hot water. Try to do this quickly to avoid the calorimeters starting at different temperatures.
4 Put a plastic lid on each calorimeter and put a thermometer through the hole in the lid.
5 Measure and record the initial temperature of each calorimeter and start the stopwatch.
6 After 1 minute, record the temperature of each calorimeter again.
7 Repeat the measurements of temperature every minute for about 10 minutes.
8 Finish the measurements and pour away the hot water when it has cooled sufficiently to handle safely.
9 Plot a graph showing the cooling curve of each calorimeter with time. You do not need to start the temperature scale at zero. Label each calorimeter's curve.

Analysing your results

1 Describe the shape of the cooling curves.
2 Calculate the temperature drop for each calorimeter.
3 Which calorimeter cooled down the most?
4 Which type of lagging insulated the best?
5 Why do you think it was important to keep the starting temperatures the same?
6 Why do you think it was important to fill each calorimeter with the same amount of water?
7 What was the purpose of the non-lagged calorimeter?
8 How could you make the results of this experiment more reliable?
9 How could you improve the procedure?
10 How would the graph change if you plotted the temperature scale starting at 0 °C?

Extension

Your teacher may show you this experiment working in reverse. If cold water is put into the calorimeters and the calorimeters are placed at equal distances from a radiant heater, which one will heat up the quickest?

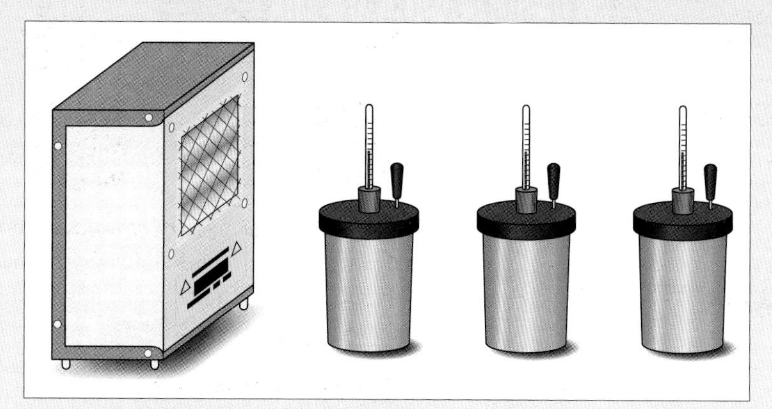

Figure 18.8 Apparatus for the insulation experiment in reverse.

continued...

PRACTICAL *contd.*

Explaining the patterns in the results

Shiny, light-coloured objects *reflect* infrared (thermal) energy really well and are *poor emitters* of thermal energy.

Dull, dark-coloured objects *absorb* infrared (thermal) energy really well and are *good emitters* of thermal energy.

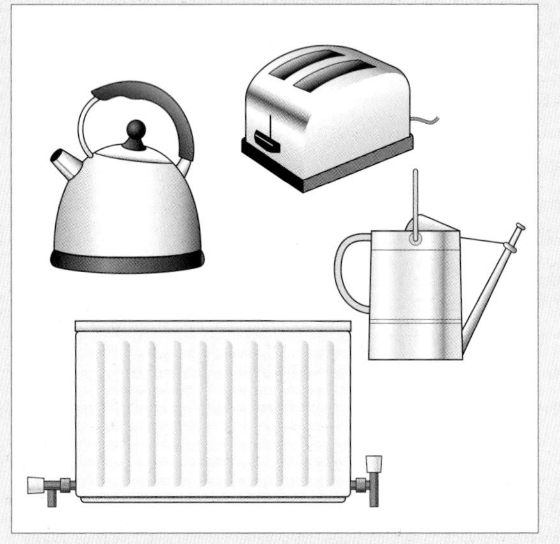

Figure 18.9 Objects that are good reflectors of thermal energy.

Figure 18.10 Objects that are good absorbers of thermal energy.

11 Use these statements to explain the results of your experiment and your teacher's demonstration.

QUESTIONS

6 Why do people who live in hot countries tend to wear light-coloured clothing?

7 Explain why solar panels are always manufactured with a dull black surface?

8 Why do shiny silver teapots keep tea warmer than dull black ones?

9 Why are the sensitive electronics of an orbiting satellite often clad in a shiny metal foil?

10 Why are marathon runners given a shiny space-blanket when they finish a race?

TASK

DESIGN A SOLAR STILL

This activity helps you with:
- ★ researching information
- ★ selecting and referencing researched information
- ★ presenting ideas graphically.

Life rafts on ships are often equipped with a device called a 'solar still' to provide drinking water. This piece of equipment uses the thermal (infrared) energy from the Sun to evaporate seawater, separating the water from the salt. These devices are often inflatable and can be towed behind the life raft. As the sun shines, water evaporates and is collected by the device.

Your task is to use your knowledge of how different surfaces reflect, absorb and emit infrared radiation to design an inflatable solar still.

Produce a cross-sectional diagram showing the design, including notes about which materials would be best for each part.

You could get some ideas for your design by using the keywords 'solar still' in an internet search engine. Don't forget to reference any information that you use.

Using infrared and microwaves

Figure 18.11 Real-time conversation across the globe.

How is it possible to have a real-time conversation with someone in Australia?

The two people in Figure 18.11 are using a enormous amount of technology to make a simple phone call. Their mobile phones are connecting to a vast network of microwave and optical fibre links that are transmitting the phone signals across the globe and back at the speed of light.

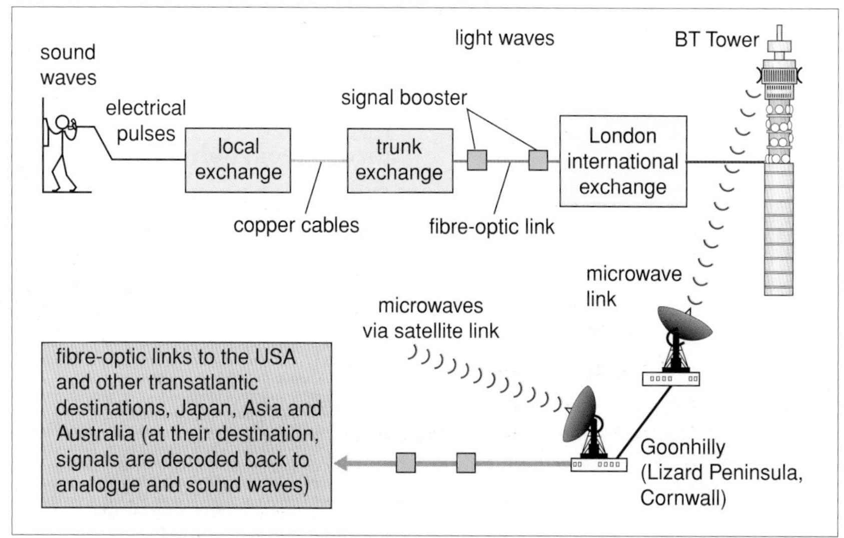

Figure 18.12 The path of an international phone call as it leaves the country.

Normal, land-line telephone calls travel down optical fibres, using infrared radiation. Optical fibres are far better at transferring information than the copper wires that used to carry long-distance phone calls. A single fibre can carry over 1.5 million telephone conversations, compared with 1000 conversations for copper wires. Most national telephone calls, faxes, internet calls, etc. pass along optical-fibre lines. The fibres can carry ten television channels (cable TV). An intercontinental optical cable carries many fibres so an enormous amount of information can be transferred. It is very cost-effective.

For long-distance telephone calls, the electrical signals are converted to digital (on/off) pulses at the exchange. The digital signal is then converted to light pulses by a laser. The infrared laser flashes at high speed. Infrared light is used because it passes through the glass optical fibres better than visible light. Repeaters boost the signal at 30 km intervals along the fibre. At the far end, another decoder converts the digital signal from the laser into a changing voltage that is then converted into sound at the telephone earpiece.

There are other advantages of optical fibres over copper wire:

- Fibre optic lines use less energy.
- They need fewer boosters.
- There is no cross-talk (interference) with adjoining cables.
- They are difficult to bug.
- Their weight is lower so they are easier to install.

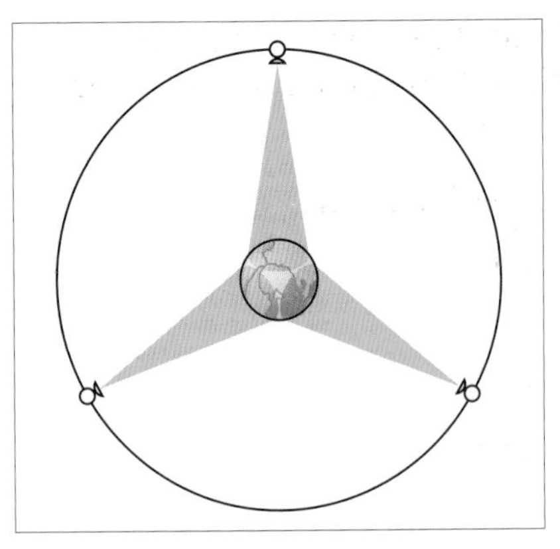

Figure 18.13 The Earth from above the North Pole; three geostationary satellites can send signals to most of the Earth.

Mobile phones use a different part of the electromagnetic spectrum – microwaves. Microwaves are wireless signals – you don't need a copper cable or an optical fibre. One disadvantage of using microwaves is that there must be a clear path between the transmitter and your television aerial or mobile phone. To cover the largest area, television and mobile phone transmitters and receivers are tall and sited on hills. The curvature of the Earth means that repeater stations have to relay the microwave signal to distant transmitters. Satellites must be used for long-distance communications around the world. Theoretically, only three satellites are needed to transmit signals around the world. In practice, more are used.

The satellites are placed in orbit at a height of 36 000 km. They orbit the Earth exactly in time with the Earth's rotation. This is called a **geosynchronous** (**geostationary**) orbit.

Here in the UK, TV, phone, fax and data signals are sent to satellites from one of three BT stations. The Madley Communications Centre near Hereford is the largest Earth station in the world, and most of the UK's satellite communications pass through it.

Figure 18.14 Satellite dishes at the Madley Communications Centre.

Optical fibres or microwaves?

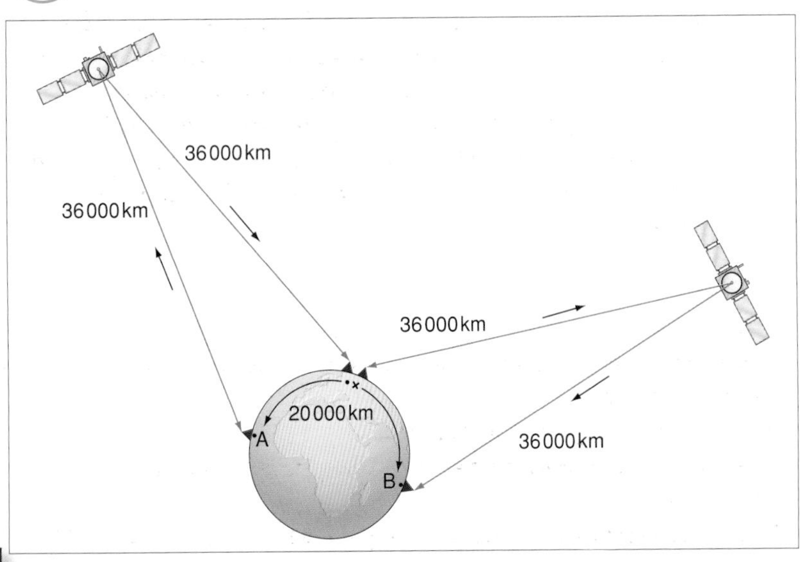

Figure 18.15 The satellite signal has much further to travel.

Both optical fibres (using the infrared part of the electromagnetic spectrum) and satellite communications (using microwaves) are used for international phone calls and TV broadcasts. It takes time for the signals to travel from an Earth station up to one of the satellites and back again (Figure 18.15). Let's compare the time delay in sending a signal from A to B.

The satellites orbit at a height of 36 000 km. Therefore the path length is 4 × 36 000 km, or 144 000 km. This is for studio-to-studio via satellite. Use the following formula:

$$\text{speed (km/s)} = \frac{\text{distance travelled (km)}}{\text{time taken (s)}}$$

Rearranging:

$$\text{time for travel} = \frac{\text{distance}}{\text{speed of light}}$$

Put the numbers in:

$$\text{time for travel} = \frac{144\,000\ \text{km}}{300\,000\ \text{km/s}}$$
$$= 0.5\ \text{s approximately}$$

An outside broadcast might increase the journey to 200 000 km, making the time for travel about 0.7 s. This time delay on news broadcast or telephone conversations will be quite noticeable. You may well have observed this effect on your television.

With optical fibres connecting the two studios, the distance travelled may be only 20 000 km:

$$\text{time delay} = \frac{\text{distance travelled}}{\text{speed of signal in glass}}$$

Put the numbers in:

$$\text{time delay} = \frac{20\,000\ \text{km}}{200\,000\ \text{km/s}}$$
$$= 0.1\ \text{s}$$

The time delay with optical fibres is only 0.1 s, which is much less noticeable.

Will optical fibres take over?

Optical fibres can handle a huge number of voice and data calls. Because of their greater information capacity, no noticeable time delay and no need for repeater stations, there is a worldwide shift towards using optical fibres for long-distance applications. However, microwaves and satellites will never be replaced by them. Microwave links often take over optical fibre traffic when the cable is being repaired.

QUESTIONS

11 **a** What type of electromagnetic radiation travels down optical fibres for communications?

b How fast will the signals travel down the fibre?

12 What type of electromagnetic radiation is used by mobile phones for communications?

13 Why do you think that communication via optical fibre is very 'cost effective'?

QUESTIONS

14 Communication satellites are normally put into geosynchronous orbit. Explain what this means with the help of a diagram.

15 Make a list of the advantages of using optical fibres rather than copper wires for communication.

16 Explain why repeater stations are needed for long-distance communication by microwaves.

17 Satellites must be used for long-distance microwave communication around the world. Draw a simple diagram to show how this is possible.

Is there a problem with mobile phones?

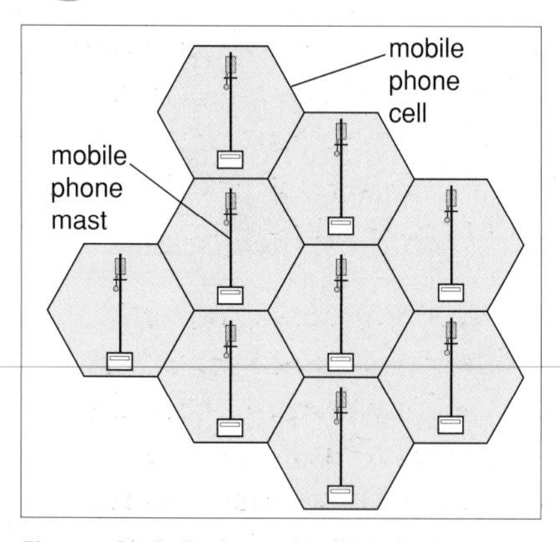

Figure 18.16 Each area is divided up into hexagonal cells with a base station. All base stations are connected to a control centre (MTSO).

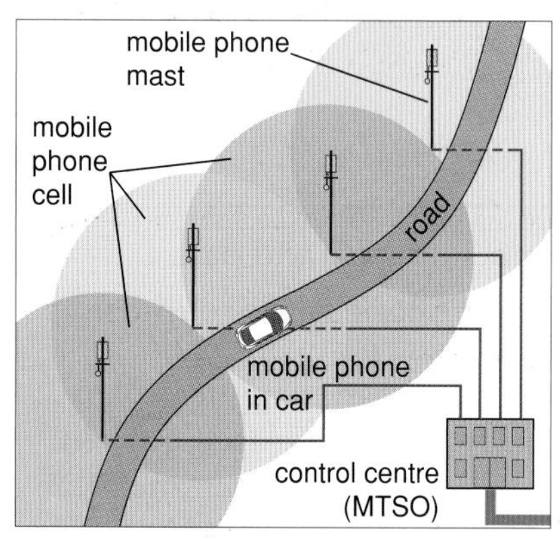

Figure 18.17 As you travel, the mobile phone signal is passed from cell to cell.

There are millions of mobile phones (called 'cell phones' in the USA) in use all over the world. Many developing countries, particularly Africa, see mobile phone systems as the only way to connect their population into the twenty-first century, and are developing these systems rather than fixed lines (copper wire or optical fibre). Mobile phones are 'radios' that can transmit and receive at the same time, using microwaves. Mobile phone companies (carriers) divide a town, city or countryside into a number of cells (Figure 18.16).

Each cell covers about 25 km^2 and contains a base station consisting of a tower and a box or small building to hold the electronic equipment. The large number of cells means that mobiles can transmit at a much lower power, typically 0.6–1.0 W, because the signal does not have to travel very far. Each carrier company has a central office in each city or district, called a mobile telephone switching office (MTSO).

This is what happens when you switch on and someone is trying to call you (Figure 18.17):

- Your phone is constantly listening for the base station code. If it cannot find one, you get a 'no service' message.
- Your phone and the base station code communicate and compare codes.
- Your phone sends a registration request. A sort of, 'I am here if anyone calls' signal.
- The MTSO keeps track of your location on its database.
- When your friend calls, the MTSO looks in its database to find which cell you are in.
- The MTSO picks a pair of microwave frequencies that your phone and the base station will use to make the call. Your phone and the base station tower switch to these frequencies.
- You are connected. You are talking by two-way radio, using microwaves.
- As you move towards the edge of your cell, the base station notices that your signal strength is getting weaker. The next base station notices your signal strength increasing. At some point your phone gets a signal from the MTSO to change frequencies. Your phone switches to the new cell.

Mobile phones and health

Some people are worried that mobile phones might affect their health. They think that because the transmitting aerial emits microwaves and has to be held close to the head, it will increase the risk of brain cancer. So far, all the research has shown that there is no proven link between radio-wave and microwave use and cancer. Radio and microwaves are at the low-energy end of the electromagnetic spectrum. They are not ionising radiations like ultraviolet and X-rays, which have been proved to cause cancer.

Although microwave ovens cause tissue heating (the microwaves make water molecules vibrate), mobile phones use very different frequencies of microwaves. These frequencies do not cause water molecule vibration and therefore do not heat up tissue. However, there may be a long-term effect. There is very little research on the long-term exposure of tissue to low levels of microwaves, particularly for the growing tissues and organs in children, where the bone of the skull is thinner. This is one of a number of areas where science cannot always give a definite answer one way or the other. A long-term international study will assess the health of around 250 000 mobile phone users, but the results are not due until 2020.

Years ago, when people first started smoking, they did not suffer any immediate effects. Now we know that smoking kills. Using mobile phones may affect us in the long term. The trouble is, if they are found to be dangerous, billions of us across the globe will be affected. If children are at an increased risk because the soft bones in their skull may let more radio waves into their brains, we may not know the effects for decades. The UK National Radiological Protection Board (NRPB) has therefore used a Precautionary Principle approach and recommended that mobile phones should not be given to children. The NRPB has also issued precautionary guidelines for the rest of the population:

Always try to:

- keep your call as short as possible
- use an earphone/microphone cable to keep the transmitter away from your head
- limit calls inside buildings (where your phone needs to transmit at higher power) and use open spaces as much as possible.

Figure 18.18 The safety of mobile phones is still under discussion.

Safety tests

Mobile phones have to be tested for radiation. The 'specific absorption rate' is a measure of the amount of energy absorbed when the microwaves enter your head. To be licensed for sale, the phone must have an SAR value of less than 2 W/kg in the UK and 1.6 W/kg in the USA. Sales brochures usually tell you all about the exciting features on their models. You should ask to see their SAR rating. They are often to be found in the small-print at the back of the catalogue.

Figure 18.19 The siting of mobile phone masts is a controversial issue.

Planning requirements for communications masts

Unlike the mobile phones themselves, the base station masts in each cell emit signals at much higher power, typically 10 to 100 times higher. Mobile phone companies cannot just put a base station aerial wherever they want – the local Planning Department must be consulted. Planning requirements can differ in the various parts of the UK. All new ground-based masts come under full planning control. Technical restrictions apply to the size, height and number of masts on a building. There are more stringent requirements in conservation and particularly scenic areas. Public consultation requirements have been increased, especially for masts under 15 m high, which originally did not require permission. School governors must be consulted on proposals for masts near schools.

There is a lot of concern about the siting of mobile phone masts.

QUESTIONS

18 Unlike two-way radios (walkie-talkies), mobile phones transmit at very low power.
- **a** How is this possible?
- **b** Why do two-way radios have to transmit at much higher power?

19 Draw a flowchart showing how a two-way mobile phone conversation between two people in different mobile phone cells occurs.

20 X-rays and ultraviolet radiation are examples of ionising radiation.
- **a** What does this mean?
- **b** What are the possible consequences of human tissue exposure to ionising radiation?

21 What is the difference between the microwaves used by a microwave oven and those used by a mobile phone?

22 How do the microwaves emitted by a microwave oven cause tissue to be heated?

23 Why might long-term exposure to microwaves be particularly harmful to children?

24 Why is it important that local people are involved with the decision to site a mobile phone mast?

25
- **a** What is meant by the Precautionary Principle?
- **b** How is the NRPB using it?

Discussion Points

1. Why do you think that developing countries such as Kenya are developing mobile phone communications systems rather than developing fixed-line systems?
2. What do you think? Are mobile phones safe or not?

TASK

MOBILE PHONES – FOR OR AGAINST?

This activity helps you with:

- ★ researching information
- ★ selecting, rewriting and drafting researched information
- ★ organising a scientific debate
- ★ speaking about scientific issues in front of other people
- ★ presenting researched information graphically using PowerPoint.

1. Research the arguments for and against the claims about:
   - **a** the health risks of mobile phone handsets
   - **b** the health risks of mobile phone masts
   - **c** planning laws or lack of them concerning the position and building of the masts.
2. Nominate a spokesperson for each side.
3. Use a PowerPoint presentation to illustrate your case.
4. Select an impartial chairperson to ensure fair play.
5. Make your presentations to the class.

The greenhouse effect

Why is **global warming** usually blamed on the **greenhouse effect**? Why is it called 'the greenhouse effect'? What's global warming got to do with growing tomatoes?

Figure 18.20 The Eden Project.

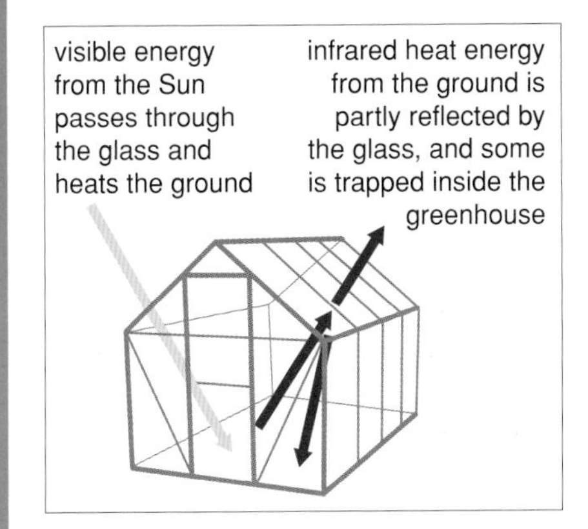

Figure 18.21 The greenhouse effect.

The Eden Project in Cornwall is home of some of the largest, and certainly the most spectacular greenhouses in the world. The largest of the greenhouses (called biomes) contains tropical plants and has an average temperature of 24 °C. The temperature is controlled by a variety of different means, but very little external energy is needed to keep the temperature high because, like all greenhouses, they let radiation from the Sun in but they don't let it escape very easily. As a result, greenhouses always seem hotter than their surrounding buildings – hence the name 'greenhouse effect'.

In a greenhouse, the glass or transparent plastic allows visible light from the Sun to pass through it. This visible light is absorbed by the ground inside the greenhouse (because it's dull and dark). The absorbed visible light gradually warms up the ground. The warm ground then emits some of its thermal energy as infrared radiation – with a longer wavelength than the visible light. Some of the infrared radiation emitted by the ground is reflected off the inside of the glass or the plastic (which acts like a partial mirror) back to the ground where it is absorbed again (and so on, and so on). The temperature inside the greenhouse therefore starts to rise.

Our planet's atmosphere acts in a similar way (see Figure 13.10, page 146).

The 'greenhouse' gases in the atmosphere, such as carbon dioxide and methane, allow visible light from the Sun through, but the infrared radiation emitted by the warm surface of the Earth is absorbed by the greenhouse gas molecules. This infrared is then re-emitted by the greenhouse gas molecules in all directions – and crucially, some of this infrared is effectively reflected back down to Earth, increasing the temperature.

The more carbon dioxide and methane in our atmosphere, the more significant the greenhouse effect becomes and the greater the rate of global warming. If we do not do something about this, world sea levels could rise by up to two metres before 2100, causing huge coastal flooding and putting the lives of hundreds of millions of people at risk.

It is quite simple – reduce greenhouse gas emissions or face the consequences.

QUESTIONS

26 What is the difference between the radiation absorbed by the ground in a greenhouse and the radiation that it emits?

27 Why is some of the radiation emitted by the ground in a greenhouse reflected back into the greenhouse?

28 In the greenhouse model of global warming:

- **a** which bit of the greenhouse is like the atmosphere
- **b** in what ways is the atmosphere different?

29 Why would increased use of fossil fuels accelerate the rate of global warming?

30 Why would deforestation of jungle areas, such as the Amazon Basin, cause an increase in the greenhouse effect?

31 Explain why a relatively small increase in temperature at the Earth's poles causes a big increase in the average sea level?

Discussion Point

Whose responsibility is global warming? What can *you* do?

Chapter summary

- Thermal radiation, also called infrared radiation, is made up of waves and is a part of the electromagnetic spectrum.
- Shiny, light-coloured surfaces are good reflectors and poor absorbers and emitters of thermal radiation. Dull, dark-coloured surfaces are good absorbers and emitters of thermal radiation.
- The temperature of an object determines the intensity and wavelength of the radiation that it emits. Hotter objects tend to emit higher-intensity, shorter-wavelength radiation. This is called black-body radiation.
- The greenhouse effect can be explained in terms of visible radiation from the Sun being absorbed by the ground and emitted as infrared radiation. This radiation is absorbed and re-emitted from the atmosphere back to Earth – contributing to global warming.
- Microwaves and infrared radiation are used for long-distance communication, via geosynchronous satellites, mobile phones and intercontinental optical fibre links.
- There are public concerns about the claimed health risks associated with mobile phone masts, some of which are unproven, but further research is needed on long-term exposure to low-levels of microwaves.

Ionising radiation

Is this the most radioactive place on the planet?

Building B30 is a large, stained, concrete edifice that stands at the centre of Sellafield, Britain's sprawling nuclear processing plant in Cumbria. Surrounded by a 3-metre-high fence that is topped with razor wire, encased in scaffolding and riddled with a maze of sagging pipes and cabling, it would never be a contender to win an architectural prize.

Yet B30 has a powerful claim to fame, albeit a disturbing one. "*It is the most hazardous industrial building in western Europe*", according to George Beveridge, Sellafield's deputy managing director.

Nor is it hard to understand why the building possesses such a fearsome reputation. Piles of old nuclear reactor parts and decaying fuel rods, much of them of unknown provenance and age, line the murky, radioactive waters of the cooling pond in the centre of B30. Down there, pieces of contaminated metal have dissolved into sludge that emits heavy and potentially lethal doses of radiation.

It is an unsettling place, though B30 is certainly not unique. There is Building B38 next door, for example. "That's the second most hazardous industrial building in Europe" said Beveridge. Here, highly radioactive cladding from reactor fuel rods is stored, also under water. And again, engineers have only a vague idea what else has been dumped in its cooling pond and left to disintegrate for the past few decades.

But the buildings, like so many other elderly edifices at Sellafield, are crumbling and engineers now face the headache of dealing with their lethal contents.

This, then, is the dark heart of Sellafield, a place where engineers and scientists are only now confronting the legacy of Britain's post-war atomic aspirations and the toxic wasteland that has been created on the Cumbrian coast. Engineers estimate that it could cost the nation up to £50bn to clean this up over the next 100 years.

Sellafield nuclear processing plant.

A power plant cooling pond.

TASK SELLAFIELD

This activity helps you with:
★ examining the arguments for and against a scientific issue important to society
★ constructing a table.

The consequences of a large-scale nuclear leak from Sellafield are almost too horrific to think about. If Building B30 were to release its contents into the Irish Sea (Sellafield is situated immediately next to the sea with the River Calder running through the site) then the radioactive waste would destroy the marine environment for hundreds of square miles. Millions of people would be in danger. The engineers estimate that the Sellafield clear-up could cost £50 billion. Can the nation afford it? £50 billion would buy 1000 hospitals or 2000 secondary schools. In 2007, the annual education budget for the UK was £77 billion and the annual budget for the NHS was £104 billion.

Construct a table showing the arguments for and against cleaning up Sellafield.

What's lurking in the cooling pond of building B30?

The materials that have come out of nuclear reactors, either straight from the reactor itself or from the surroundings, are highly radioactive. This means that they contain atoms that give out dangerous ionising radiation. Radioactivity is a naturally occurring (and also manmade) physical phenomenon. The nuclei of some atoms are **unstable**. This means they have too much energy and need to lose some of their energy to become more stable. There are three ways that the nuclei of most radioactive atoms can do this. They can emit (give out):

- alpha (α) particles
- beta (β) particles
- gamma (γ) rays.

The emitted particles or rays take energy away from the atom's nucleus, making it more stable. These particles are called **nuclear radiation**.

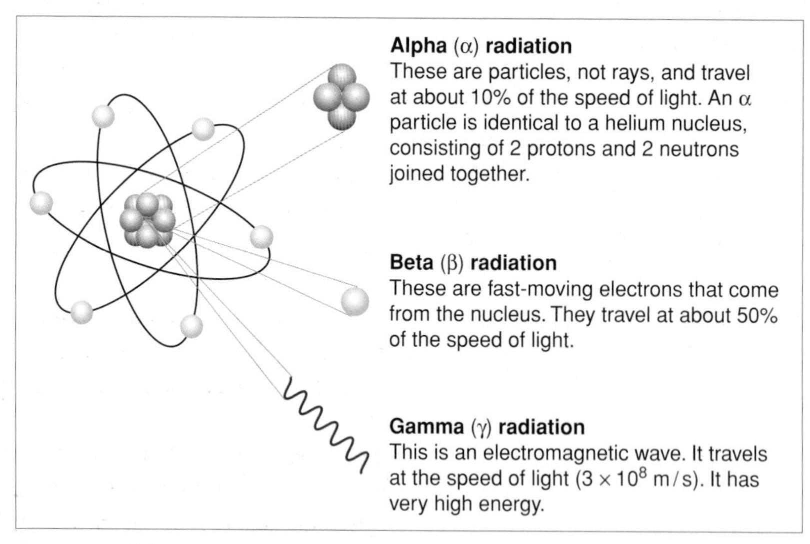

Figure 19.1 Alpha, beta and gamma radiation.

The problem is that the energy emitted as alpha, beta or gamma radiation moves out and away from the atoms. If human tissue is in the way, the energy will harm or kill the cells of that tissue. The radiation can ionise the cell (making it charged), killing it directly, or it can alter the DNA of the cell, causing it to mutate, forming cancers or genetic abnormalities if the cells are sex cells.

Gamma rays are part of the electromagnetic spectrum, like ultraviolet light and X-rays. Ultraviolet light and X-rays are also ionising and can cause cells to die or mutate. Radiation emitted by radioactive substances, ultraviolet light and X-rays are all called **ionising radiation**.

Alpha radiation is the most ionising form of radiation (about 20 times more than all the others). Beta radiation, gamma radiation, X-rays and ultraviolet light all have approximately the same ionising effect on the body.

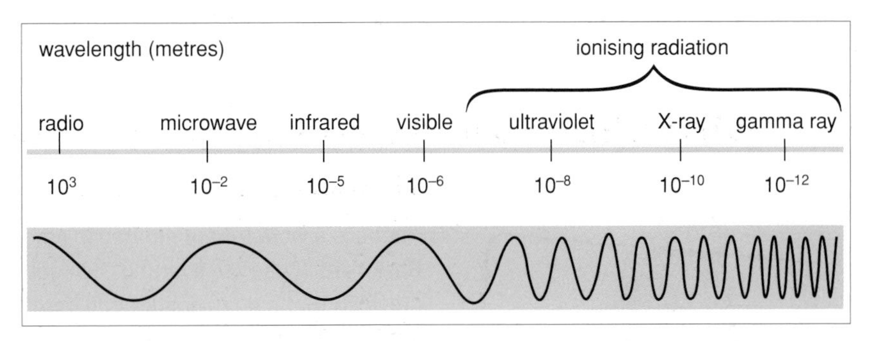

Figure 19.2 The electromagnetic spectrum.

Figure 19.3 Children near Chernobyl were born with physical abnormalities.

QUESTIONS

1. Why are some atoms radioactive?
2. What are the three ways that nuclei can become more stable?
3. Radon is a radioactive gas that emits alpha radiation. It can be inhaled into the lungs. Explain what can happen to cells in the lungs if a person breathes in radon gas.
4. Following the nuclear reactor explosion at Chernobyl in the Ukraine in 1986, many babies have been born in the Ukraine and Belarus with genetic abnormalities. Explain what might have happened in the cells of the parents of these children who have been born with physical abnormalities.
5. Beta radiation consists of high-energy electrons emitted from the nuclei of radioactive atoms. Ultraviolet light is part of the electromagnetic spectrum of waves. In what way are beta radiation and UV light similar?
6. Why is alpha radiation more dangerous than ultraviolet light?

Radioactive materials are stored in water, surrounded by lots of concrete and sometimes lead shielding. The radiation is absorbed by these materials rather than by people – making it considerably safer.

Sellafield plans to use robots to dredge the ponds in B30 and B38, and then encase the solid radioactive sludge in blocks of glass – a process known as 'vitrification'. They will then store the glass blocks deep underground where the surrounding rocks will absorb the radiation. This process will take decades and the radioactive material will stay radioactive for millions of years.

QUESTIONS

7 Why does Sellafield plan to use robots to dredge the cooling and storage ponds in Building B30?

8 Why will radioactive sludge be safer if it is encased in glass?

Discussion Point

Why do you think that underground storage is the long-term 'preferred option' for the storage of nuclear waste? Are there other options?

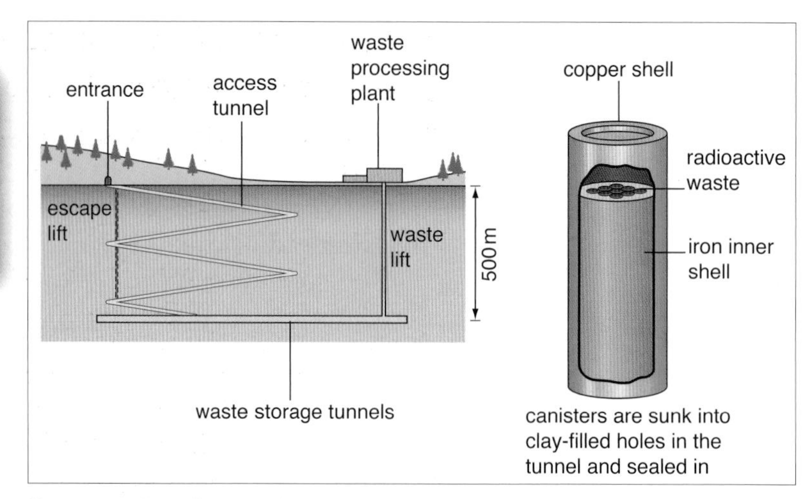

Figure 19.4 Deep disposal of radioactive waste – the Finnish model.

TASK

ARE YOU A NIMBY?

This activity helps you with:

- ★ conducting a survey
- ★ analysing the results of a survey
- ★ presenting the finding of a survey graphically.

Nimby is a word that originated in America. It is an acronym of **N**ot **I**n **M**y **B**ack **Y**ard. If you are a nimby then you are generally in support of something, provided that it does not directly impact on you. In this case, most scientists and the general public would agree that nuclear waste is best stored securely deep underground. But would you want a nuclear waste facility built under your house? If you don't want a facility built near you, but you want it somewhere else, then you are a nimby. In this task you need to devise a test of nimbyness. You need to devise a scale to rate a person's nimby attitude. For example, you could make a simple numerical scale with 'raving nimby' at one end, 'non-nimby' at the other end and a range of values in between (1 to 5 is usually a good range). You could then devise a series of scenarios that could be put to people to test their nimbyness. Here are a few examples:

- Your neighbour wants to convert his garden into a chicken run to produce free-range eggs.
- The local council want to put a pedestrian crossing with lights in front of your house.
- A mobile phone company want to erect a phone mast next to your house.
- The local water company want to build a small sewerage works near your house.
- Your neighbour wants to install a 15-m-high wind turbine in her garden.
- An energy company want to build a nuclear power station 3 miles from your house.
- The Government wants to build a nuclear waste storage facility deep underground beneath your house.

You could use these examples or develop some of your own.

1 Ask a range of people (friends, family, teachers) what they think about each scenario. Get them to use your rating system.
2 Record their values and add up their answers to give an overall nimby rating.
3 If you also record the person's age and gender you can see if there are any patterns in nimbyness.
   a Are young people less nimby than older people?
   b Are males more nimby than females?
   c Does nimbyness depend on the potential harm caused by the issue?
4 Present your findings graphically.

What sort of radioactive materials might be in the sludge at the bottom of the ponds?

Radioactive waste from nuclear reactors contains many highly radioactive elements. One of these elements is **uranium-235**. Uranium-235 is the main atom involved with producing the energy in a nuclear reactor. About 0.7% of naturally occurring uranium is made up of uranium-235 (most is uranium-238). The fuel rods used in a nuclear reactor undergo a special process that enriches them, increasing the amount of uranium-235 up to about 5%.

Uranium-235 remains radioactive for millions of years. In fact, it takes about 703 800 000 years for a sample of uranium-235 to become half as radioactive. It takes about five times this value, about 3 500 000 000 years, before the radioactivity returns to a value similar to naturally occurring background radiation. About 0.8% of a used (or spent) fuel rod is uranium-235. Radioactive waste from inside a nuclear reactor will need to be stored for a very, very long time to keep it safe.

Radioactivity is also used in hospitals to treat cancers and to image the body. One of the radioactive atoms used is radium-226. Radium-226 remains radioactive for about 16 000 years, so very special arrangements have to be made to store the used radioactive materials. Most radioactive waste from hospitals ends up at Sellafield for processing – adding to the accumulated radiation.

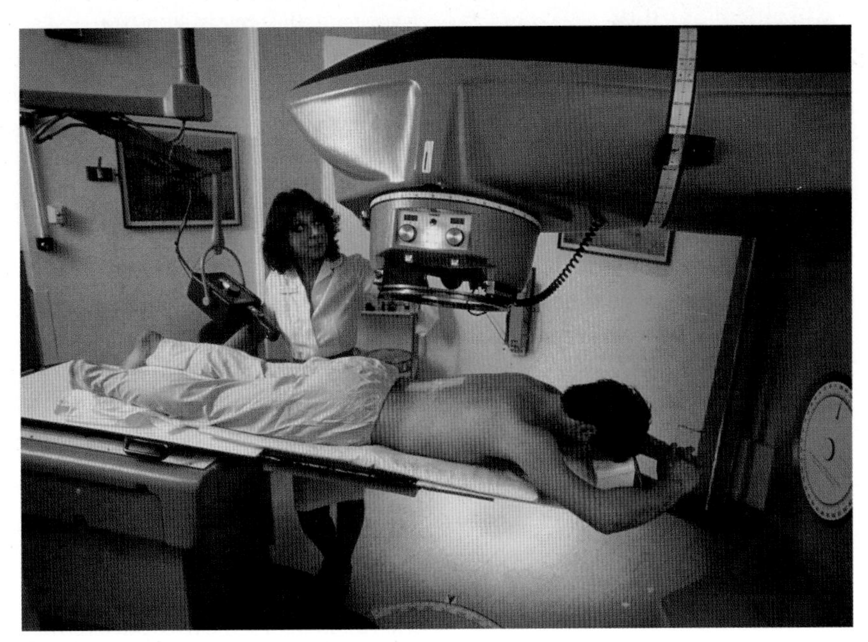

Figure 19.5 A patient undergoing radiotherapy to treat a tumour.

TASK NUCLEAR WASTE AND HUMAN EVOLUTION

This activity helps you with:
- ★ plotting data graphically
- ★ comparing two 'unconnected' phenomena
- ★ considering the consequences of current scientific actions for future generations.

The decay time of radioactive atoms is compared using a measurement called **half-life**. The half-life of a radioactive substance is the time it takes for the activity of a sample to halve. The radioactive atoms in Table 19.1 are all found in spent nuclear fuel, and have very long half-lives:

Table 19.1 Half-lives of some radioactive atoms.

Radioactive atom	Half-life (millions of years)
Technetium-99	0.211
Tin-126	0.230
Selenium-79	0.327
Zirconium-93	1.53
Caesium-135	2.3
Palladium-107	6.5
Iodine-129	15.7

Your task is to compare the half-lives of these radioactive substances to a timeline of human evolution.

Table 19.2 Some major human evolution events.

Event	Time (millions of years ago)
Great Apes appear	15
Orang-utan ancestor	13
Gorilla ancestor	10
Chimpanzee ancestor	7
Ardepithicus (first bipedal ancestor)	4.4
Australopithecus	3.6
Homo habilis	2.5
Homo erectus	1.8
Neanderthal man	0.6
Homo sapiens	0.2

1 Plot a chart of the half-lives of the radioactive atoms present in spent nuclear fuel. Use the time axis of the chart to plot the significant events in the human species evolution.
2 How do the two charts compare?

Discussion Point

The whole of human evolution from the common ancestor of the Great Apes has occurred in the time of *one* half-life of Iodine-129. Radioactive waste will remain radioactive for a very, very long time. What are the consequences of this for humanity? What steps will need to be taken to ensure the long-term security of this waste?

How do we monitor radiation?

The company responsible for the nuclear processing facility at Sellafield in Cumbria is called Sellafield Ltd. It is responsible to the Government for making sure that the facility is safe, and that all the radiation is confined within the site and none of it is leaking into the local environment. It is required to monitor, measure and record the radiation in and around the site over time. To do this it uses sensitive **Geiger counters** that can pick up and detect the different types of radiation.

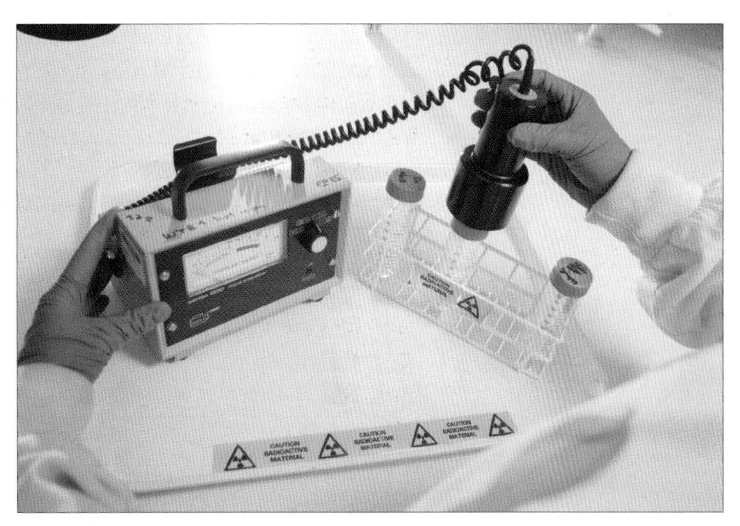

Figure 19.6 A Geiger counter.

Geiger counters will pick up radiation coming from any source, so it is important to know the level of **background radiation** and how much is coming from potential leaks from Sellafield. Background radiation is all around us. It comes naturally from our environment and from artificial (manmade) sources. Figure 19.7 shows (on average) the different sources of background radiation.

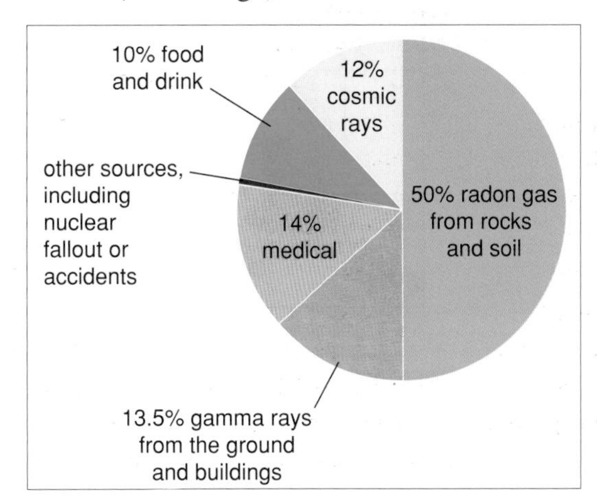

Figure 19.7 Sources of background radiation.

Most of the background radiation comes from naturally occurring sources, primarily from the ground, rocks and from space. Artificial background radiation mostly comes from medical sources, predominantly as a result of the medical and dental examinations using X-rays.

The vast bulk of the background radiation that we receive comes from the radioactive element **radon**, emitted from rocks and the soil. Some rocks are much more radioactive than others. Granite is a particularly radioactive rock because it contains uranium. The uranium in granite decays, eventually producing radon. As radon is a gas, it can escape from the granite and is easily breathed in by humans. The radon enters our lungs where it can decay, and the alpha particles emitted by the decaying radon are absorbed by the cells lining the lungs, causing the cells to die or mutate (forming cancers).

The map in Figure 19.8 shows the risk of exposure to radon across England and Wales. The magnified portion of the map shows the radon risk in and around the Sellafield plant in Cumbria.

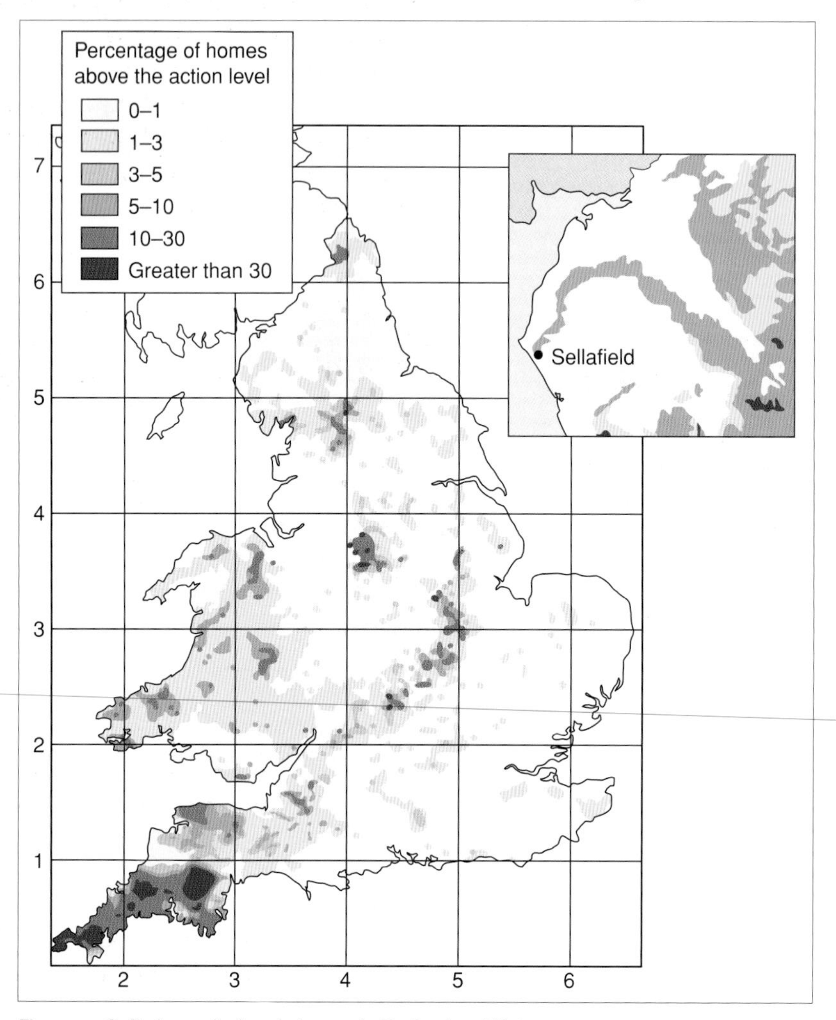

Figure 19.8 Radon emissions in homes in England and Wales.

If you want to find out how much you are at risk from radon, you can get a large-scale map of where you live from:

www.ukradon.org/map.php?map=englandwales

If scientists want to measure the effects of radioactivity they have to take the level of background radiation into account – it has to be subtracted from the measured values. Once the background level has been subtracted, any remaining radioactivity is due to other factors, such as leakage from the nuclear storage facilities at Sellafield.

Radiation dose (how much radiation we receive) is measured in **sieverts** (**Sv**). It is a big unit and a dose of 1 Sv is a big dose of radiation. In practice, we use the **millisievert** (**mSv**) which is one-thousandth of a sievert. The average annual dose received from radon throughout the UK is 1.3 mSv, but in places like Cornwall where there is a lot of granite, the background radiation due to radon can be as high as 6.4 mSv – nearly five times higher!

Monitoring dose

In the UK, the company tasked with monitoring the radiation dose received by the general population is called the Health Protection Authority (HPA). The charts in Figure 19.9 compare the average annual radiation dose for three averaged groups of people. The first group is the whole UK population, the second group is people living in Cornwall and the final group is people living around Sellafield in Cumbria. (The effect of exposure from medical examinations has been removed to make it a fair comparison.)

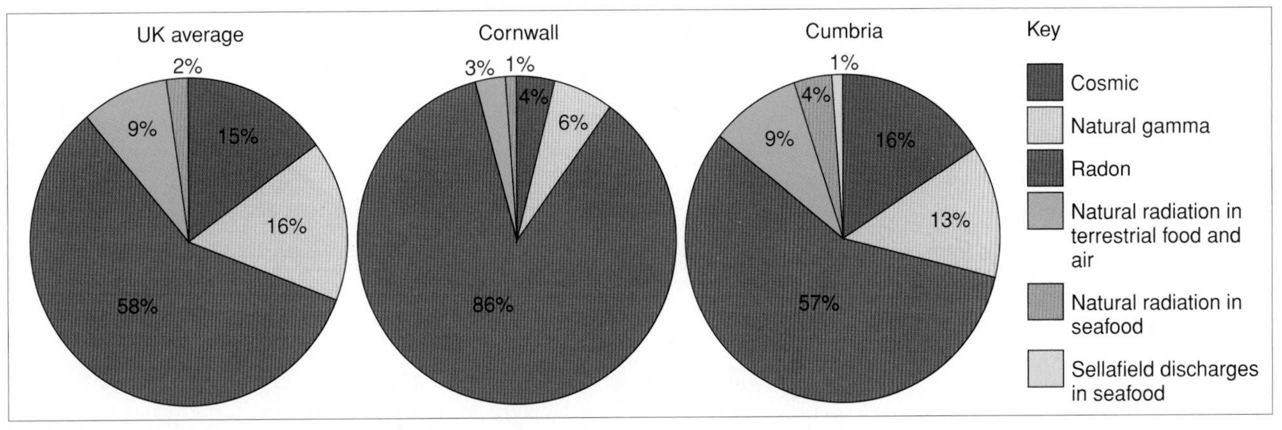

Figure 19.9 Sources of radiation in the UK, Cumbria and Cornwall.

The charts show that only 1% of the annual average radiation dose in Cumbria is attributed to the Sellafield nuclear processing facility.

QUESTIONS

9 What are the two main groups of sources of background radiation?

10 On average, across the UK, how much background dose (in mSv) is received from space?

11 Which source of background radiation contributes the largest dose on average across the UK?

12 Why is the background dose due to radon different in Cornwall and Cumbria?

13 Why is the background dose from medical examinations (X-rays) usually taken out of the comparisons between different locations?

14 Most of the background radiation dose that we receive through food and drink comes from consumption of seafood (fish and shellfish mostly). Why do you think that seafood generally contains a higher background dose than fruit, vegetables and meat?

What's the problem with radon?

The problem with radon is it's a gas. When it is emitted by rocks like granite it can enter our lungs. Outside, this isn't really a problem, but radon can build up inside houses, particularly if the houses have poor under-floor ventilation, or, the houses are made of granite (as some older house are in places like Cornwall).

QUESTION

15 Most deaths (per 1000 people) occur due to cancers. From the graphs in Figure 19.11 and Figure 19.12:

a What is the total number of non-cancer related deaths per 1000 people?

b How many times higher is the risk of dying from any form of cancer than the risk of dying from background radon?

Discussion Point

About 100 000 cosmic rays from outer space will pass through your body every hour. You get a bigger dose when you are in an aeroplane because there is less air between you and space to absorb them. A 1 h flight will increase your dose by about 0.005 mSv. Do frequent fliers such as pilots and cabin crew have to worry about their increased background radiation due to cosmic rays?

The diagrams in Figure 19.10 show how easily radon can enter a house, but also how easily it is to ventilate the house of radon by adding extra vents and by fitting radon stacks and sumps to new-build houses.

The UK Government has advised that a level above 200 Bq/m^3 requires remedial action. One **becquerel (Bq)** is equivalent to one disintegration (decay of one unstable atom) per second.
Figures 19.11 and 19.12 show that the lifetime risk from exposure to average levels of radon is very small. You are three times more likely to die as a result of an accident in your home. But, as the graphs show, as the concentration of radon rises so do the risks. The diagram shows the risks for non-smokers. If you smoke 15 cigarettes a day, you increase your risk 10 times.

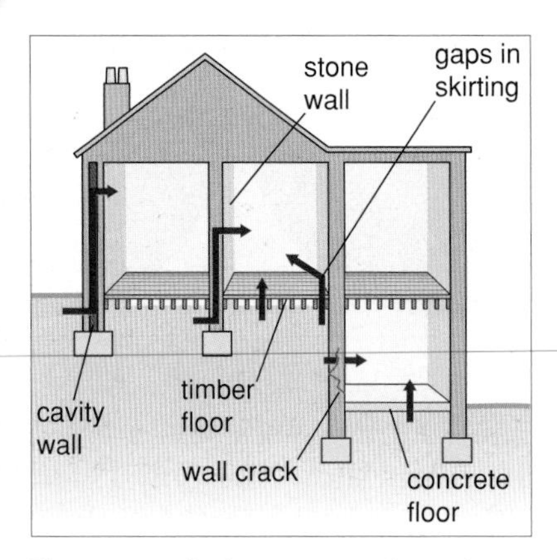

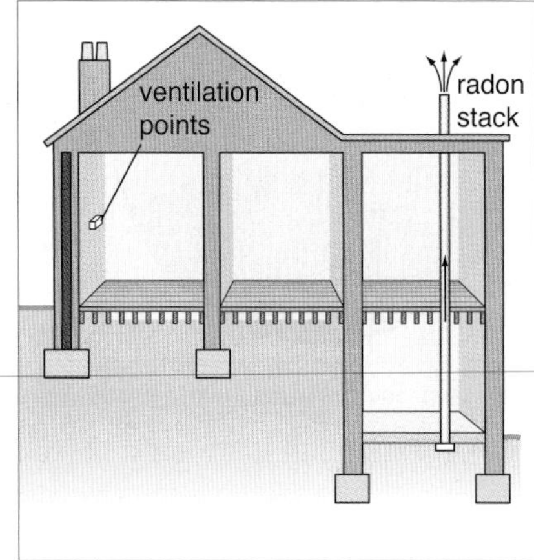

Figure 19.10 Radon can enter a house by many different routes.

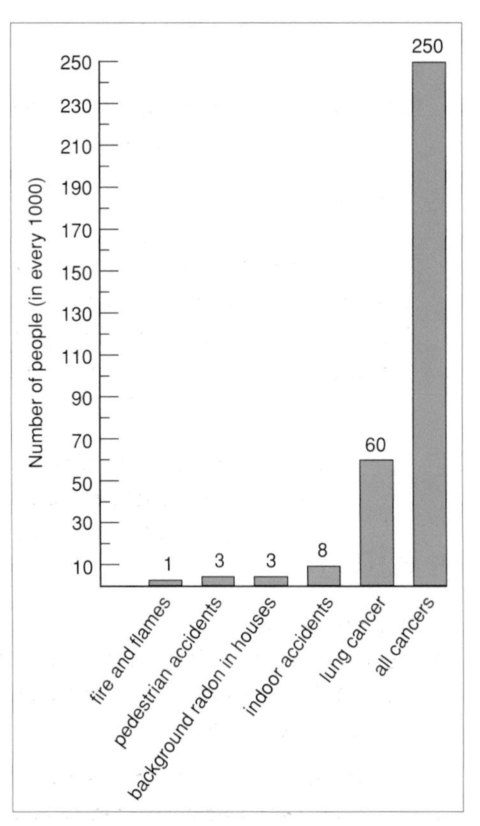

Figure 19.11 Lifetime risks of death from common causes (UK average for smokers and non-smokers).

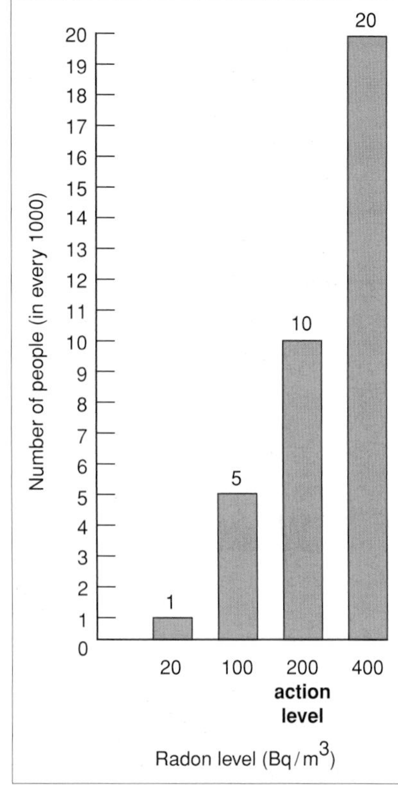

Figure 19.12 Lifetime risks of lung cancer from radon (for non-smokers).

Is Sellafield safe?

The answer is yes, and no! The sheer amount of radioactive fuel and waste poses a real threat. If the radiation were to leak in significant amounts into the local environment it would be a natural disaster. Sellafield does not have a particularly high natural background radiation level (it's about 3.5 times smaller than Cornwall's average) and only 1% of that background is down to the processing plant. So Sellafield is doing a good job in shielding the local environment from the lethal levels of radiation contained within it. Places like Building B30, that hold huge amounts of radioactive waste, contain the radioactivity within the building. So how is that done?

PRACTICAL INVESTIGATING THE ABSORPTION OF RADIOACTIVITY

This activity helps you with:

★ observing a demonstration and making observations
★ using a scientific computer simulation/model
★ investigating a scientific effect
★ forming conclusions based on experimental and/or simulation observations.

Your teacher can demonstrate to you the different absorption properties of alpha, beta and gamma radiation. In schools and colleges, students under 16 years of age are not allowed to do experiments involving ionising radiation.

You can download a simulation programme that will demonstrate the same thing to you virtually. Some good simulations you might like to try are:
http://visualsimulations.co.uk/software.php?program=radiationlab
www.furryelephant.com/player.php?subject=physics&jumpTo=re/2Ms16

Risk assessment

- **This experiment must be demonstrated by a teacher.**
- **Your teacher will provide you with a risk assessment for this activity.**

Apparatus

* radioactive sources (α, β and γ)
* Geiger–Müller tube
* rate meter
* tweezers
* sheet of card
* sheets of aluminium of different thicknesses
* sheets of lead of different thicknesses

Procedure

Your teacher will set up the apparatus as shown in Figure 19.13 and measure the background radiation count.

Alpha radiation

The source (americium-241) emits α radiation only.

1 Your teacher will place the source close to the Geiger–Müller tube with tweezers and measure the count rate.

2 Using tweezers, your teacher will place a piece of card between the α radiation source and the tube, and measure the count rate.

Beta radiation

The source (strontium-90) emits β radiation only.

Your teacher will repeat the demonstration with the β radiation and will try to stop the radiation, first with a card, and then with sheets of aluminium of increasing thickness.

Gamma radiation

Using a source that emits γ radiation only, such as cobalt-60, your teacher will repeat the experiment and try to stop the radiation with card, aluminium and finally sheets of lead.

If your school has a radium source you can use the experiment to show that radium emits alpha, beta and gamma radiation.

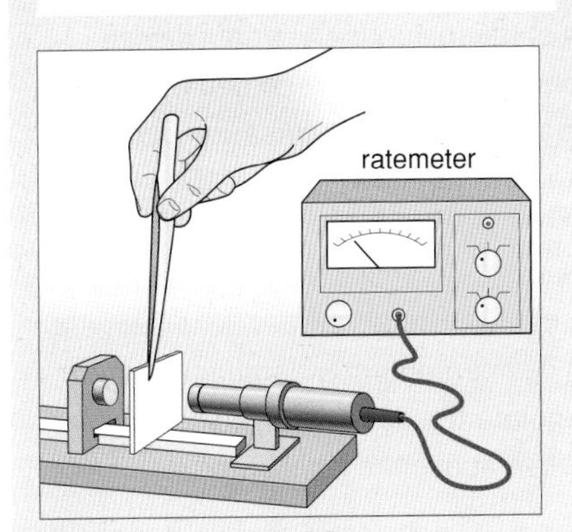

Figure 19.13 Testing to see which materials will stop radiation.

The demonstrations on the previous page show that each type of radiation has a different penetration power. Gamma radiation is the most penetrating, followed by beta radiation, with alpha radiation being the least penetrating. Figure 19.14 summarises this.

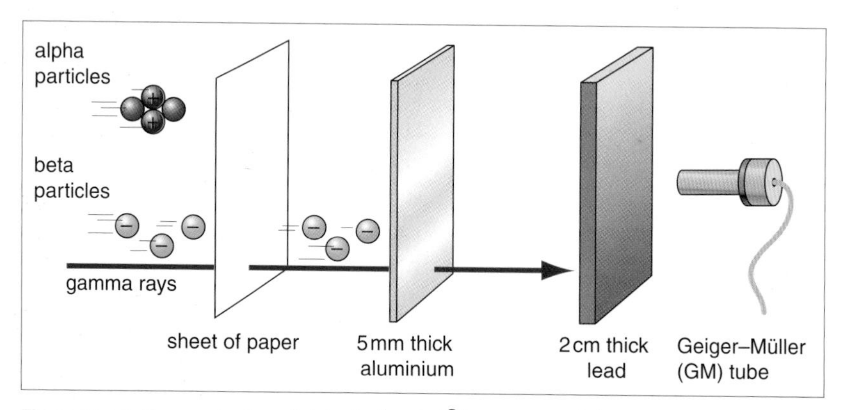

Figure 19.14 The penetration of alpha (α), beta(β) and gamma(γ) radiation.

RADIATION DOSIMETERS

This activity helps you with:

★ analysing a practical application of science
★ communicating scientific ideas using diagrams
★ designing an electronic device
★ explaining how ICT can be used in connection with science to improve a process.

The different penetration powers of alpha, beta and gamma radiation are put to use in hospitals, nuclear power stations and in other industries that use radiation as a way of measuring the dose received by workers. A device called a **dosimeter badge** is used to measure the amount of exposure to the different forms of radiation. One design of dosimeter badge is shown in Figure 19.15.

The photographic film is labelled and shown in black on the diagram. It is covered with a very thin light-proof case. When ionising radiation hits the film it exposes the film, causing it to be fogged. If more radiation hits the film, the 'fog' becomes denser. The film from individual badges is periodically examined and the exposure fogging is measured. The film is partially covered with two 'windows': one window is made out of a thin piece of aluminium and the other is made out of aluminium behind a thin piece of lead.

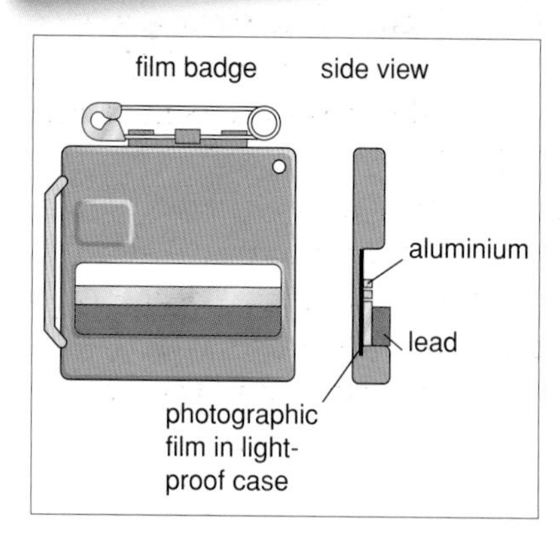

19.15 Dosimeter film badge.

1. Using a suitable diagram, and your knowledge of the penetrating power of the different types of radiation (alpha, beta and gamma), explain how this badge can measure the wearer's exposure to each type of radiation.
2. Design an electronic device for measuring the density of the fog on the film when it is periodically examined. Draw a suitable diagram for your device next to your diagram of the dosimeter, showing how it detects the different types of radiation.
3. Explain what the advantages would be of having your electronic device connected to a computer with a suitable database that recorded the density of the fog on the film for each radiation worker in an organisation *over time*. How could a system like this be used to automatically monitor the radiation dose of each worker?
4. The Health and Safety Executive states that the legal dose limit for any person (over 18 years) working with radiation is 20 mSv per year. Suggest how your dosimeter badge/measuring device/computer database system could be used by an organisation to regulate the dose received by workers and ensure that no worker received a dose above the legal limit.

Storing radiation at Sellafield

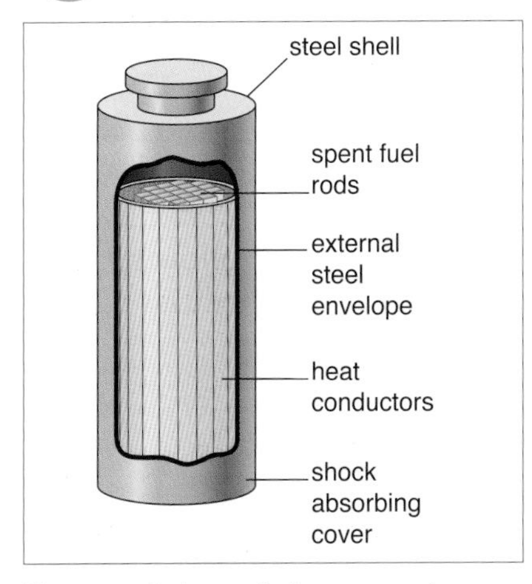

Figure 19.16 A spent fuel storage canister.

At Sellafield they use this knowledge of the penetrating power of the different types of radiation to contain the radiation within buildings. A variety of different shielding methods is used to prevent high levels of radiation from escaping.

Figure 19.17 A storage pond.

The spent fuel rods from reactors, containing highly radioactive fuel waste, are loaded into steel canisters. The canisters are designed to be very strong and to conduct away the heat still being generated by the fuel rods. The steel material absorbs a large proportion of the radiation emitted from the spent fuel. The steel canisters are then stored inside a large 'cooling pond' of water. The water cools the rods and also absorbs more of the radiation. The cooling pond is contained within a building with thick reinforced-concrete walls and ceilings. The building is therefore strong, and the concrete absorbs the radiation even more. Plutonium-239 and uranium-235, two of the main radioactive decay products in spent nuclear fuel, are alpha and gamma radiation emitters.

The alpha radiation is absorbed by the fuel rod casing, and the gamma radiation is mostly absorbed by the steel–water–concrete combination. Other radioactive elements contained within the spent fuel cells are beta radiation emitters, and the beta radiation is absorbed by the steel canisters and the water.

The future for Sellafield and Building B30

Sellafield is a place designed to reprocess nuclear waste. It was never designed as a long-term storage facility for the waste. Sellafield successfully manages the 'temporary' storage of 67% (by volume) of the UK's radioactive waste materials. Across the UK, the amount of radioactive waste continues to build up. In 2006 an assessment was made by CoRWM, the Government's Committee on Radioactive Waste Management. It estimated the following volumes of radioactive waste were stored in various facilities across the country:

- high-level waste – 2000 cubic metres
- intermediate-level waste – 350 000 cubic metres
- low-level waste – 30 000 cubic metres
- spent fuel – 10 000 cubic metres
- plutonium – 4300 cubic metres
- uranium – 75 000 cubic metres

CoRWM's recommendation was that long-term storage of radioactive waste should be developed deep underground. It estimated that approximately one third of the UK has geology suitable for the long-term storage of nuclear waste.

QUESTIONS

16 How is the highly ionising alpha radiation from radioactive waste shielded at Sellafield?

17 Why are the radioactive waste storage canisters made out of steel?

18 What is the purpose of the water inside the storage ponds?

19 Why do you think that the buildings are made out of thick reinforced concrete, not standard bricks?

20 Why might it be impractical to shield the entire structures in lead, to substantially reduce the emission of gamma radiation from the buildings?

21 The storage buildings are serviced only by robots. Why do you think this is a good idea?

22 An Olympic swimming pool needs to be 50 m long × 25 m wide × 2 m deep. What is the volume of (water in) an Olympic swimming pool? How many Olympic-sized swimming pools would be needed to store the UK's radioactive waste?

23 Eventually, the radioactive waste will need to be transported to its underground storage facility. Describe five steps that Sellafield Ltd will have to take in order to protect the public from the radiation as the waste is moved – particularly if the underground storage facility is in a different part of the country.

Discussion Point

Hand's up anybody who wants a nuclear waste storage facility at the bottom of their garden! Do you?

Chapter summary

- The term 'radiation' is used to describe both electromagnetic waves and energy given out by radioactive materials.
- Substances that are radioactive can emit alpha (α), beta (β) and gamma (γ) radiation.
- Gamma radiation is also part of the family of waves called the electromagnetic spectrum – like ultraviolet light and X-rays. Gamma rays have the shortest wavelengths and highest energies.
- Alpha (α), beta (β) and gamma (γ) radiation, ultraviolet light and X-rays are all types of ionising radiation.
- Ionising radiation is able to interact with atoms and damage cells because of the energy it carries.
- The waste materials from nuclear power stations and nuclear medicine are radioactive; some of them will remain radioactive for thousands of years.
- Experiments and/or ICT simulations of experiments can be used to investigate the penetrating power of nuclear radiation.
- When measurements of radiation are taken, an allowance for background radiation must be made.
- Alpha, beta and gamma radiation have different penetrating powers. Alpha radiation is absorbed by a thin sheet of paper, beta radiation is absorbed by a few mm of aluminium or Perspex, but gamma radiation is only absorbed by a few centimetres of lead.
- The differences in the penetrating power of alpha, beta and gamma radiation determine their potential for harm. Alpha radiation is easily absorbed but is the most ionising. Gamma rays are very penetrating but are about 20 times less ionising than alpha radiation.
- Radioactive waste is stored in a series of containment systems. Steel canisters, water, concrete and lead are all used to shield the environment from harmful doses of radiation. The radiation produced by the waste is absorbed by the different types and thicknesses of containment.
- The long-term solution to the storage of radioactive waste is deep underground, where the harmful radiation can be absorbed by the surrounding rocks.
- Background radiation is all around us and comes from natural or artificial (manmade) sources.
- Natural sources of background radiation include radon from rocks, gamma rays from the ground and buildings, cosmic rays from space and radiation contained in food and drink.
- Artificial sources of background radiation include X-rays from medical examinations and nuclear fallout from weapons tests or accidents.
- Most (between 50 and 90%) of our background radiation comes from radon gas (depending on where you live).
- Places like Cornwall, where there is a lot of granite rock, have higher levels of radon gas because the granite contains uranium that decays (eventually) to radon.

20 Space

How big is space?

Space is big. You just won't believe how vastly, hugely, mind-bogglingly big it is. I mean, you may think it's a long way down the road to the chemist's, but that's just peanuts to space.

Douglas Adams (1952–2001),
The Hitchhiker's Guide to the Galaxy

The Hitchhiker's Guide to the Galaxy was originally a Radio 4 programme, first broadcast in 1978. Douglas Adams, the author, wanted to get over to listeners (and subsequent readers and viewers) the concept that space is so big, it's difficult for human beings to get their heads around it. In order for us to start to think about space we even have to start inventing new units, like light years and parsecs, in order to cope with the truly massive scale of the numbers.

The best way to truly appreciate the huge awesomeness of space is to start on a 'small' scale and gradually get bigger – each change of scale clinging on to the previous one. By building up a 'local' picture of space we can slowly get a feel for the overall, bigger picture.

What does our 'local' patch of space look like?

Figure 20.1 Earth and its moon from space.

Our home planet, Earth, is a relatively small rocky planet, situated in the 'Goldilocks Zone' of our local star, the Sun. The Goldilocks Zone of a star is the orbital range of distance around a star where water would be liquid on an Earth-like planet and Earth-like life would be possible.

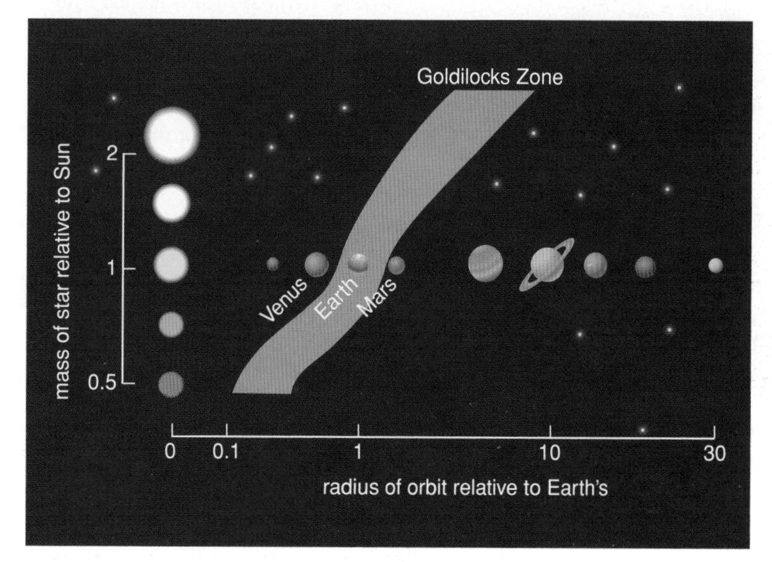

Figure 20.2 The Goldilocks Zone.

Discussion Point

Why do you think that the 'habitable zone' is usually referred to as the 'Goldilocks Zone'? What other factors do you think would be important for a planet or moon that was inside a star's Goldilocks Zone if it were to be a suitable place for a human 'colony'?

"This porridge is too hot," Goldilocks exclaimed.
So she tasted the porridge from the second bowl.
"This porridge is too cold."
So she tasted the last bowl of porridge.
"Ahhh, this porridge is just right!" she said happily.
And she ate it all up.

From the nursery rhyme *Goldilocks and the Three Bears*

TASK

GOLDILOCKS AND THE THREE PLANETS!

This activity helps you with:

- ★ reading an article adapted from a scientific magazine
- ★ writing your opinions on good scientific method
- ★ doing estimation-type calculations
- ★ using data to imagine conditions on different planets.

The following task is adapted from an article published in the *New Scientist* magazine.

Found: first rocky exoplanet that could host life

Astronomers have found the first alien world that could support life on its surface. It is both at the right distance from its star to potentially harbour liquid water and probably has a rocky composition like Earth.

The planet orbits a dim red dwarf star 20 light years from Earth called Gliese 581. Four planets were already known around the star, with two lying near the inner and outer edges of the habitable zone, where liquid water – and therefore potentially life – could exist on its surface.

One of those, which travels on a 13-day orbit, seems too hot for liquid water. The other, on a 67-day orbit, may be just warm enough for liquid water. Opinions may continue to swing back and forth because it is hovering right near the outer edge (of the Goldilocks Zone).

The newly found 'Goldilocks' planet, called Gliese 581 g, lies in between the hot and cold ones. Steven Vogt of the University of California, Santa Cruz, and Paul Butler of the Carnegie Institution of Washington, DC, used the 10-metre Keck I telescope in Hawaii to measure the wobbles of the parent star in response to gravitational tugs from its planets. They combined their data with measurements published by Michel Mayor of Geneva Observatory and his colleagues using a 3.6-metre telescope at the European Southern Observatory in Chile.

Rocky super-Earth

The wobbles revealed two previously undiscovered planets around the star, for a total of six. One is about seven times the mass of Earth and in a 433-day orbit – much too far from its star to support liquid water. The other, Gliese 581 g, lies in the Goldilocks Zone and has a 37-day orbit. Its mass is between 3.1 and 4.3 times that of Earth. Its relatively low mass means it should be made mostly of rock, like Earth.

Twilight zone

Conditions on the planet would be very different from those on Earth. The host star is a low-mass red dwarf that is just 1 per cent as bright as the Sun. Because it puts out so little light and warmth, its Goldilocks Zone lies much closer in than does the Sun's. At such tight distances, planets in the zone experience strong gravitational tugs from the star that probably slow their rotation over time, until they become 'locked' with one side always facing the star, just as the Moon always keeps the same face pointed towards Earth. That would mean perpetual daylight on one side of the planet and permanent shadow on the other. A first approximation suggests the temperature would be 71 °C on the day side and −34 °C on the night side, though winds could soften the differences by redistributing heat around the planet. Travelling from one side of the planet to the other, there would be a range of intermediate temperatures. The most comfortable place on this planet would be along the terminator, the line between light and dark, you basically see the star sitting on the horizon – you see an eternal sunrise or sunset.

First of many

The discovery suggests habitable planets must be common, with 10 to 20 per cent of red dwarfs and Sun-like stars boasting them. If you take the number of stars in our galaxy – a few hundred billion – and multiply them by 10 or 20 per cent, you end up with 20 or 40 billion potentially habitable planets in the Milky Way alone. Although the new planet is in the habitable zone, we are unlikely to find out whether it is actually inhabited anytime soon. One way to find out would be to measure the planet's light spectrum, which could reveal molecular oxygen or other possible signs of life in its atmosphere. But the overwhelming glare from its parent star makes it impossible to do that with current instruments. New planets will soon be discovered that can be observed with existing telescope technology.

continued...

TASK contd.

Discussion Point

The Drake Equation – in 1961 Frank Drake came up with an equation to estimate the number of civilisations that might exist in our galaxy, that we might be able to communicate with. Check it out at:
www.pbs.org/wgbh/nova/origins/drake.html
http://library.thinkquest.org/C003763/index.php?page=origin09
How many civilisations did you come up with?

Questions

1 What is the 'Goldilocks Zone'?
2 What do you think would be the general features of a Goldilocks planet?
3 How was Gliese 581 g 'discovered'?
4 Why do you think that the observations and measurements of Gliese 581 g were carried out by two independent sets of astronomers working in two different observatories? Why do you think this is 'good science'?
5 Why do you think that the mass of Gliese 581 g is given as 'between 3.1 and 4.3' times that of the Earth?
6 What do you think the gravity would be like on Gliese 581 g compared with Earth? How would this affect the everyday lives of human astronauts living on Gliese 581 g, if they were to get there?
7 How do you think that astronauts would adapt to life on a planet that was 'locked' with one side always facing the star?
8 M31, the Andromeda Galaxy, is a nearby galaxy to the Milky Way. It is bigger than the Milky Way. Observations made by the Spitzer Space Telescope estimated that M31 contains 1 trillion (1×10^{12} or 1 000 000 000 000) stars. If only 10% of stars have Goldilocks Planets, how many such planets could there be in the Andromeda Galaxy?
9 Why are we unlikely to find out if there is life on Gliese 581 g?

Relative units – comparing distances in the Solar System

You can see that the Goldilocks Zone depends on the mass of the star, and that the units used in Table 20.1 are given as 'relative' to the Earth and the Sun. This means the mass of the Sun is 1 and the radius of the Earth's orbit is 1. For a larger (and hotter) star, the Goldilocks Zone would be further away from the star. In the Solar System, astronomers usually use a relative scale.

Table 20.1 Some 'relative' values with their actual values in SI units.

Relative unit	Actual value and SI unit
The mass of the Earth, $M_{\oplus}$ The average radius of the Earth, $R_{\oplus}$	$M_{\oplus} = 6 \times 10^{24}$ kg $R_{\oplus} = 6\,371\,000$ m $= 6.371 \times 10^{6}$ m
The (average) distance from the Earth to the Sun, called 1 astronomical unit (1 AU)	1 AU = 149 598 000 000 m (1.5×10^{11} m)
The mass of the Sun, $M_{\odot}$ = 1 solar mass The radius of the Sun, $R_{\odot}$ = 1 solar radius	$M_{\odot} = 2 \times 10^{30}$ kg $= 333\,333\ M_{\oplus}$ $R_{\odot} = 7 \times 10^{8}$ m $= 0.0046$ AU

Inside the Solar System, the best units to use are relative ones. Distances are usually given in AU and masses in $M_{\oplus}$.

Discussion Point

Pluto has now been 'downgraded' to a dwarf planet by the International Astronomical Union. Find out why. A lot of people have been very unhappy about this decision. What do you think?

Table 20.2 Data on the planets of the Solar System.

Planet	Symbol	Average orbit radius (in AU)	Orbital period (in Earth Years)	Average radius, (in $R_{\oplus}$)	Mass, (in $M_{\oplus}$)
Mercury	☿	0.39	0.24	0.38	0.06
Venus	♀	0.72	0.62	0.95	0.82
Earth	⊕	1.0	1.0	1.0	1.0
Mars	♂	1.5	1.9	0.53	0.11
Jupiter	♃	5.2	12	11	320
Saturn	♄	9.6	29	9.5	95
Uranus	⛢	19	84	4.0	15
Neptune	♆	30	170	3.9	17

QUESTIONS

1 What are the actual values (in SI units) of the following?
  a the average orbit radius of Mercury
  b the average radius of Jupiter
  c the mass of Neptune
2 Draw a graph of orbital period (y-axis) against average orbit radius (x-axis). Make sure you put a title on your graph and label each axis. Draw a best-fit line through your data points.
3 The best-fit line is not straight. What is the pattern in your data?
4 The dwarf planet Pluto has an orbital period of 248 Earth years. Using your graph, predict the average orbit radius of Pluto.

PRACTICAL MAKING A SCALE MODEL OF THE SOLAR SYSTEM

This activity helps you with:
- ★ working as a team
- ★ making a scale model.

Neptune
Uranus
Pluto
Saturn
Jupiter
Mars
Venus
Earth
Mercury
Sun

Figure 20.3 The relative sizes of the Sun and its planets.

continued...

Risk assessment

- **Your teacher will provide you with a suitable risk assessment for this activity.**

Apparatus

* school playing field (650 m long)
* 20 cm diameter ball (the Sun)
* 2 cress seeds (Mars and Mercury)
* 2 peppercorns (Earth and Venus)
* 23 mm diameter ball of plasticine (Jupiter)
* 18 mm diameter ball of plasticine (Saturn)
* 2 × 7 mm balls of plasticine (Neptune and Uranus)
* a trundle wheel
* sticky tape (optional)

Procedure

Work in a team of three or four. You may find it easier to stick all the 'planets' to the top of tent-pegs so that they are easier to stick in the ground. You could make your own 'flags' (like sand-castle flags), using the same scale.
If your school playing fields are not 650 m long, you will need to adapt the scale using an online Solar System calculator such as:

www.exploratorium.edu/ronh/solar_system/
http://thinkzone.wlonk.com/Space/SolarSystemModel.htm
www.smileyxtra.co.uk/etdistance/

To lay your scale model of the Solar System out, you need to start at one end of your playing field with the 20 cm diameter ball, representing the Sun. To scale, the planets are then the following distances from the Sun:

Planet	Mercury	Venus	Earth	Mars	Jupiter	Saturn	Uranus	Neptune
Distance (m)	8	16	21	33	112	205	412	647

The truly awesome scale of our Solar System then becomes apparent!

Extension

Work in pairs for this activity.

Drawing a scale map of the Solar System on paper is rather difficult. You could use toilet roll (about 30 m long) but even so, using the same scale for the diameter of the planets as their orbital radius would mean that Mercury would be 0.02 mm in diameter and the Earth, 0.06 mm, i.e. just tiny dots. So, you could use two scales, one for the orbit radius and one for the diameter.

Use an online Solar System calculator to make a scaled map of the Solar System that will fit on two sheets of A3 paper stuck side by side in landscape orientation. You will have to experiment with the orbital radius scale and the diameter scale so that Neptune fits on the two sheets, and Mercury, Venus, Earth and Mars are not overlapping. You could use scaled colour images of the planets to stick on your map.

How big is our Solar System?

The answer to this question depends on what we consider to be *in* our Solar System. There are two ways of looking at this. The first way is by thinking about the particles of matter that stream away from the Sun, called the Solar Wind, and the particles of matter that are coming towards us from nearby stars, called the Stellar Wind. The point where the force of the Solar Wind acting away from the Sun balances the force of the Stellar Wind coming in towards the Sun, is called the **Heliopause**. This happens at about 100 AU from the Sun – about three times further away from the Sun than Neptune.

The second way to think about this is by considering gravity. The Sun's gravitational field extends a very long way into space, but there comes a point between the Sun and its nearby stars where the Sun's pull of gravity is less than that of the nearby stars. This happens at about 125 000 AU – over 4000 times further away than Neptune. (About the same distance as London to Moscow on our school field model!)

If we take the larger of the two definitions, then the Solar System consists of:

- one star (the Sun)
- eight planets (MVEMJSUN)
- 146 moons (as of October 2010, according to the International Astronomical Union). A moon is a natural satellite of a planet, like our Moon. The largest moon in the Solar System is Ganymede, a moon of Jupiter
- one asteroid belt (between Mars and Jupiter – of which the largest known asteroid, Ceres, is actually classified as a dwarf planet, like Pluto)
- five dwarf planets (Ceres, Pluto, Eris, Makemake and Haumea)
- many short-period and long-period comets (such as Halley's comet)
- objects called centaurs (which are unstable minor planets, a cross between asteroids and comets, of which the largest known centaur is Chariklo, which is only 260 km in diameter)
- interplanetary dust
- the Solar Wind
- a cloud of dust and ice called the Oort Cloud, which is thought to be a sort of 'comet nursery'. The Oort Cloud extends from about 2000 AU out to about 50 000 AU.

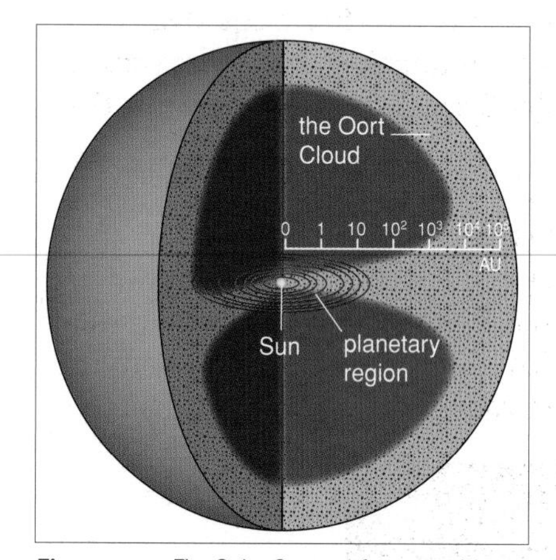

Figure 20.4 The Solar System from outside.

Figure 20.4 shows what our Solar System looks like as seen from outer space.

The next scale up is our group of stars – our galaxy, The Milky Way. Figure 20.5 shows a map of The Milky Way as seen from above.

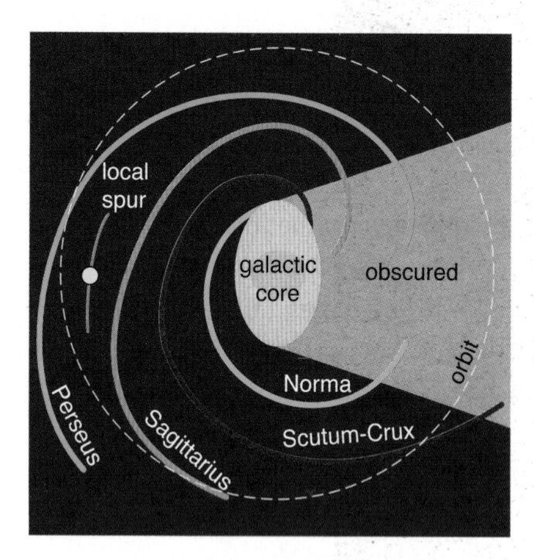

Figure 20.5 The Milky Way from above.

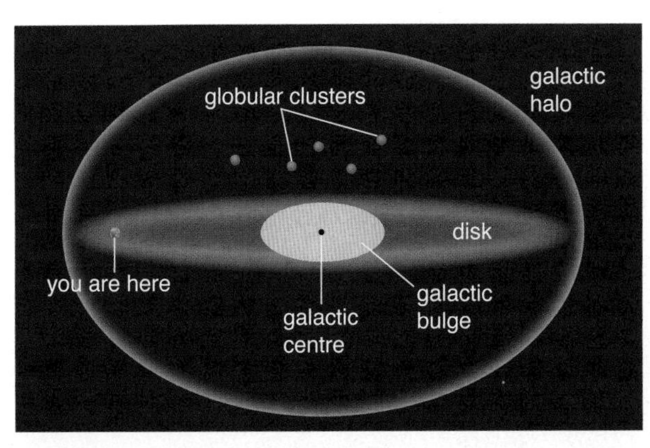

Figure 20.6 The Milky Way from the side.

Figure 20.6 is a diagram of the Milky Way drawn from the side.

The Milky Way is a very big place. The Astronomical Unit (AU), which we use to compare distances within the Solar System, is too small. The unit we now need to use is the light year (ly).

1 light year (1 ly) is defined as the distance that light travels in 1 year. The speed of light has been measured as 300 000 000 m/s.

1 year contains 365¼ days; each day has 24 hours; each hour has 60 minutes; each minute has 60 seconds.

$$\text{So, 1 year} = 365.25 \times 24 \times 60 \times 60 = 31\,557\,600 \text{ seconds}$$

$$\text{So, 1 light year (ly)} = 300\,000\,000 \times 31\,557\,600$$

$$= 9\,467\,280\,000\,000\,000 \text{ m}$$

If the Solar System is 250 000 AU in diameter, this is equal to 37 500 000 000 000 000 m or 4 ly – or put another way, light would take 4 years to travel across the Solar System from one edge to the opposite edge.

The Milky Way galaxy is 100 000 ly across and the Sun is about 27 000 ly from the galactic centre (just over half-way out). Our nearest star, Proxima Centauri, is a mere 4.2 ly away. Travelling at the speed of light, it would take 4.2 years to reach it. (Proxima Centauri is part of a small group of stars called Alpha Centauri.) The map in Figure 20.7 plots the nearest stars to the Sun (closer than 14 ly).

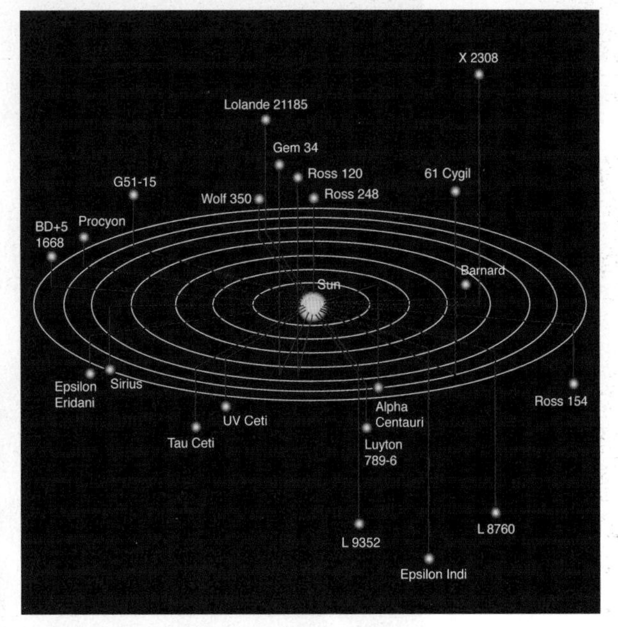

Figure 20.7 A map of stars closest to the Sun.

Figure 20.8 Charles Messier.

Our galaxy is part of a 'Local Group' of galaxies, consisting of the Milky Way, another large spiral galaxy called M31 or the Andromeda Galaxy (the 'M' stands for Messier object – a series of deep-space objects first catalogued by the French astronomer Charles Messier and published in 1774), another spiral galaxy called Triangulum, M33, and a whole series of small 'dwarf' galaxies (Figure 20.9).

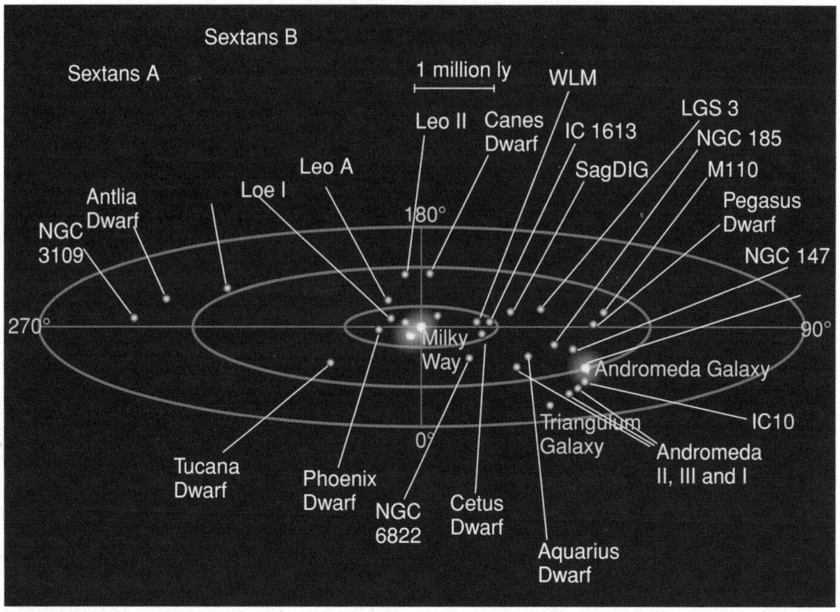

Figure 20.9 Our Local Group of galaxies is 10 million ly in diameter – about 100 times wider than our own galaxy.

It then turns out that our Local Group is part of a supercluster of galaxy groups called the Virgo Supercluster, which is 110 million light years in diameter (11 times bigger than our Local Group and over 1000 times bigger than the Milky Way).

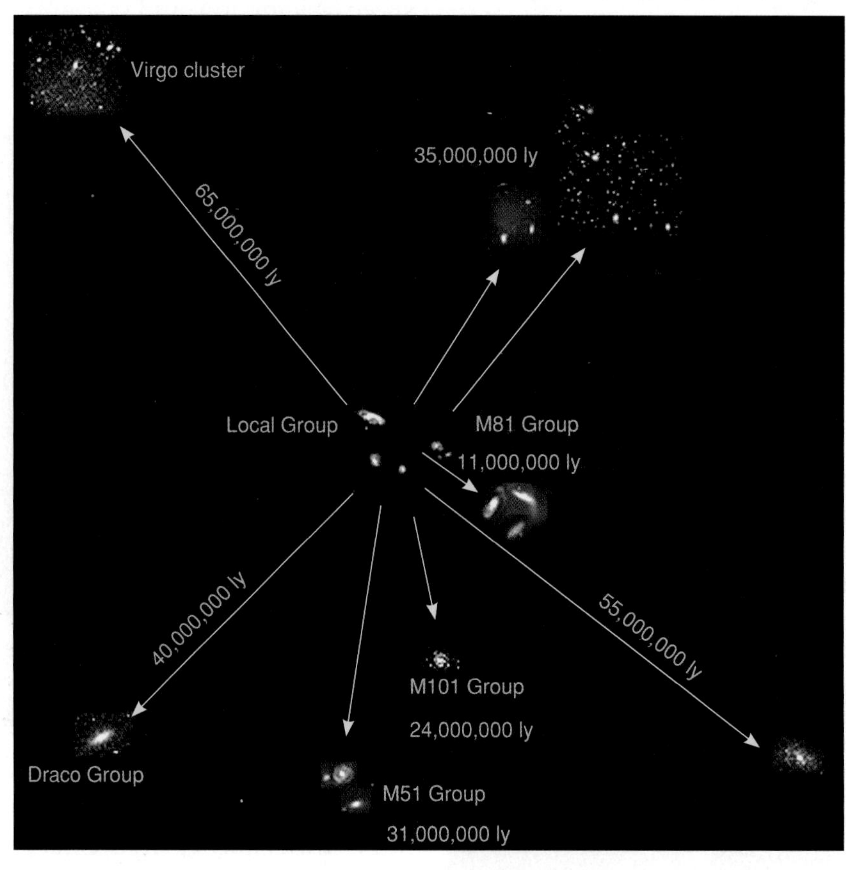

Figure 20.10 The Virgo supercluster.

TASK HOW MUCH BIGGER?

This activity helps you with:
★ communicating ideas as a diagram
★ designing a diagram
★ calculating scale.

How much bigger is each object from the last? Starting with the Solar System, followed by the Milky Way, the Local Group and the Virgo Supercluster, draw a diagram showing how much bigger each object is from the last as a scale factor (e.g. ×10). You could use Figure 20.11 as a template, or you could print out your own pictures and make another design.

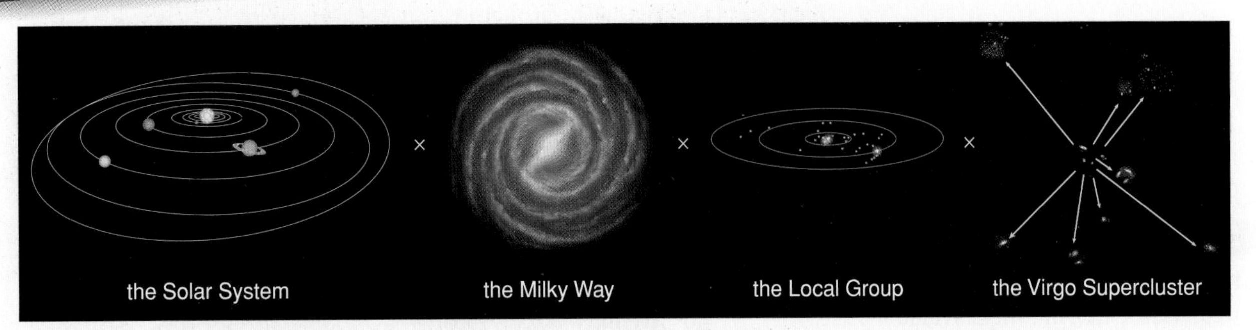

Figure 20.11

QUESTIONS

5 a Why do you think that there are two definitions of our Solar System?
  b What are the main objects in our Solar System?
6 What is the Heliopause?
7 a What is a moon?
  b The planets Jupiter and Saturn have the most moons. Why do you think that they have more moons than other planets?
8 Explain what is meant by the following:
  a a dwarf planet
  b the Oort Cloud
  c a galaxy
  d a supercluster
9 a What is a light year (ly)?
  b How many astronomical units (AU) are there in 1 ly?
10 Why is The Andromeda Galaxy called M31?

Chapter summary

- The Solar System consists of one star – the Sun, eight planets, several dwarf planets and many moons.
- 'Life' is possible inside a band of space near to a star called the habitable or Goldilocks Zone.
- Our local patch of space is called the Solar System, which is inside a galaxy called the Milky Way, which is part of a group of galaxies called the Local Group, which in turn is part of a cluster of 'groups' called the Virgo Supercluster.
- There are a range of distance scales needed when discussing the Universe: from the scale of planets and the Solar System, where comparisons to the Earth and the Sun are best; to the the Milky Way galaxy and the observable Universe, where the distance that light travels in 1 year, called a light year, is the best unit to use.

21 The Universe

Where does space end?

Perhaps this question should be phrased in different ways. Does space go on for ever? Is the Universe infinite? Or for that matter, what is the Universe? The last question is perhaps easier to answer than the others. The Universe is: all space; all time; all matter and all energy – simple really!

Discussion Point

Is the Universe any bigger than this? Well a little bit. The Universe is 13.7 billion years old, so the 'edge' of the Universe is only 0.6 billion light years further away than GRB 090423 (another 4% of the overall size of the Universe). But that's it. No more. Or is it? The visible Universe (the Universe that we can observe with the electromagnetic spectrum) can be thought of as a sphere with a diameter approximately 28 billion light years (approximately 2 × 13.7 billion ly), but as for what's beyond? Or what went before? Who knows?

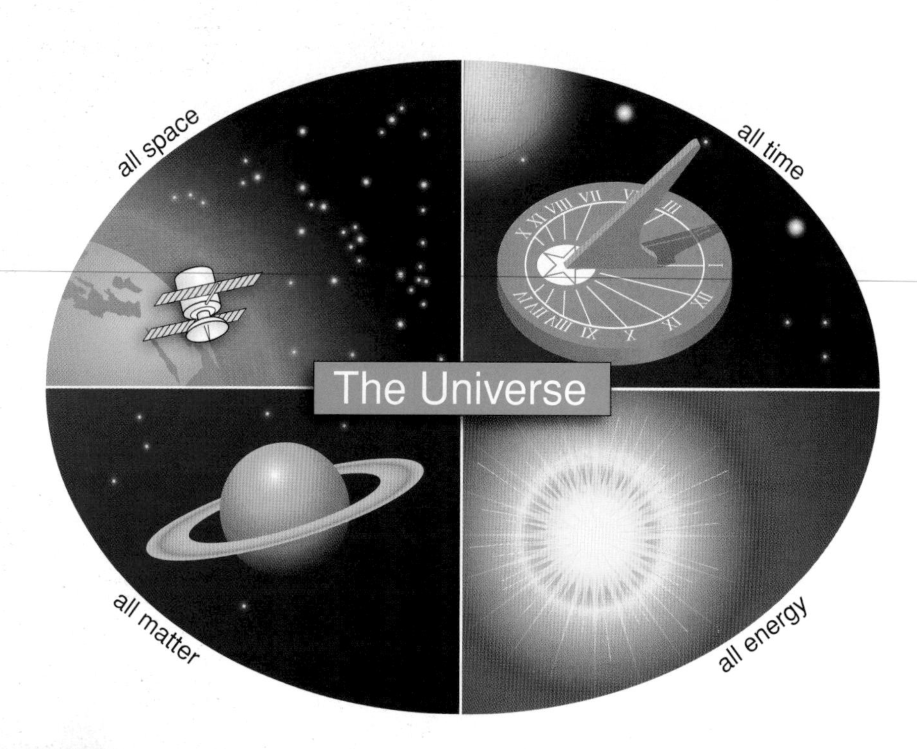

So if the Universe is everything, how big is everything? When astronomers use the largest, most powerful telescopes in the world, they can see objects that are a very, very long way away. The furthest object ever imaged is GRB 090423, a gamma-ray burst object measured to be 13.1 billion light years away.

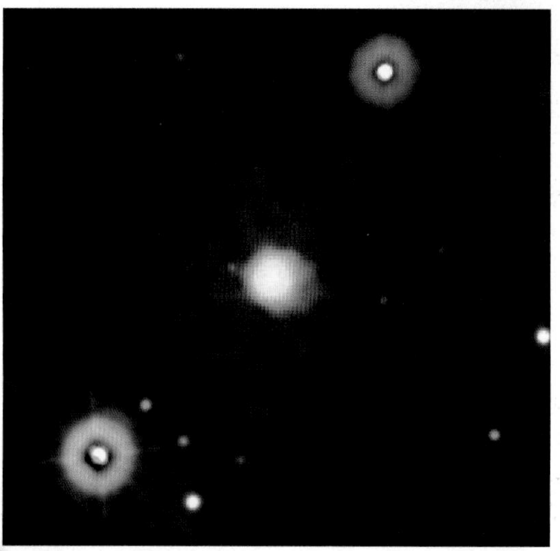

Figure 21.1 The furthest object ever imaged – GRB090423.

1 How big is the Universe?
2 If you were writing your address to someone else who lived just the other side of the Universe, how would you write your address, so that they could send you a postcard?
3 How long would it take an e-postcard travelling at the speed of light to reach you from the other side of the Universe?

TASK THE EDGE OF THE UNIVERSE?

This activity helps you with:
★ evaluating scientific claims based on critical analysis of data
★ seeing how a scientific theory developed over time
★ processing, analysing and interpreting secondary data.
★ drawing evidence-based conclusions.

Figure 21.2 The Hubble telescope.

Figure 21.3 The Hubble Ultra Deep Field.

Is this the most important scientific photo ever taken? During the months of September 2003 to January 2004, the Hubble Space Telescope's cameras were directed at a patch of apparently black, starless space and the cameras' apertures were left open for a time lasting just over 11 days. When the resulting images were processed and added together the fantastic image shown in Figure 21.3 was formed. It is known as the Hubble Ultra Deep Field.

continued...

This 'blank' patch of sky is in fact teeming with over 10 000 galaxies, each containing over a hundred billion stars. Each of these specks and smudges is a whole galaxy!

The size of space surveyed in this picture is also fantastic. Imagine looking at space through a 2.5 m long drinking straw – that's how big the patch of space in this photo is.

Many of these galaxies are so far away from us that the light reaching us from them has taken billions of years to arrive. We are looking at them as they were shortly after the Big Bang that produced the Universe, 13.7 billion years ago.

In this task you will be supplied with two sets of data, taken by astronomers over the last 100 years. The data give you the distance of a galaxy away from Earth and its speed away from us. Dataset 1 (Table 21.1) was used by the American astronomer Edwin Hubble in 1929 (Figure 21.4) and Dataset 2 (Table 21.2) is a collection of modern data using observations and measurements of distant exploding supernovae (huge exploding stars).

Figure 21.4 Edwin Hubble.

You can plot the data either by hand on graph paper, or using a spreadsheet program such as Excel.

Table 21.1 Dataset 1—Edwin Hubble's 1929 data.

Galaxy distance from Earth, *d* (light years, ly)	Galaxy speed away from Earth, *v* (km/s)
10	170
150	200
170	290
210	200
270	300
300	650
310	150
340	920
370	500
480	500
580	960
660	500
670	800
680	1 090

continued...

1 Plot Hubble's data (for nearby galaxies) on a graph with 'Distance from Earth, *d* (light years)' on the *x*-axis and 'Speed away from Earth, *v* (km/s)' on the *y*-axis.
2 How would you describe the pattern in these data? (***Hint***: Do the data all sit on a straight line or is it a curve? Is there a general pattern/trend? Are the data widespread or close together?
3 Draw a best-fit straight line through the data. Your line must start at the origin (0,0). This best-fit line must go through the middle of the pattern of the data.
4 Can you draw any other lines showing patterns in these data?

Edwin Hubble concluded that these data showed that the Universe was expanding. The further away the star was, the faster it was moving. He also said that there was a direct mathematical relationship between the two quantities, $v = H_0 \times d$, or if you double the distance away from Earth, you double the speed of the star. The value H_0 became known as the Hubble Constant, and is the gradient (or slope) of the line; Hubble calculated the value as 500 km/s/Mpc. (1 Mpc = 1 megaparsec = 3.2×10^6 ly.) This relationship became known as the Hubble Law.

5 Do you agree with Hubble's conclusions?
6 How strongly do you think Hubble felt about this conclusion?
7 Do you think other astronomers were in general agreement with him, or do you think some might have been sceptical?

Almost as soon as Hubble published his data, other astronomers realised that if the Universe was expanding, there must have been a time when the Universe was much, much smaller, in fact the Universe must have started in one place, at one time in the past. This must have been a fantastic explosion – the creation of the Universe – and this became known as The Big Bang.

Table 21.2 Dataset 2 – Modern supernova data.

Distance of supernova from Earth, *d*, (Mpc)	Speed of supernova away from Earth, *v* (km/s)
60	4100
80	5400
100	7200
120	7900
140	9000
160	12000
180	13700
200	14800
220	15000
240	16900
260	18400
280	19000
300	21600
320	23600
400	26500
420	30600

continued...

8 Plot the supernova data on a graph with 'Distance from Earth, *d* (Mpc)' on the *x*-axis and 'Speed away from Earth, *v* (km/s)' on the *y*-axis.
9 How would you describe the pattern in these data? (***Hint***: Do these data all sit on a straight line or is it a curve? Is there a general pattern/trend? Are the data widespread or close together?
10 Draw a best-fit straight line through this data. Your line must start at the origin (0,0). This best-fit line must go through the middle of the pattern of the data.
11 Can you draw any other lines showing patterns in these data?
12 **Extension:** Calculate the gradient of this best-fit line, in km/s/Mpc.
13 Do these data confirm or deny Hubble's 1929 conclusions?
14 Do you think modern astronomers are confident or sceptical about Hubble's conclusions?
15 Describe the difference between the two datasets.

Astronomers now have many more points plotted on this graph, using very accurate and powerful telescopes, including the Hubble Space Telescope. The latest value of the Hubble Constant from the Hubble Space Telescope measurements in 2009 gives $H_0 = 74.2 \pm 3.6$ km/s/Mpc. Working backwards, this gives the age of the Universe as 13.75 ± 0.17 billion years old, so the edge of the Universe could be 13.75 billion light years away!

16 Working forwards, what do you think that these data mean about the future fate of the Universe?

QUESTIONS

4 What does the Hubble Ultra Deep Field photo in Figure 21.3 show?
5 The Hubble Space Telescope made 400 orbits of the Earth during the time of the observations and made 800 exposures – or two every orbit.
   a Why couldn't the Hubble Space Telescope just point at the same point in the sky continuously for 11 days?
   b If all the exposures were the same length of time, how long was each exposure?
6 a How many stars could there be in the photo of the Ultra Deep Field?
   b Why is it that when we are looking at the Ultra Deep Field photo, we are effectively looking back in time?
   c Why is it that Hubble's peers – the other astronomers working on similar problems in 1929 – found it difficult to accept Hubble's Law?
7 a Using Hubble's dataset (Table 21.1), estimate the maximum and minimum speeds of a galaxy 400 ly (light years) from Earth.
   b Why is there such a range of speeds?
8 Dataset 2 in Table 21.2, the data from modern observations of supernovae, shows a very linear (straight line) trend. Where on this graph would Hubble's data from 1929 sit?
9 Why are modern astronomers very confident about Hubble's law?
10 If the Universe is 13.75 billion years old, then light from the furthest objects in the Universe (that were created just after the Big Bang) will have travelled 4297 Mpc. Using Hubble's Law, $v = H_0 \times d$, and a Hubble Constant of 74.2 km/s/Mpc, how fast will the furthest objects in the Universe be travelling?

Discussion Point

Why do you think that The Hubble Ultra Deep Field photo is one of the most important photos ever taken?

How did Hubble measure the speed of galaxies?

It was Isaac Newton who first realised that the cheap 'fairground' curio called a prism was splitting sunlight up into its constituent colours. He published his ideas about light in his 1704 book called '*Optiks*'.

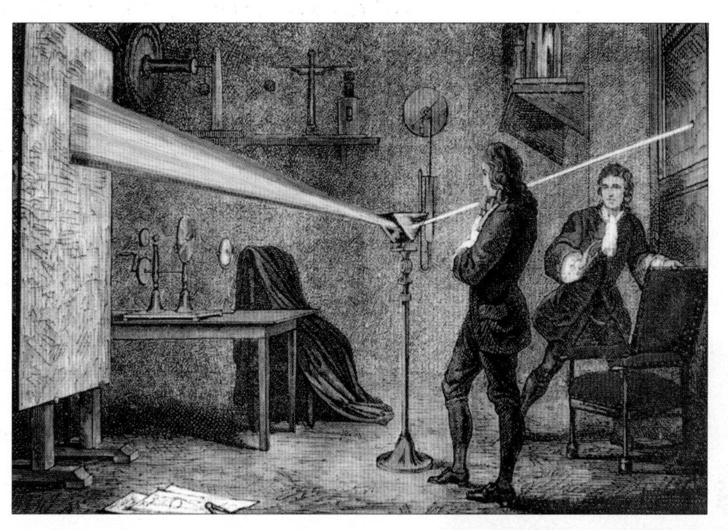

Figure 21.5 Isaac Newton, with his prism.

Newton used prisms to study the light from lots of different sources. *Optiks* is a book describing Newton's thoughts about what light and colour were. One hundred years later Joseph von Fraunhofer (the 'father of modern spectroscopy'), discovered that the continuous spectrum of colour produced by light from the Sun in fact contained over 700 tiny black lines, later called the 'Fraunhofer Spectrum'.

Figure 21.6 Joseph von Fraunhofer.

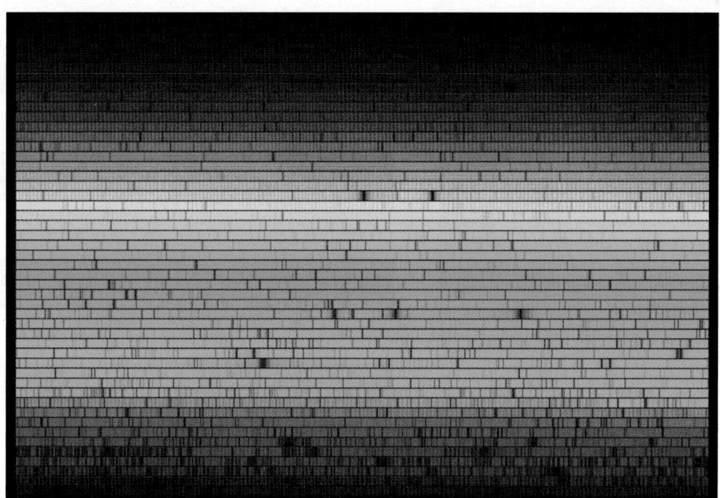

Figure 21.7 The Fraunhofer Spectrum.

Figure 21.8 Gustav Kirchoff.

In 1859, Fraunhofer's spectrum was finally explained by Gustav Kirchoff and Robert Bunsen (of Bunsen burner fame). Kirchoff and Bunsen discovered that when different elements were vaporised in Bunsen's burner flame they emitted light. They then used a prism device to study the spectra of the different elements. They discovered that each element had its own signature spectrum of light. These are called **emission** spectra.

Kirchoff also found that when he passed the light from different elements through a gas of that element (for example, shining hydrogen spectrum light through hydrogen gas) the gas absorbed the colours of the spectrum.

Kirchoff realised that the black lines on Fraunhofer's spectrum of the Sun were produced by the elements that made up the Sun. Kirchoff and Fraunhofer had discovered a way of identifying different elements on stars far away from Earth. **Stellar spectroscopy** had been born.

On 18 August 1868, during an expedition to Norway to observe a solar eclipse, Sir Norman Lockyer discovered an unusually prominent yellow spectral line in the spectrum of a solar flare observed during the totality of the eclipse. The *yellow* line corresponds to a colour with a wavelength of 588 nm (5.88×10^{-7} m). At the time no known element produced a spectral line with this colour and wavelength. Lockyer suggested that this line corresponded to a new element that he called helium, after the Greek word 'helios', meaning 'sun'.

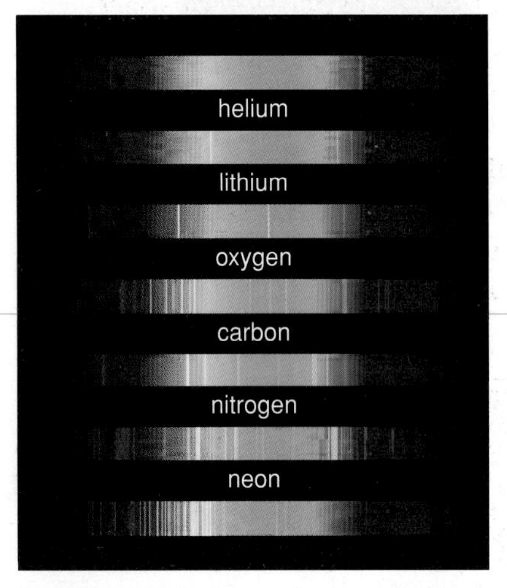

Figure 21.9 Emission spectra.

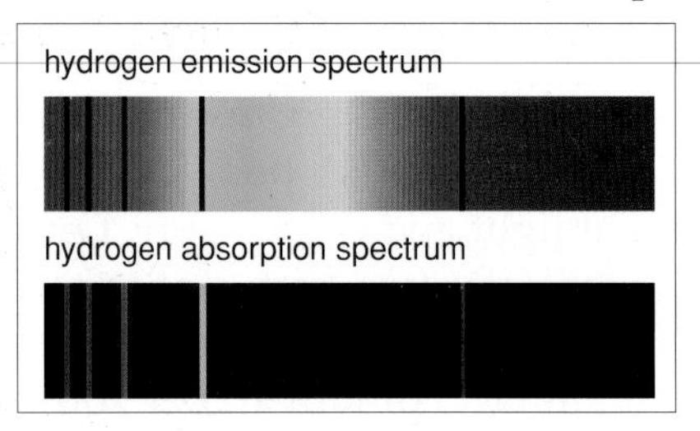

Figure 21.10

Figure 21.11 Sir Norman Lockyer.

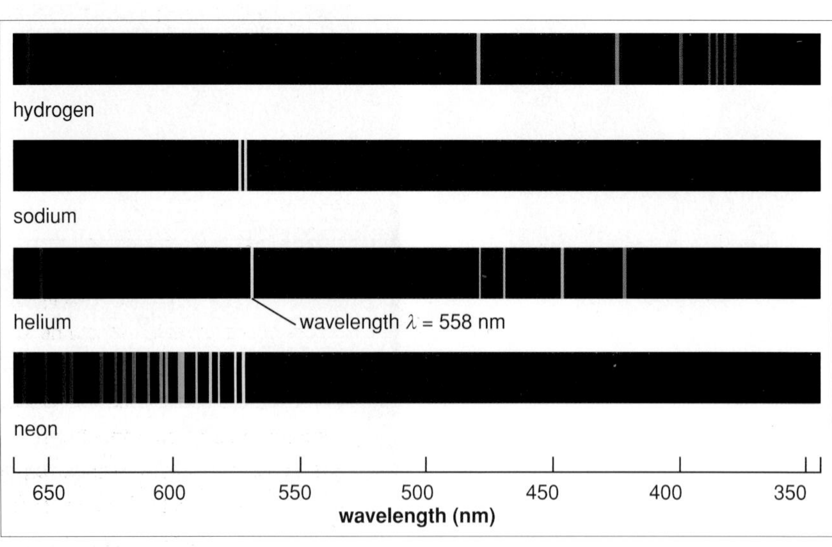

Figure 21.12 Absorption spectra of hydrogen, sodium, helium and neon.

Helium was finally isolated and identified in the laboratory in 1878 by William Ramsey, and Lockyer's stellar spectroscopy techniques have been used ever since to study the chemical composition of stars.

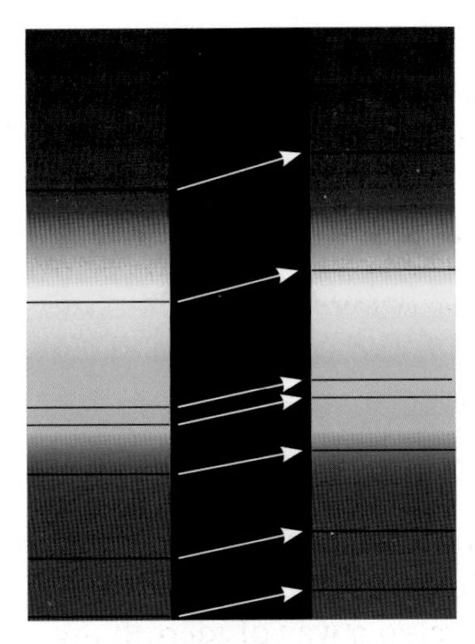

Figure 21.13 Redshift.

Soon after Fraunhoffer's discovery of spectroscopy, a French physicist called Hippolyte Fizeau discovered that the spectral lines of some stars appeared to be 'shifted' towards higher wavelength, i.e. they became slightly more 'red'. The patterns always stayed the same but each of the different spectral lines appeared to move by the same amount towards the red end of the visible spectrum. This effect became known as **redshift**.

Fizeau assumed that the shift in the lines was due to the star moving away from Earth at speed, and in 1868 the British astronomer William Huggins was the first person to use measurements of redshift to determine the speed of another star moving away from Earth.

It was Edwin Hubble, though, who first realised that his measurements of the redshift of other galaxies far away could be explained, not only by the relative movement of those galaxies away from our own, but by the expansion of space. This is now known as **cosmological redshift**. Figure 21.14 shows a model of cosmological redshift. The Universe is represented by the surface of a balloon and a wave of light is drawn on the surface of the balloon before it is inflated. As the balloon is inflated, the surface of the balloon (the Universe) expands and the wave of light is stretched, increasing the wavelength of the light. If the wavelength increases it becomes more red, or it redshifts.

Figure 21.14 Cosmological redshift.

QUESTIONS

11 What does a prism do to white light?

12 One hundred years is a long time in science and technology. Why do you think that it was Joseph von Fraunhofer who observed the spectral lines in the Sun's spectrum rather than Newton?

13 a What is an emission spectrum?
   b How is it different from an absorption spectrum?

14 a How did Norman Lockyer discover helium? Helium is a noble gas and is very inert – it does not react with other chemicals very much.
   b Why do you think it took 10 years to confirm Lockyer's discovery?

15 Explain how the emission spectra of elements here on Earth can be used to determine the chemical composition of stars from their spectra.

16 a What is redshift?
   b Explain the difference between Hippolyte Fizeau's explanation of redshift and Edwin Hubble's explanation.

Discussion Point

How can a balloon be used to model cosmological redshift? What is good about the model? Where does the model break down?

PRACTICAL MAKING AND MEASURING A MODEL OF THE UNIVERSE

This activity helps you with:
★ making a scientific model
★ analysing a scientific model
★ discussing the similarities and differences between a scientific model and 'the real thing'
★ taking measurements.

Figure 21.15

Risk assessment

- **Your teacher will provide you with a suitable risk assessment for this activity.**

Apparatus

* thick rubber band
* pair of scissors
* 4 sticky-backed stars
* ruler

Procedure

1 Make a cut in the rubber band and lay the rubber band out on the table.
2 Draw a wave in pen all the way down one side of the rubber band. Try to keep the wavelength constant – you could use the ruler to help do this.
3 Stick the stars to the rubber band in increasing distances away from one end. Label the stars alpha (α), beta (β), gamma (γ) and delta (δ).
4 Hold the rubber band at either end and stretch it.
5 Observe and record what happens to the wavelength of the wave that you have drawn on the rubber band.
6 Observe and record what happens to the 'interstellar distance' between each star.
7 Turn the rubber band around and repeat the stretch and the observations. Does it matter where you are on the rubber band to make these observations?
8 Mount the rubber band vertically between two clamps on a stand. Tighten one clamp across each end of the rubber band. Stretch your model and secure the bosses holding the clamps.
9 Copy and complete the following table of measurements made with your model of the Universe:

Measurement	Before stretching	Stretched
Length of the rubber band		
Number of complete waves on the rubber band (the wave number)		
Wavelength of the waves		
Distance between stars α and β		
Distance between stars β and γ		
Distance between stars γ and δ		

Discussion Point

Explain the similarities between this model and the real Universe. How good do you think the model is? Is this model any different from the balloon model?

Analysing your results

1 What happens to the stars as the rubber band is stretched?
2 Explain why, using this model, stars don't get stretched as the Universe expands.
3 What is the relationship between the increase in the wavelength on the rubber band and the length of the rubber band as it is stretched?
4 What is equivalent to cosmological redshift on this model?

The Big Bang

The measurements made by Hubble were the starting point of The Big Bang model of the Universe. 13.75 billion years ago the Universe (all space, time, mass and energy) came into being as the result of a huge explosion. Ever since that time the Universe has been expanding, and this can be measured by the cosmological redshift and the Hubble Equation, $v = H_0 \times d$. At the time of the Big Bang though, huge amounts of energy must have been created in the form of gamma rays. What's happened to them? Cosmological redshift will have stretched these gamma rays, and as the Universe has expanded, their wavelengths will have got longer and longer. In 13.75 billion years, these gamma rays will have been stretched so much that they should have wavelengths equivalent to those of microwaves. If we look hard enough, the Universe should be filled with these characteristic microwaves, now called the **Cosmic Microwave Background Radiation** or **CMBR**.

The CMBR was actually discovered by accident in 1964. Two physicists working for the Bell Telephone Company near New York – Arno Penzias and Robert Wilson – were actually searching for radiation from space that could harm satellites used for communications. What they found was the remnant of the radiation produced by the Big Bang.

Figure 21.16 Arno Penzias and Robert Wilson.

The CMBR discovered by Penzias and Wilson had exactly the right wavelengths predicted by the Big Bang model. The WMAP satellite has made a map of the CMBR (Figure 21.17).

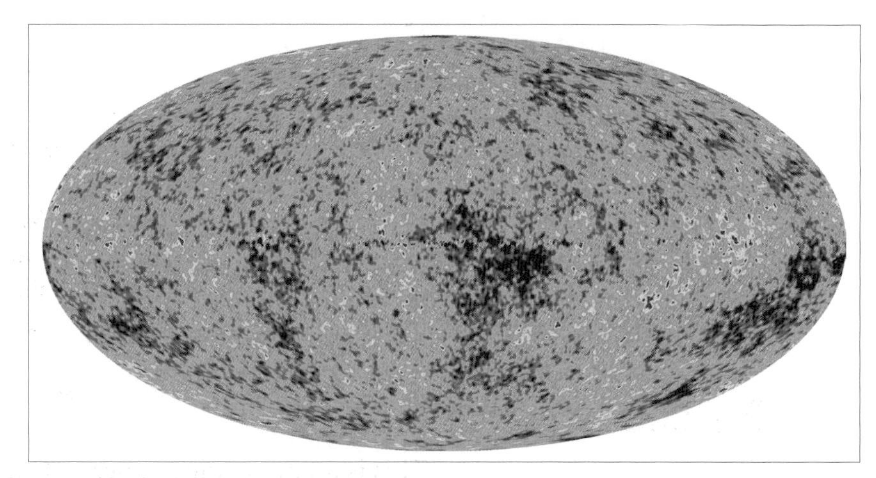

Figure 21.17 Cosmic Microwave Background Radiation mapped out by the WMAP satellite.

The different colours on the map represent small changes in the intensity of the CMBR. Without these small changes, matter would not have clumped together to form stars and galaxies. Figure 21.18 shows the evolution of the Universe from The Big Bang to the observations of the WMAP satellite.

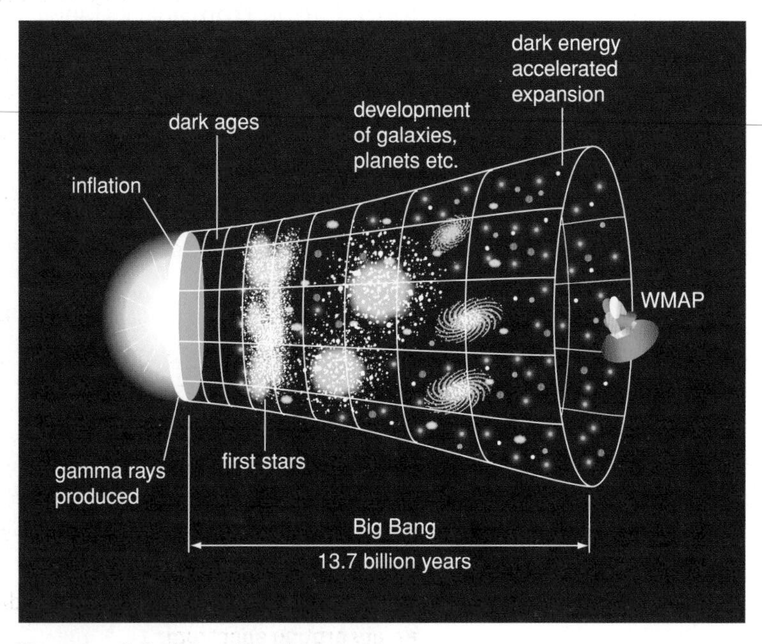

Figure 21.18 Evolution of the Universe.

The Big Bang is an amazing theory. It was first proposed by a Belgian physicist called Georges Lemaitre in 1927, and has developed over the last 80 years or so. What makes it such a good theory is its basis in observation and evidence. Without cosmological redshift and the CMBR, the Big Bang theory would only be an interesting theoretical exercise in physics. The theory does have its problems though: there doesn't seem to be enough matter in the Universe (leading to a theory of 'Dark Matter'); and no one is quite sure how it started, and whether there was anything before the Big Bang. Is the Universe a 'one-off' or are there more Universes? Were there Universes before *our* Universe?

Discussion Point

Quite a lot of theoretical cosmlogists now think that there was definitely 'something' before the Big Bang. What could the 'somethings' be?

QUESTIONS

17 What is the CMBR?

18 Why do you think that the discovery of the CMBR is considered to be an 'accident'?

19 What does the WMAP 'map' of the CMBR show us?

20 Describe the evolution of the Universe.

21 a Why is the Big Bang such a good theory?

b What is a 'theory'? Why are physicists working on new theories of the Universe?

The end?

This is an interesting concept. Current observational evidence suggests that the Universe will go on expanding forever, but that's a very long time, and who knows...perhaps another, better theory might just come along instead!

As Douglas Adams wrote, in *The Hitchhikers Guide to the Galaxy*:

> Some time ago a group of hyper-intelligent pan dimensional beings decided to finally answer the great question of Life, The Universe and Everything.
> To this end they built an incredibly powerful computer, Deep Thought. After the great computer programme had run (a very quick seven and a half million years) the answer was announced.
> The Ultimate answer to Life, the Universe and Everything is...
> (You're not going to like it...)
> Is...
>
> **42**
>
> Which suggests that what you really need to know is 'What was the Question?'

Chapter summary

- Atoms of a gas absorb light at specific wavelengths, which are characteristic of the elements in the gas.
- You can use data about the spectra of different elements to identify gases from an absorption spectrum.
- Scientists in the nineteenth century were able to reveal the chemical composition of stars by studying the absorption lines in their spectra.
- Edwin Hubble's measurements on the spectra of distant galaxies revealed that the wavelengths of the absorption lines are increased and that this 'cosmological redshift' increases with increasing distance.
- The cosmological redshift of the radiation emitted by stars and galaxies is due to the expansion of the Universe since the radiation was emitted.
- The Big Bang theory of the origin of the Universe predicted the existence of background radiation, which was subsequently detected accidentally in the 1960s, and that the Cosmic Microwave Background Radiation (CMBR) is the redshifted remnant of radiation from the origin of the Universe.
- Cosmological redshift and the Cosmic Microwave Background Radiation have provided the evidence for the establishment of the Big Bang theory of the origin of the Universe.

Index